D1058478

The Breast

Morphology, Physiology, and Lactation

THE BREAST

Morphology, Physiology, and Lactation

HELMUTH VORHERR

Departments of Obstetrics–Gynecology and Pharmacology
School of Medicine
The University of New Mexico
Albuquerque, New Mexico

ACADEMIC PRESS New York San Francisco London 1974

A Subsidiary of Harcourt Brace Jovanovich, Publishers

ACADEMIC PRESS, INC.
111 Fifth Avenue, New York, New York 10003

United Kingdom Edition published by
ACADEMIC PRESS, INC. (LONDON) LTD.
24/28 Oval Road, London NW1

Library of Congress Cataloging in Publication Data

Vorherr, Helmuth.
The breast: morphology, physiology, and lactation.

Bibliography: p.
1. Breast. 2. Breast–Diseases. 3. Lactation.
I. Title. [DNLM: 1. Breast. WP800 V953b 1974]
QP188.M3V67 612.6'64 74-1635
ISBN 0–12–728050–2

PRINTED IN THE UNITED STATES OF AMERICA

To My Beloved Wife, Ute, in Appreciation of Her Many Invaluable Contributions

Contents

Preface

In this book, an effort has been made to assemble essential information available on mammary morphology and function related to endocrine physiology, as well as on pathophysiologic disorders such as galactorrhea. It should prove a welcome addition to medical students, physiologists, endocrinologists, pharmacologists, basic and clinical investigators, physicians, and clinicians.

The milk glands of mammals are vital for the nursing, rearing, and survival of the newborn, thus assuring preservation of the species. It appears that nursing, besides its beneficial effects on the infant and on postpartum uterine involution through suckling-induced neurohypophysial oxytocin release, may offer some protection against the development of breast cancer in the later life of a breast-feeding mother.

In order to understand the morphology and physiology of the breast, the relationship between mammary gland function and the various endocrine organs has to be taken into consideration. Knowledge of the many hormones influencing the structure and function of the breast enables us to relate cyclic endocrine ovarian changes to symptoms of premenstrual mammary tension, parenchymal alterations, and breast neoplasia.

Since breast malignancies are the most commonly found form of cancer in women, it is essential that we understand the multifaceted interrelationship between mammary morphology and physiology and endocrine glands such as hypophysis, ovaries, and adrenals for therapeutic management.

Physiologic breast changes occurring in pregnancy and lactation are induced by the combined action of pituitary, ovarian, placental, thyroid, parathyroid, adrenocortical, and pancreatic hormones. A better understanding of the mechanical and hormonal factors involved with the onset and maintenance of lactation provides tools for the diagnosis and treatment of abnormalities such as hypogalactia and galactorrhea. In puerperas who do not wish to breast-feed, suppression of lactation is instituted. Accordingly, the understanding of the various endocrine mechanisms leading to the postpartum mammary changes for lactation provides a basis for medicamentous inhibition of lactation.

Physiologic progressive and regressive mammary changes during a woman's life cycle are remarkable, and can be compared only with those occurring in the uterus and ovaries with which they are closely interrelated. I hope this book will contribute to a better understanding of the morphology and physiology of the breast in relation to mammary changes induced by hormonal activities of various endocrine glands.

I am indebted to my secretary, Mrs. Virginia Jones, for her dedicated and skillful assistance during preparation of this book. Also, I would like to express my sincere appreciation to Mrs. Eva Baerwald for her editorial contributions.

HELMUTH VORHEER

The Breast

Morphology, Physiology, and Lactation

CHAPTER I

Development of the Female Breast

A. EMBRYONIC AND FETAL MAMMARY DEVELOPMENT

Around the fifth week the 2- to 4-cell-layered primitive milk streak ("galactic band") develops in the human embryo. This thickened epithelial band is derived from the ectoderm and runs along each side of the lateral embryonic trunk extending from the axilla to the groin (Fig. 1). During the sixth and seventh weeks of intrauterine life, the epithelial band widens to form a 4- to 6-cell layer, the mammary ridge, in the thoracic area, which later becomes the site of the definite mammary glands; simultaneously, the other parts of the milk streak involute. In case of incomplete regression of the primitive mammary streak or dispersion of its cells, accessory mammary tissues (gland tissues, nipples, or both) develop along the mammary streak or other parts of the body (Fig. 1). Accessory mammary tissues are observed in 2–6% of females, and in some an additional nipple is mistaken for a pigment nevus. Later in life these supernumerary mammary tissues may cause discomfort during pregnancy and lactation by increasing in size and secreting milk, especially when they are located in the axilla or groin (Fig. 1). In the 7-week-old embryo, the round, epithelial, ectodermal mammary anlage, situated medially from the root of the upper extremity, becomes thicker (milk hill stage) and then starts to invaginate the underlying mesenchymal tissue (disc stage, Fig. 2a). Beginning with the seventh and extending into the eighth embryonic week, the mammary anlage grows tridimensionally leading to a globular protrusion of the overlying skin (globular stage). At the eighth week, cells of the epi-

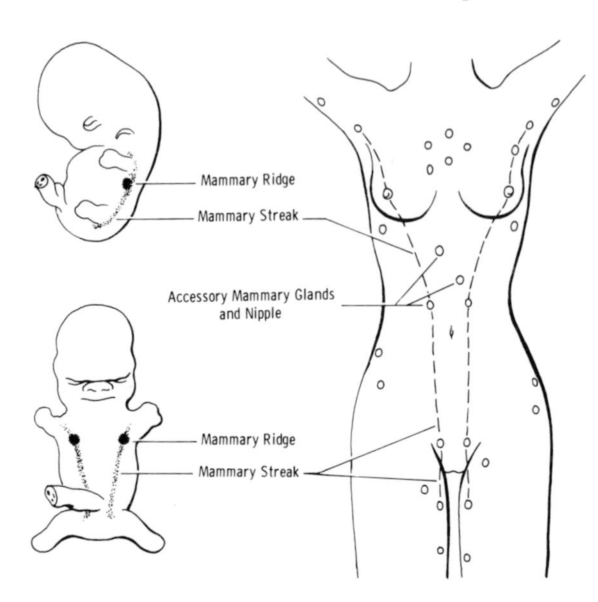

Fig. 1. Development of embryonic mammary tissues. In the 5-week-old human embryo the 2- to 4-cell-layered ectodermal mammary streak develops bilaterally on the embryonic trunk, extending from the axilla to the groin. Around the seventh week of intrauterine life the 4- to 6-cell-layered mammary ridge is formed in the thoracic area, which later becomes the site of the definite mammary glands (left side of figure). With development of the mammary ridge, the other parts of the mammary streak involute. Through inadequate regression or dispersion of ectodermal cells of the mammary streak, accessory mammary tissues may develop along the area of the former embryonic streak (milk line) or on other parts of the body (right side of figure).

thelial globular mammary anlage begin to descend and to invade the underlying mesenchyme. Consequently, at the tenth and fourteenth week, the overlying skin is no longer protruding and forms a nipple groove, and the inward growing primary epithelial anlage changes from a globular into a cone-shape (cone stage). Between the twelfth and sixteenth week, further descent and mesenchymal invasion by the tip of the epithelial cone-shaped cylinder take place, and the overlying skin becomes slightly more indented. Underneath the invaginated skin, mesenchymal cells differentiate into smooth muscle fibers to form the smooth musculature of the nipple and areola mammae. Owing to continued basal proliferation and mesenchymal invasion around the tenth and fourteenth week of gestation, the upper part of the primary mammary anlage is drawn out to form a neck-shaped thinning. Thereafter, epithelial buds begin to develop at the tip of the invading parenchymal tissue (budding stage) subsequently showing distinct furrow-

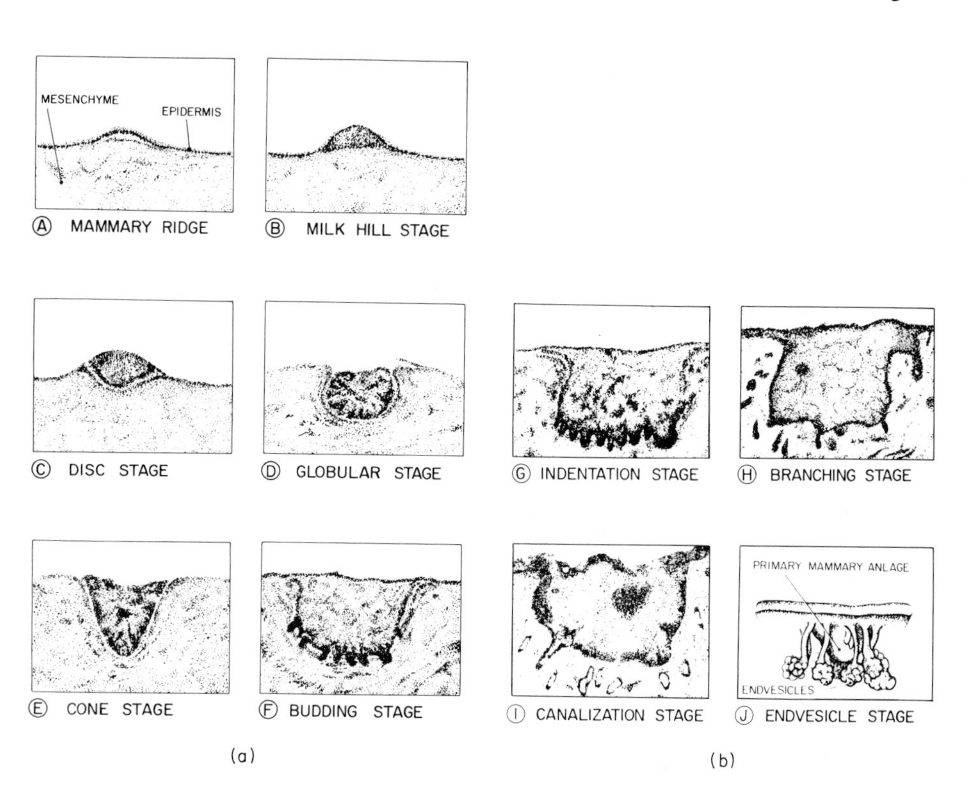

Fig. 2. Development of the human embryonic mammary gland and the phases of fetal mammary parenchymal differentiation (illustrated by Norviel based on drawings in Dabelow, 1957). (a) The development of the mammary ridge (embryonic milk line) is observed at the fifth week of intrauterine life. Through thickening of the ectodermal mammary ridge at the sixth embryonic week the milk hill stage is reached. Subepidermal descent of parenchymal cells of the primary milk hill sets in at the seventh embryonic week, leading to the mammary disc stage. Tridimensional proliferation of descending parenchymal cells of the primary mammary anlage at the seventh to eighth embryonic week leads to the globular stage. Further inward growth of parenchymal elements at the ninth embryonic week, with regression in the protrusion of the overlying skin, accomplishes the cone stage. Progressing epithelial inward growth at the tenth to twelfth week, causing skin indentation and neck-line thinning of superficial parenchymal parts, induces the budding stage. (b) Development of basal epithelial foldings producing a lobular notchlike appearance of the epithelial–mesenchymal border at the twelfth to thirteenth week (indentation stage). Subsequently, basal epithelial tissue sprouts appear, introducing the secondary mammary anlage between weeks 13 and 20 of pregnancy (branching stage). Canalization of secondary epithelial sprouts then leads to formation of milk ducts between 20 and 32 weeks of gestation (canalization stage). Peripheral parenchymal differentiation into lobular-alveolar structures (endvesicles) occurs between weeks 32 and 40 of pregnancy (endvesicle stage).

ing (indentation stage). In the 15-week-old fetus, secondary tissue sprouts develop, leading to further branching into 15–25 epithelial strips (branching stage). With the branching stage around the sixteenth week of fetal development, the phase of the secondary mammary anlage is reached, and anlagen for hair, sebaceous, and sweat glands appear simultaneously in adjacent subcutaneous tissues. Only sebaceous glands, however, develop, whereas anlagen for hair and sweat glands regress. It appears that phylogenetically the mammary parenchyma has evolved from sweat glands. In addition, special apocrine and holocrine glands, the Montgomery glands, develop around the secondary mammary anlage. These early stages of mammary development are independent of specific hormonal effects; whereas, during the last trimester of pregnancy, placental sex steroids enter the fetal circulation and induce pronounced breast stimulation (for more details see Section C). At above 20–24 weeks of gestation, the secondary dichotomic epithelial mammary sprouts begin to canalize (canalization stage, Fig. 2b). After the sixth fetal month, the secondary mammary branches become progressively canalized. Near term, about 15–25 tubules (mammary ducts), into which sebaceous glands merge near their epidermal surface, are formed in the fetal mammary gland. The formation of glandular lumina begins at the peripheral parenchymal tissue sprouts, proceeds centrally and superficially toward the epithelial tissues of the primary mammary gland anlage, and is completed around the eighth month of intrauterine life. Simultaneously, with progressing epithelial canalization, increased vascularization and formation of connective tissue and fat are observed in the peripheral parts of the mammary gland. From the ninth to tenth fetal months, the mass of mammary gland tissue increases fourfold, and a parenchymal differentiation into lobular-alveolar structures, resembling vesicles, takes place peripherally (endvesicle stage). These monolayered epithelial "endvesicles" contain colostrum and are surrounded by connective tissue and fat. At this stage of fetal breast development, the smooth musculature of the areola mammae is established in its typical circular-elliptic structure. The hair- and gland-free area of the areola mammae becomes increasingly pigmented, and 15–25 milk ducts (ductus excretorii) merge in its indented center, the location of the future nipple to be formed in puberty. Postnatally, the peripheral, colostrum-filled endvesicles (alveoli) regress to ductal structures. At birth the epidermis of the areola mammae with its central dimple is thickened, and the squamous epithelium extends to the infundibular part of the milk ducts into which sebaceous glands discharge. The infundibular ducts may also harbor hair anlage, but hair does not develop (Dabelow, 1957; Bässler, 1970).

B. INTRAUTERINE DEVELOPMENT OF THE MESODERMAL COMPONENTS OF THE MAMMARY GLAND: VASCULAR APPARATUS, CONNECTIVE TISSUE, SMOOTH MUSCULATURE, AND FAT

1. Mammary Vasculature

During the early stage of embryonic development, the thickened epithelial mammary knot begins to stimulate a growth reaction of the underlying mesenchymal tissue. Thus, a condensation zone of mesenchymal cells is observed around the primary epithelial anlage; the proliferating mammary epithelial elements grow into and through this zone. In the 7-week-old embryo, erythroblasts and primitive blood vessels begin to differentiate from mesenchymal cells. In the 9- to 10-week-old embryo (budding stage), fine mammary capillaries appear, and at 12 weeks numerous capillaries are formed and embedded in connective tissue. At 12–13 weeks of gestation the vascular network shows a concentric distribution around the primary mammary anlage similar to the connective tissue. Accordingly, three vascular zones around the primary mammary anlage are defined: (1) Inner vascular zone, containing the smallest vessels (they function as a mesenchymal mantle being in close contact with the basal part of the primary epithelial mammary anlage). (2) Intermediate vascular zone, located in a fiber-rich second connective tissue envelope around the parenchymal tissue. (3) Outer vascular zone, consisting of larger vessels forming a polygonal frame around the two inner zones (Fig. 3).

At the fetal age of 16 weeks the vascular mammary development is basically established, and the three blood vessel layers begin to provide blood supply: the outer subepithelial zone supplies the corium, the intermediary zone supplies the deeper layers of the cutis, and the inner vascular zone nourishes parenchymal, connective, and fat tissues. As mammary parenchymal proliferation progresses into the deeper zones of subcutaneous tissues, the epithelial sprouts are increasingly accompanied by blood vessels. At this stage of fetal development mammary blood vessels invade and penetrate the surrounding connective tissue capsule to reach the proliferating parenchymal tissue and supply it with nutrients. Subsequently, the proliferating vasculature is joined by connective tissue and becomes located within connective tissue septa dividing the proliferating glandular elements.

Toward the end of pregnancy the fetal mammary blood vessels form a perialveolar network, and the mammary ducts are supplied by capillaries

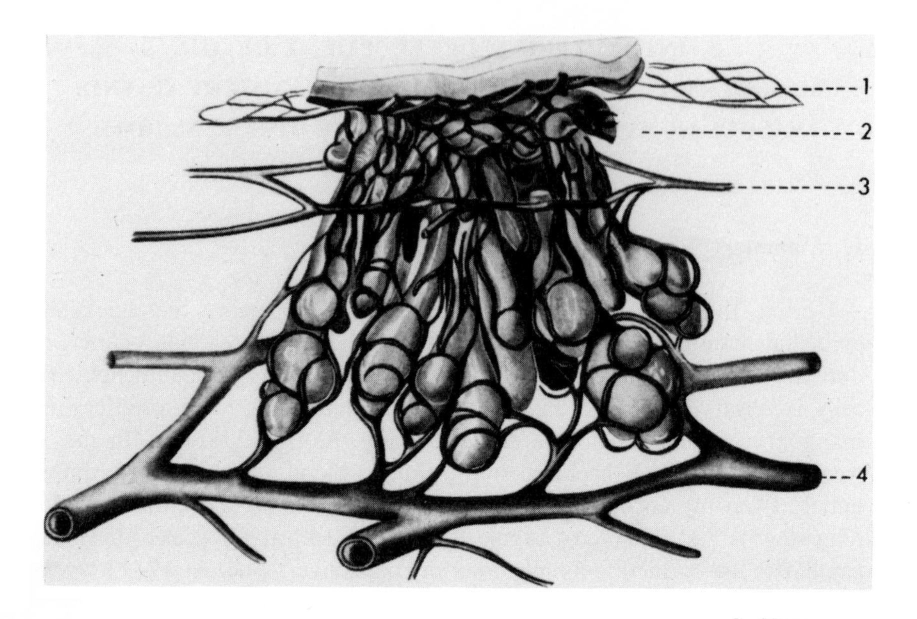

Fig. 3. Scheme of vascular development of the fetal mammary gland (from Dabelow, 1957, by kind permission of the author and publisher). In the 16- to 20-week-old fetus a concentrically layered vascular mammary arrangement can be recognized. An inner subepithelial vascular network (1) surrounds the central epithelial hill, or primary mammary anlage (2). This is followed by an intermediate zone of larger blood vessels situated in the deeper layers of the corium (3). The peripheral vascular zone contains the largest vessels and forms a meshwork with vascular branches growing toward the peripheral parenchymal mammary structures and fat (4).

from longitudinal branches, which anastomose among each other, thus producing a cylindrical vascular envelope around the milk ducts (Dabelow, 1957).

2. Mammary Connective Tissue, Mesodermal Smooth Musculature, and Fat

The mammary connective tissue apparatus supports the glandular elements. It serves as a carrier for blood vessels and as a fiber grid for the embedded smooth musculature of nipple and areola mammae.

The connective tissue apparatus extending between skin and pectoral fascia consists of three layers:

1. A superficial layer of connective tissue is attached to the overlying corium and surrounds the primary mammary anlage and the ductus excre-

torii. This superficial connective tissue layer covers the whole area of the areola mammae, including the mesenchymal condensation zone of the smooth musculature of the areola mammae and the future nipple.

2. The intermediate zone of connective tissue surrounds the mammary ducts and it extends into peripheral connective tissue septa, which divide the endvesicles (alveoli) of the fetal mammary gland.

3. The inner connective tissue layer envelops and supplies the deepest epithelial cellular elements. Strandlike tips of connective tissue septa are attached to glandular elements and to fat located in the deeper zones of the breast. Around the time of birth and in the early postpartal days the newborn's mammary connective tissue displays considerable vascularization and fullness of blood capillaries. At this time the connective tissue also contains cellular elements such as erythroblasts, lymphocyte-like cells, larger round nuclear cells, myelocytes, leukocytes, mast cells, and histiocytes.

Mamilla (nipple) and areola mammae are free of terminal hair; they are rich in connective tissue fibers forming a diagonal rhomboid grid in which areolar glands (Montgomery glands, sebaceous glands) and rudimentary hair anlage are embedded. The connective tissue fibers located in the areola show the same circular-elliptic course as the smooth muscle fibers of that area. Accordingly, connective tissue and smooth muscle fiber bundles reveal a polar crossing at three and nine o'clock at the periphery of the areola mammae. The areolar smooth musculature is divided into inner, intermediate, and outer muscular layers (Figs. 4 and 5). The inner part of the smooth musculature surrounds the infundibular region of the mamillary milk ducts in a basketlike manner. The inner muscular fibers continue into the stronger intermediate layer, which is covered by an outer layer of smooth musculature situated around the ducts of the Montgomery glands. The inner basketlike layer as well as the outer circular-elliptic fibers merge into the connective tissue fiber grid of the surrounding skin.

Mammary fat is most likely formed by fibrillar connective tissue in which some of the cells have lost their fiber-forming capacity and have assumed the function of fat storage. Fat serves as matrix tissue for future parenchymal mammary growth (Dabelow, 1957; Langman, 1969).

C. THE MAMMARY GLAND OF THE NEWBORN AND THE CHILD

During pregnancy the abundantly secreted luteal and placental sex steroids partially enter the fetal circulation. Thus, the fetal mammary tissues become stimulated, and a hemispheroidal elevation of the mammary discs

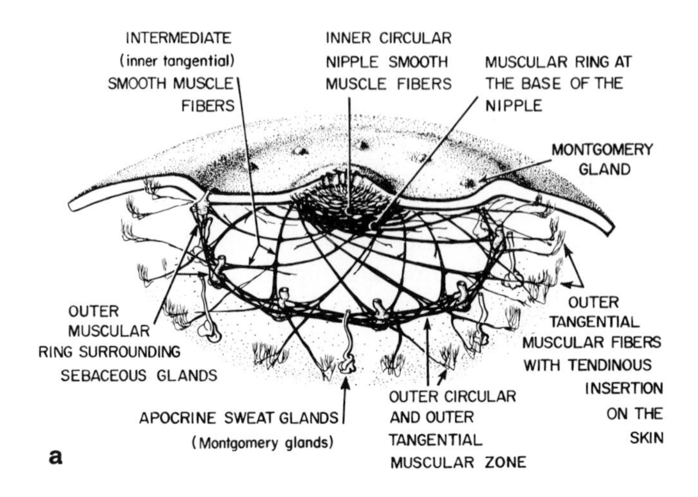

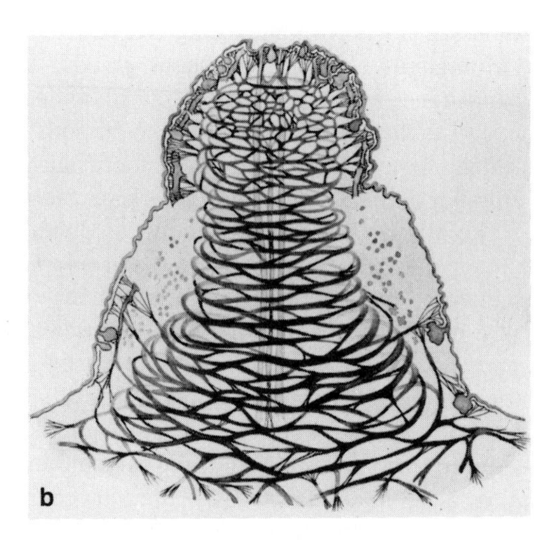

Fig. 4. Scheme for the smooth musculature arrangement of areola mammae and nipple (part a illustrated by Norviel based on Dabelow, 1957; part b reproduced from Dabelow, 1957). (a) Smooth musculature arrangement in the nonerected nipple. The smooth musculature of areola mammae and nipple represents a muscular-elastic fiber system. As revealed by thick tissue slices, the muscular system is arranged in a cone-like manner, with a rather spread-out basis. The muscle fibers around the nipple show a circular course, and from the nipple basis inner tangential, stronger fibers (intermediate muscle layer) leave to form a circumferential fiber meshwork at the periphery of the tissue fibers, thus, creating counterdirected double spirals before inserting as tendinous fibers on papillae of the superseated corium. This fiber arrangement allows a concentric diminution of the size of the areola mammae as necessary

may be observed in neonates of both sexes. In histologic section the peripheral lobuli of the newborn's mammary gland are enlarged, and the 1- to 2-layered glandular epithelium of the alveoli is secretorily active (Table 1 and Fig. 6). Accordingly, colostral milk ("witches' milk") can be expressed from the nipple of 80–90% of newborns of either sex, 2–3 days postpartum. This mammary colostral secretion of the newborn is caused by withdrawal of sex steroids, allowing prolactin to stimulate mammary alveolar epithelium for milk secretion, and reaches its maximum within 4–7 days postpartum. It subsides within 3–4 weeks when the newborn's pituitary prolactin secretion declines and glandular tissues, which were prenatally stimulated by maternal sex steroids, involute. These regressional necrobiotic cellular processes are minor compared to the involutional changes of the adult lactating mammae (Dabelow, 1957).

During the neonatal and early childhood period the endvesicles of the newborn's mammary gland are changed into ductular structures by some longitudinal growth and dichotomic branching. Throughout childhood, the rudimentary mammary tissues are quiescent, the mamilla does not protrude, and only slight branching of the primary mammary ducts, lined with a flattened epithelium, is observed.

D. MAMMARY GLAND DURING PUBERTY AND ADOLESCENCE

Puberty in the female sets in at the age of 10–12 years at the time of somatic maturation of the hypothalamus. Because of the incipient function of the hypothalamic gonadotropin-releasing hormones, the secretion of gonadotropins (FSH–LH) from the basophilic cells of the anterior pituitary gland is initiated. In turn, the released gonadotropins (FSH mainly) stimulate ovarian primordial follicles to mature into antral and, eventually, into Graafian follicles. The growing follicles secrete estrogens (17β-

---

for nipple erection and the act of suckling (see also Fig. 5). (b) Smooth musculature arrangement of the erected nipple. In the erected nipple the circular muscular fiber arrangement is accentuated, especially at the nipple base. Tangential fibers are emerging from the basal muscular ring, forming an outer muscular ring and then ending as tendinous insertions on corial papillae. This elastic tendinous skin insertion of smooth muscle fibers is also observed within the nipple. The function of the musculo-elastic system of areola mammae and nipple lies (a) in the erection of the mamilla, which is supported by dilation of the rich arteriovenous anastomotic network of the areola mammae and nipple base; and (b) in securing lactiferous ducts, sebaceous and sweat glands, and blood vessel of nipple and areola mammae against undue pressure, stretching, and tearing during the act of suckling.

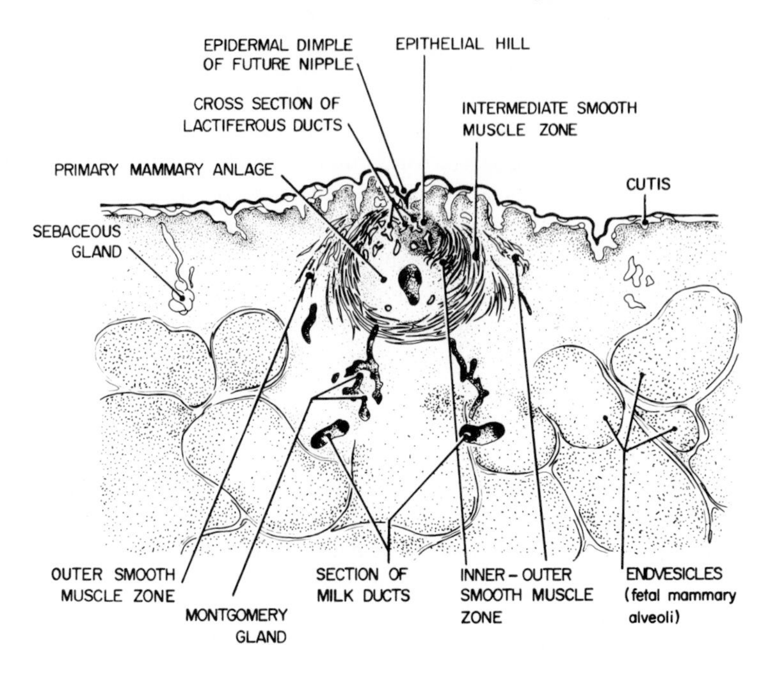

Fig. 5. Arrangement of the areola mammae smooth musculature—sagittal section of fetal mammary gland and preparation of 250 μm thick tissue slices stained with Resorcin Fuchsine (illustrated by Norviel, based on Dabelow, 1957). This drawing represents the morphology and topography of the areola mammae smooth musculature and the epidermal dimple of the future nipple area of the newborn human female mammary gland. This type of sectioning, as well as surface-near horizontal flat sections, show that circular-elliptic smooth muscle and connective tissue fibers surround the parenchymal structures. The smooth musculature of the areola mammae can be divided into three different layers, which are connected with each other and with the connective tissue fiber grid of the overlying skin. The inner muscular layer consists of gracile fibers enveloping the infundibular part of the lactiferous milk ducts in a basketlike manner. An intermediate zone of fewer thicker and longer muscular fibers functions as a tissue bridge, connecting the inner space with the outer smooth muscle zone via a relative muscular-free tissue space. The outer layer is composed of fibers with circular-elliptic course, lying around the ducts of Montgomery glands and impinging on the overlying fiber grid of the corial connective tissue. This smooth musculature arrangement of the newborn is basically the same as the muscular structure of nipple and areola mammae of the adult female breast.

estradiol mainly), which induce growth and maturation of the tissues of the genital organs and the breasts. According to recent investigations (Rifkind *et al.,* 1967; Yen *et al.,* 1969), using radioimmunoassays for determining hypophysial gonadotropins in urine and serum, the basophilic cells

TABLE 1

Mammary Gland of the Newborn

1. Stimulation of fetal mammary gland through transplacental sex steroids leading to ductular-alveolar-lobular proliferation
2. Secretion of colostrum in the newborn 4–7 days after birth and lasting 3–4 weeks due to withdrawal of placental sex steroids, allowing pituitary prolactin to act on mammary alveolar epithelium
3. Regression of hemispheroidal neonatal breast development into a small mammary disk of childhood occurring within 3–4 weeks postpartum

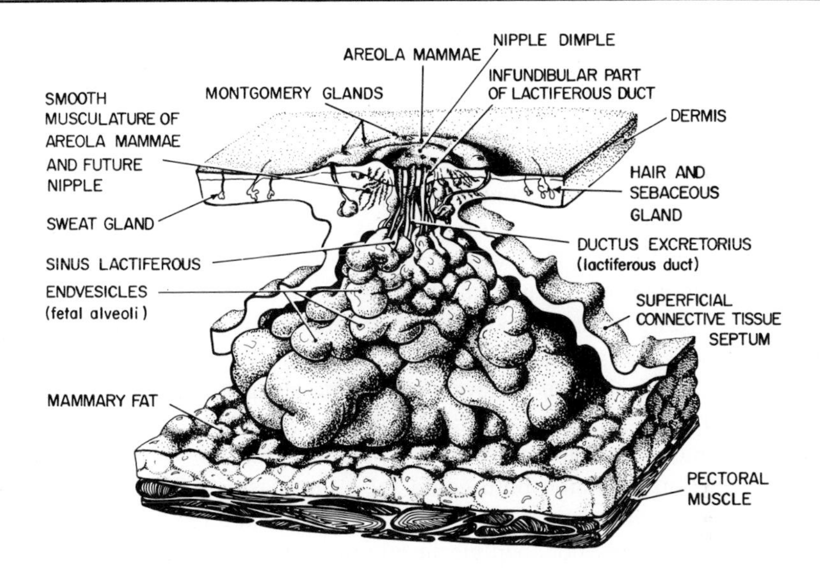

Fig. 6. The mammary gland of the newborn—schematic reconstruction from 300 μm thick slices stained with Alum Carmine, Resorcin Fuchsine, and Azan (illustrated by Norviel, based on Dabelow, 1957). At birth, the presence of colostrum-filled endvesicles (alveoli) governs the picture. The peripheral dichotomic ductular system merges into 15–25 larger ducts, running toward the indented surface of the areola mammae. Below the mammary dimple they form lactiferous sinuses and continue as ductus excretorii (lactiferous ducts) to reach the skin surface via the funnel-shaped ductular endings situated in the epidermal dimple of the future nipple. Also, the smooth musculature of areola mammae and future nipple is established. The areola mammae contains Montgomery glands, which are considered to be rudimentary milk ducts; neither sweat glands nor hair are found in the area of the areola mammae. The parenchymal elements are divided by a capillary-rich connective tissue, which also separates the mammary fat into small lobuli. A continuous subcutaneous connective tissue septum carrying blood vessels superfically and peripherally envelops the mammary gland and centrally continues into the fascia of the pectoral muscle.

of the anterior pituitary gland seem to secrete some FSH and LH before the onset of puberty. It is not clear, however, why the hypothalamic–anterior pituitary function increases distinctly at puberty. It appears that the hypothalamus becomes increasingly sensitive with progressing somatic maturation of the whole organism, and, thus, its function intensifies and certain inhibitory influences are removed. Because anterior pituitary FSH and LH are excreted during childhood, it is assumed that the hypothalamus secretes, albeit in minor amounts, gonadotropin-releasing hormones, which reach the pituitary gland through the portal veins. Recently, in animal studies, the existence of a hypothalamic hormone stimulating adenohypophysial gonadotropin synthesis has been suggested (Corbin and Milmore, 1971). Thus, the hypothalamus may possess a dual function in regulating not only release but also the synthesis of anterior pituitary gonadotropins. It is not understood to what extent at puberty the higher cortical centers of the brain contribute to hypothalamic stimulation of increased secretion of gonadotropin-releasing hormones. In 1969 the releasing hormone for TSH was chemically identified as a tripeptide containing the amino acids glutamine-histidine-proline (Schally *et al.,* 1969). In recent investigations a hypothalamic gonadotropin (FSH–LH) releasing hormone with the amino acid sequence glutamine-histidine-tryptophan-serine-tyrosine-glycine-leucine-arginine-proline-glycinamide could be identified (Schally *et al.,* 1971). The natural porcine polypeptide and the synthesized one shared the same biological properties; within 5–30 minutes after their i.v. or i.m. injection both could raise the plasma FSH and LH content in women with unovulatory cycles (Schally *et al.,* 1971; Winkelmann, 1971). Whether this nonapeptide in addition to its releasing activity also stimulates gonadotropin synthesis is not understood. Thus, it appears that the same hypothalamic nonapeptide, termed LH-releasing hormone, regulates release of both FSH and LH (White, 1970).

After the onset of puberty and during the period of adolescence, a phase of unbalanced hypothalamoadenohypophysial function is often observed. Consequently, in early puberty, the maturation processes of primordial follicles usually do not result in ovulation with formation of a corpus luteum. Therefore, only a small amount of progesterone is available during the first 1–2 years postmenarche, and, thus, the breast is stimulated predominantly by estrogens. As soon as ovulatory cycles begin to occur, progesterone is secreted from the corpus luteum, in addition to estrogens. Estrogens stimulate the growth of mammary ducts to elongate and reduplicate their epithelial lining (Fig. 7); they also induce proliferation of the terminal ductular parts, resulting in the formation of tissue sprouts and lobular buds for the future breast lobules. As a result of the effect of es-

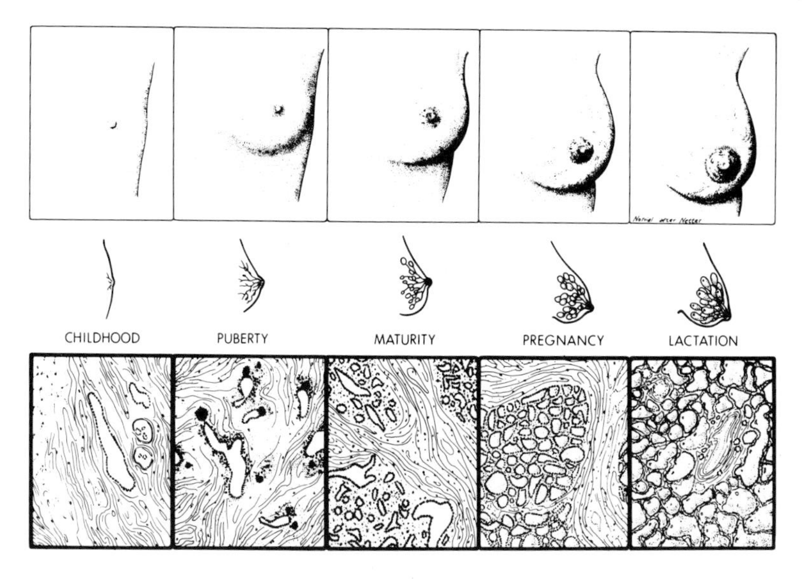

Fig. 7. The development of the female breast from childhood to maturity—mammary changes during pregnancy and lactation (from Vorherr, 1972b, by kind permission of the publisher). The breast of the female child consists of a few small rudimentary ducts lined with a flattened epithelium. With the onset of puberty, the mammary ductal tissues are stimulated by estrogens: they sprout, branch, and form terminal glandular tissue buds for the future development of alveoli and lobules. Ovulation and availability of progesterone from the corpus luteum together with estrogens produce the typical ductular-lobular-alveolar structure of the mature breast. In addition, both sex steroids induce growth of mammary connective tissue and deposition of fat, which contributes to the shape of the mature female breast. During pregnancy the ductular-lobular-alveolar mammary development is further stimulated and increased by enhanced amounts of luteal and placental sex steroids, as well as prolactin, placental lactogen and other maternal metabolic hormones. During pregnancy, areola mammae and nipple show more pigmentation and the Montgomery glands become more apparent. From midpregnancy on, alveoli and smaller milk ducts may contain some colostrum. From the beginning of pregnancy until term, the breasts, together, increase about 1½ pounds in weight and become rather firm and tight. Postpartum withdrawal of placental sex steroids triggers the onset of lactation with prolactin-induced milk synthesis and alveolar milk secretion. Thus, ⅓ of the breast volume during lactation may be due to the milk secreted into alveoli and smaller milk ducts.

trogens, the volume and elasticity of the periductal connective tissues increase; also, mammary vascularization and fat deposition become enhanced. In summary, estrogens are responsible for the first and major increase in mammary tissue in the young preadolescent female. The development of the breast into its adolescent size, firmness, and lobular-alveolar structure occurs under the combined influence of estrogens and progester-

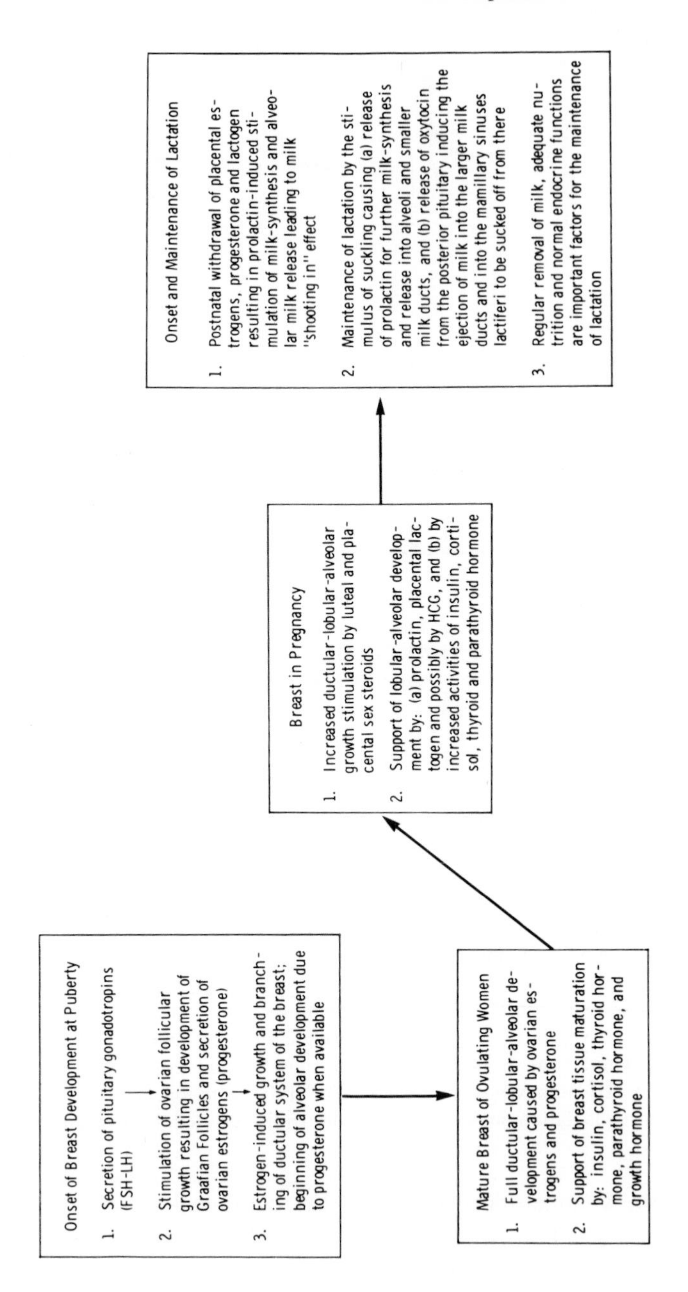

See legend on opposite page.

one. Also, permanent and typically dark-brown pigmentation of the mammary areolae and nipples is only observed when both female sex steroids are operative (Fig. 7 and scheme of Fig. 8) (Dabelow, 1957; Netter, 1965; Vorherr, 1968; Harley, 1969; Schwenk, 1969). In experiments with female immature rats, given either estrogens or progesterone or both sex steroids together over a period of 10 days, a pronounced ductular, sparsely ductular, and full ductular-lobular-alveolar development of the mammary glandular tissues, respectively, were induced (H. Vorherr, unpublished observations, 1971).

What are the various stages of breast development leading to maturity? At thelarche, at the age of 8–10 years, the mammary glandular tissue adjacent and beneath the areola mammae starts to grow, and on each breast a "mammary bud," the size of a cherry, develops; these developments are brought about primarily by estrogens. With the beginning of mammary epithelial tissue proliferation, fat is increasingly deposited and functions as a matrix for the growth of parenchymal elements (Table 2). This early stage of breast development is called "mamma-areolata," and it is recognized by increased tissue succulence and visible elevation of areola mammae and nipples. These initial changes of the breast's form and size encompass a time span of 2–3 years. The first recognizable enlargement of

---

Fig. 8. Development of female breast from puberty to lactation. Puberty of the female begins at the age of 10–12 years with the onset of hypothalamic maturation and the secretion of a hypothalamic gonadotropin-releasing hormone, which probably stimulates anterior pituitary gonadotropin synthesis and/or causes its release. The stimulation of ovarian primordial follicles by pituitary gonadotropins induces growth of primordial follicles and their maturation into Graafian follicles. The growth and development of breast and genital tissues are initiated by the availability of ovarian estrogens from maturing follicles and of progesterone, as soon as ovulation takes place. Whereas estrogens stimulate mainly the development of mammary epithelial ducts, both estrogens and progesterone are responsible for the alveolar growth and final ductular-lobular-alveolar structure of the mature female breast. During pregnancy, hormonal stimulation by estrogen and progesterone is achieved through increased luteal and placental sex steroid secretion. Placental lactogen and chorionic gonadotropin may also contribute to breast stimulation in pregnancy. Prolactin, insulin, cortisol, thyroid hormone, parathyroid hormone, and growth hormone provide additional support for optimal glandular mammary development and alveolar epithelial differentiation in connection with the upcoming process of lactation. In the second trimester some colostrum milk may already be found in mammary alveoli and pituitary prolactin appears to be responsible. With postnatal withdrawal of placental lactogen and sex steroids, prolactin-induced lactation is triggered. The baby's suckling brings about spurt release of prolactin (milk secretion) and secretion of oxytocin (milk ejection) through afferent neural impulses.

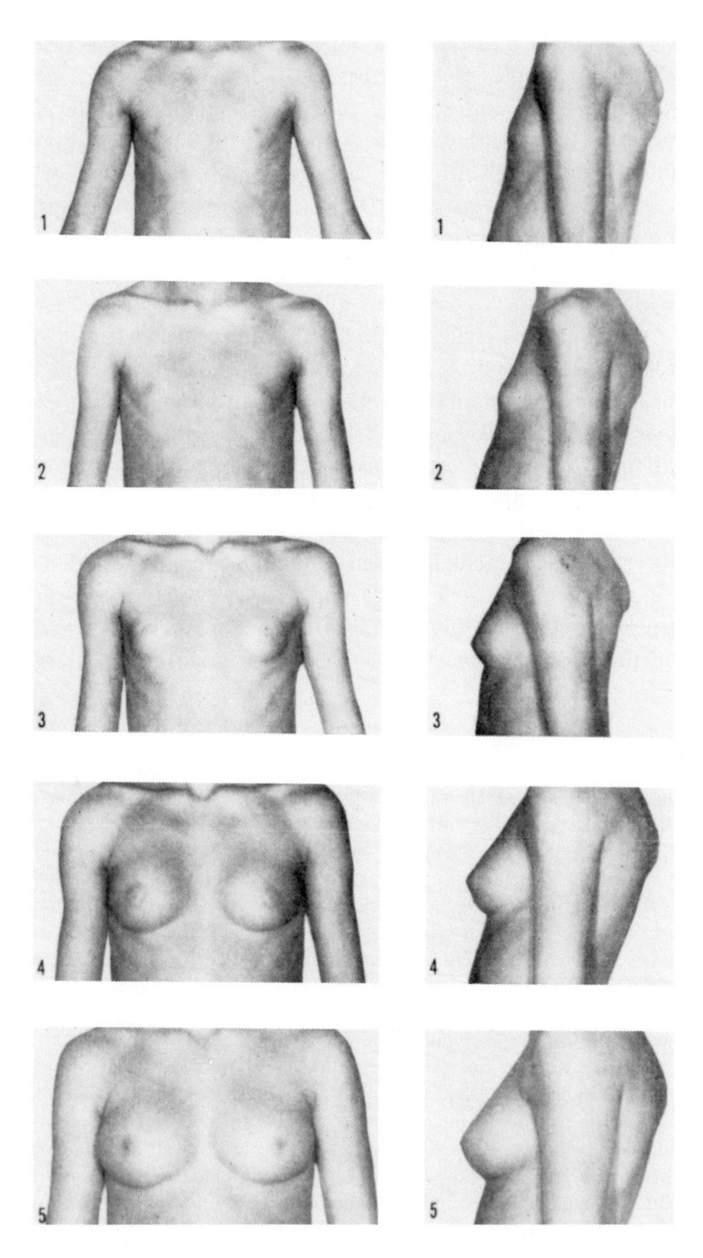

Fig. 9. The five phases of breast development from puberty to adulthood (from Tanner, 1962, by kind permission of the author and publisher). (1) In the prepuberal girl only elevation of the mamilla is recognized. (2) At the age of 10, under the influence of ovarian estrogens, the mammary tissues adjacent and beneath the areola

TABLE 2

Development of Mammary Gland from Puberty to Adolescence

1. Horizontal discoidal tissue growth and vertical spheroidal breast tissue development at onset of puberty due to ovarian estrogens. Dichotomic sprouting of primary and secondary mammary ducts, formation of connective tissue and fat
2. Mammary fat functions as matrix tissue allowing epithelial glandular proliferation
3. Mammary connective tissue provides the grid for vascular supply and parenchymal growth
4. Successive breast development from onset of puberty to adolescence:
   a. Slight elevation of mamilla
   b. Beginning of mamillary growth and sprouting of subareolar ductular breast tissues
   c. Progressing mamillary growth and beginning of nipple elevation. Visible development of globular breast tissue
   d. Distinct elevation of nipple and areola mammae from the globular-shaped breast
   e. Development of the mature dome-shaped breast under the combined influence of ovarian estrogens and progesterone. Mamilla is protruding; whereas the areola mammae levels with the contours of the whole breast

the breast disc usually occurs at the age of 10 years, and, thereafter, breast maturation proceeds. The puberal breast develops in a horizontal disclike fashion and in a vertical direction with dichotomic sprouting of primary and secondary milk ducts. Thus, the mammary ductal system of a nonmenstruating puberal girl appears to be more branched than that of a female 2–3 years after menarche because more connective tissue and fat have been deposited in the breasts at that stage. In the adolescent female the mammary ducts grow larger and continue to differentiate into lobular-alveolar structures. The evolution of the breast until maturity can be divided into five phases (Tanner, 1962) (Fig. 7 and 9):

---

area begin to grow, leading to an increased diameter of the areola ("mamma areolata") and to the formation of a "breast hill" ("mammary bud," "primary mount") around the age of 12 years. (3) Between years 12 to 14, due to increasing production of ovarian follicular estrogens, mammary growth becomes more intense and development of the nipple is apparent; there is further elevation of breast and mamilla but no distinct separation of their contours (intermediate stage). (4) Between years 14 to 15 the development of subareolar mammary tissues is accelerated, leading to elevation of nipple and areola mammae ("secondary mound") distinct from the globular shape of the rest of the breast ("stage of primary mamma"). (5) After the age of 15, due to ovarian follicular estrogens and luteal estrogens and progesterone, the definite shape of the mature breast is gradually formed. In the adult woman, only the mamilla protrudes whereas the areola mammae recedes to the general contour of the breast ("mature stage").

Phase 1: Preadolescent elevation of the mamilla (nipple) only

Phase 2: Beginning of mamillar growth and sprouting of subareolar ductular breast tissue leading to the formation and protrusion of a small "breast hill" and to an increase of the diameter of the areola mammae

Phase 3: Further growth of the nipple with intensified budding of breast tissues; the contours of areola and the whole breast remain unchanged

Phase 4: Progressing growth of the breast; accelerated development of subareolar tissues leading to elevation of nipple and areola, distinct from the globular shape of the breast

Phase 5: Development of the shape of the mature breast owing to estrogens and progesterone and other metabolic hormones (Fig. 8); in the adolescent breast only the mamilla protrudes whereas the areola mammae remains flush with the contours of the whole breast

E. CONGENITAL BREAST ANOMALIES

Accessory mammary glands (hypermastia) are phylogenetic remnants of the embryonic mammary ridge and are rather frequently observed (Fig. 1). Those supernumerous mammary glands may be located not only along the embryonic milk line but also around the urogenital region, on buttocks, and on the back. Aberration of breast tissue may involve corpus mammae, areola mammae, and nipple. Defective embryonic mammary development may result in complete absence of breast tissue (amastia). Mammary aplasia is a condition in which breast tissue is lacking; yet areola mammae and nipple are visible albeit they may sometimes be present only as small pigmented specks. In cases of hyperthelia or polythelia (in 1–7% of women) abundant, more or less developed, nipples are observed, but mammary tissue is lacking. Conversely, hyperadenia indicates the existence of mammary tissue without nipples. Frequently, a more or less pronounced difference in the size of right and left breast is noticeable.

(*Note:* In the following, the male mammary gland is discussed briefly for the sake of comparison to the female breast.)

At birth a slight transitory breast enlargement is observed in both sexes due to the effect of maternal transplacental sex steroids on fetal mammary tissue. Until puberty the mammary glands of both sexes are alike in structure. In 60–70% of males, increased mammary growth observed around puberty can be attributed to the unbalanced sex steroid metabolism in the juvenile organism, whereby transient increases of serum estradiol and/or an abnormally high estradiol/testosterone ratio is observed. It appears that such an unbalanced sex steroid metabolism leads to increased conversion of androgens into estrogens. Increased plasma sex-hormone-binding globulin levels leading to a decline in free plasma testosterone may also be a cause for pubertal gynecomastia. This condition, however, subsides within the following 3 years when synthesis, plasma binding, and metabolism of male sex steroids normalize. At senescence, enlargement (proliferation of ductular structures) of breast tissues occurs in some males; it may be caused by a dysfunction of androgen production or testosterone metabolism with preponderance of estrogen activity.

The corpus mammae in the adult male measures about 2 cm in diameter and 4 mm in thickness. The parenchyma consists of short rudimentary ducts only, with no true acini. Thus, the male breast represents a small structure composed of fat and connective tissue and interspersed with a few small ducts, which may be reduced to cords or epithelial strands. Therefore, the glandular body appears whitish in color and is of hard consistency. Blood supply, innervation, and lymphatics are similar to that of the female breast. Areola and nipple of the male breast are amply supplied with free nerves and tactile corpuscles of Meissner (stratum papillare), end bulbs of Krause (stratum papillare), tactile elements of Vater-Pacini (subcutis), and Ruffini-like endings (subcutis) (for the female breast see Chapter II). The topography of the male mammary gland is the same as that of the female gland (nipple is positioned at fourth intercostal space, 12 cm from the median line). In profile the male breast shows hardly any elevation above the surrounding tissues, except at the nipple (3–4 mm in height). Areola and nipple are pigmented; the areola has a diameter of 2–3 cm. Areolar tubercles of Morgagni (orifices of Montgomery glands) may be recognized, and areolar and nipple smooth musculature is present. Also, in the male, hypermastia occurs (Kopsch, 1955; Dabelow, 1957).

CHAPTER II

Morphology of the Mature Female Breast

A. GROSS ANATOMY OF THE BREAST

Stimulated primarily by ovarian estrogens and progesterone, the breast tissues mature and display a characteristic ductular-lobular-alveolar structure (Fig. 10). This process of maturation is supported by other hormones such as insulin, cortisol, thyroid hormone, parathyroid hormone, growth hormone, and probably prolactin. It has been suggested that pituitary prolactin may be secreted increasingly during the second half of the cycle in menstruating women (Simkin and Goodart, 1960), but this could not be confirmed by recent investigations using a radioimmunoassay for prolactin (Jaffe, 1971).

The differentiation of breast tissues as found in the female adult mammary gland is accomplished around the age of 14–15 years. However, the mammary lobular-acinar development continues up to 30 years due to cyclic ovarian sex steroid stimulation (Dabelow, 1957). Under these sustained hormonal influences, the growth of connective tissue and the deposition of fat are maintained, contributing to the appearance of the mature female breast (Figs. 7 and 10).

1. Topographical Anatomy and Morphology of the Mature Breast

The adolescent breast is dome-shaped, and the areola mammae, a circular pigmented skin area of 2–3 cm in diameter, is located in its center (Table 3). Generally, the taller the woman the more cranially the breasts

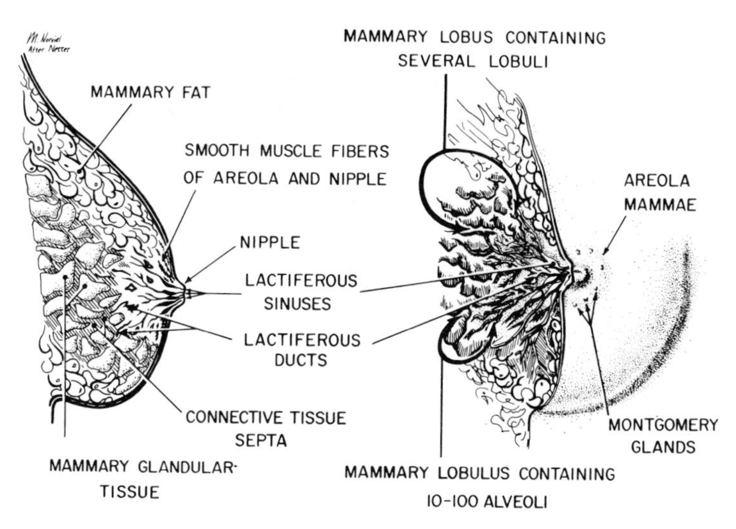

Fig. 10. Morphology of the mature female breast (from Vorherr, 1972b). The adolescent female breast (sagittal section on left side of the picture; partly frontal section and view on right side of picture) consists of 15–25 lobi, which are subdivided into lobuli and alveoli. During lactation, milk is secreted into the lumina of alveoli and smaller milk ducts. Due to the effect of endogenous oxytocin released during suckling, milk is ejected into the larger milk ducts and lactiferous sinuses from which it is removed by the suckling infant. The constitutionally and hormonally regulated volume of mammary glandular and connective tissues as well as of fat, are major factors for the size and appearance of the mature breast. Subareolar tissues and nipples contain smooth muscle fibers inducing erection of the nipples.

are located at the thorax. The sulcus between the two breasts is called "sinus mammarium" (bosom). Small ducts from "aberrant" or "accessory" mammary glands (large sebaceous glands), also called "Montgomery glands," open into the area of the areola mammae (Fig. 11). They are different from areolar subcutaneous sweat (partly of apocrine type) and free sebaceous gland. Montgomery glands are free of terminal hair, only papillas of lanugo hair may be present; usually no other hair exists within the area of the areola mammae and nipple. A significant increase in the size of Montgomery (large sebaceous glands) and other sebaceous glands of the areola mammae is observed during pregnancy and lactation (lubrication of nipple). Miniature ducts of Montgomery glands open into small sinuses of Morgagni's tubercles in the epidermis of the areola, and these ducts secrete milk during lactation. Montgomery glands probably represent an intermediate stage between sweat and mammary glands and are stimulated in the fetus by transplacental sex steroids; these changes regress within 3–4 neonatal weeks. At pubescence the growth of Montgomery

TABLE 3

Topographic Anatomy and Morphology of the Mature Breast

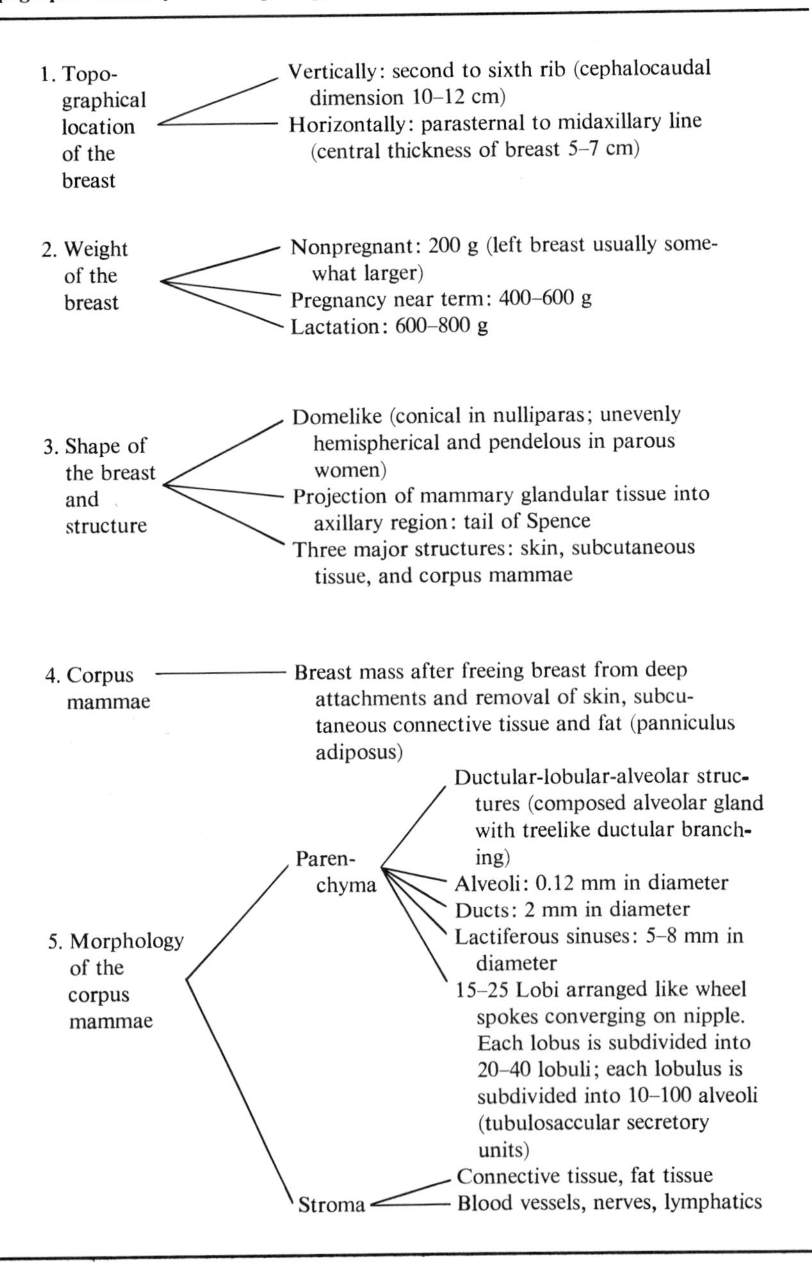

1. Topographical location of the breast		Vertically: second to sixth rib (cephalocaudal dimension 10–12 cm)
		Horizontally: parasternal to midaxillary line (central thickness of breast 5–7 cm)
2. Weight of the breast		Nonpregnant: 200 g (left breast usually somewhat larger)
		Pregnancy near term: 400–600 g
		Lactation: 600–800 g
3. Shape of the breast and structure		Domelike (conical in nulliparas; unevenly hemispherical and pendelous in parous women)
		Projection of mammary glandular tissue into axillary region: tail of Spence
		Three major structures: skin, subcutaneous tissue, and corpus mammae
4. Corpus mammae		Breast mass after freeing breast from deep attachments and removal of skin, subcutaneous connective tissue and fat (panniculus adiposus)
5. Morphology of the corpus mammae	Parenchyma	Ductular-lobular-alveolar structures (composed alveolar gland with treelike ductular branching)
		Alveoli: 0.12 mm in diameter
		Ducts: 2 mm in diameter
		Lactiferous sinuses: 5–8 mm in diameter
		15–25 Lobi arranged like wheel spokes converging on nipple. Each lobus is subdivided into 20–40 lobuli; each lobulus is subdivided into 10–100 alveoli (tubulosaccular secretory units)
	Stroma	Connective tissue, fat tissue
		Blood vessels, nerves, lymphatics

TABLE 3 *(continued)*

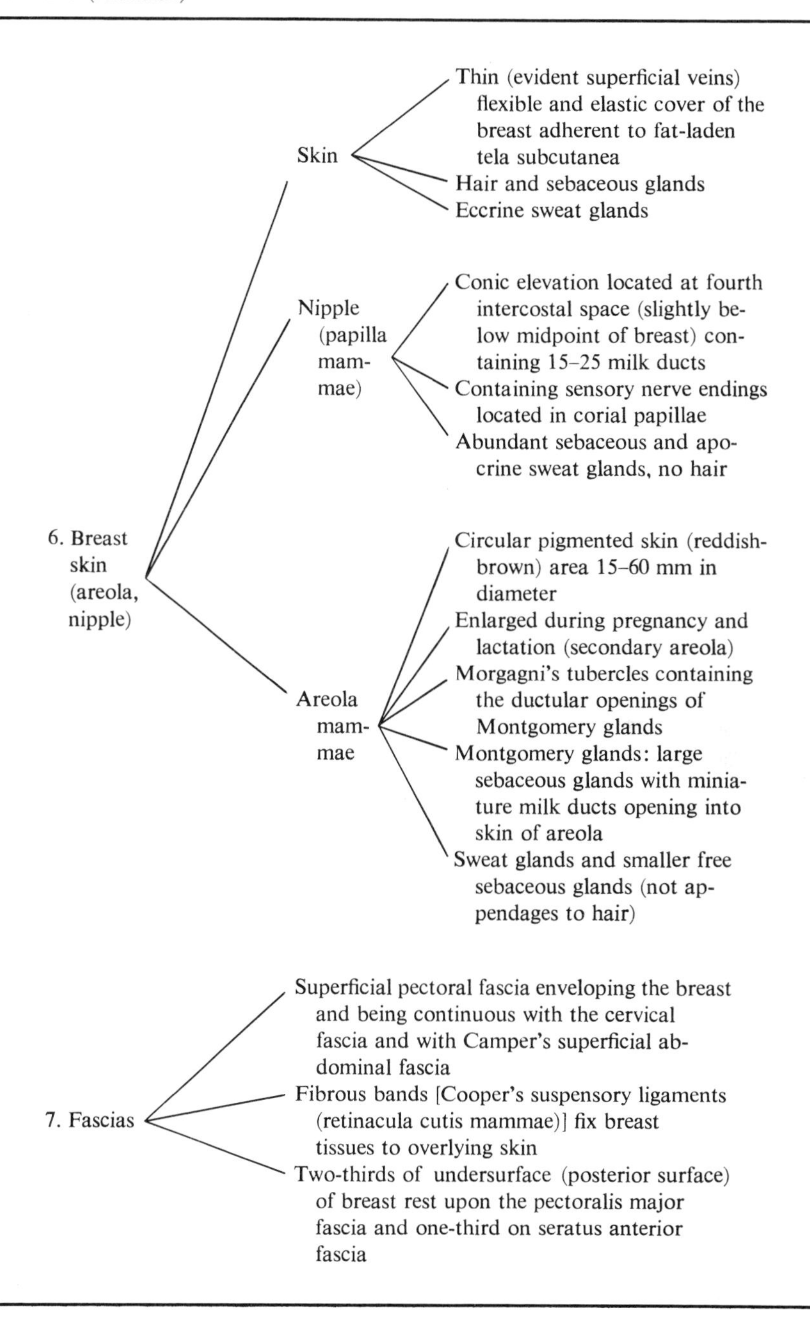

Structure	Component	Features
6. Breast skin (areola, nipple)	Skin	Thin (evident superficial veins) flexible and elastic cover of the breast adherent to fat-laden tela subcutanea
		Hair and sebaceous glands
		Eccrine sweat glands
	Nipple (papilla mammae)	Conic elevation located at fourth intercostal space (slightly below midpoint of breast) containing 15–25 milk ducts
		Containing sensory nerve endings located in corial papillae
		Abundant sebaceous and apocrine sweat glands, no hair
	Areola mammae	Circular pigmented skin (reddish-brown) area 15–60 mm in diameter
		Enlarged during pregnancy and lactation (secondary areola)
		Morgagni's tubercles containing the ductular openings of Montgomery glands
		Montgomery glands: large sebaceous glands with miniature milk ducts opening into skin of areola
		Sweat glands and smaller free sebaceous glands (not appendages to hair)
7. Fascias		Superficial pectoral fascia enveloping the breast and being continuous with the cervical fascia and with Camper's superficial abdominal fascia
		Fibrous bands [Cooper's suspensory ligaments (retinacula cutis mammae)] fix breast tissues to overlying skin
		Two-thirds of undersurface (posterior surface) of breast rest upon the pectoralis major fascia and one-third on seratus anterior fascia

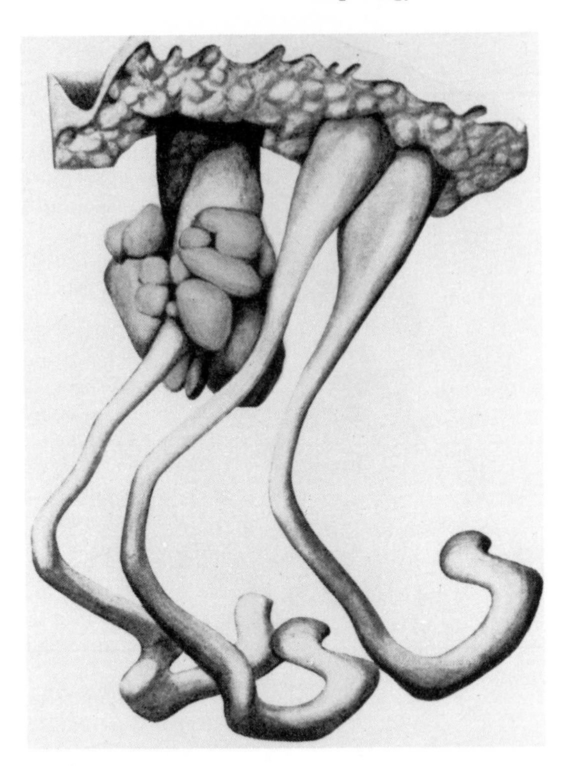

Fig. 11. Apocrine Montgomery glands, their connection with holocrine, sebaceous glands and hair follicles (from Dabelow, 1957). Montgomery glands are phylogenetically derived from apocrine sweat glands. Before ending on the skin surface in the form of a funnel-shaped dilated terminal duct, the Montgomery glands are more or less intensely connected with sebaceous glands and hair papillae opening into the infundibular part of the Montgomery ducts; this is shown in the left of the three depicted Montgomery glands. Maternal Montgomery and sebaceous glands undergo hypertrophy during pregnancy, and also in the newborn the Montgomery glands are hormonally stimulated. The physiologic function of these glands is not clear. It has been suggested that apocrine mammary glands are analogous to the accessory tissue of sexual glands with the function of a scent organ. Also lubrication of areola mammae and nipple surface has to be considered as a functional aspect. Whereas Montgomery glands undergo postmenopausal involution, sebaceous glands seem not to be affected by these regressional processes.

glands is stimulated again. After the menopause the Montgomery glands undergo senile involution, whereas the sebaceous glands show no signs of regression. The corium of the areola mammae lacks fat, but it contains smooth muscle and collagenous and elastic connective tissue fibers in radial and circular arrangements. The contraction of smooth muscle fibers causes erection of the nipple. This process of erection is supported by local venous stasis and hyperemia because mamilla and areola mammae are rich in arteriovenous anastomoses. Nipple erection is induced by tactile sensory or autonomic sympathetic stimuli. The dermis of the nipple and areola mammae contains a large number of multibranched free nerve fiber endings, Ruffini-like bodies, and end bulbs of Krause. Thus, through stimulation of these sensory nerve endings, nipple erection and afferent neural and efferent humoral hypothalamo-pituitary reflex mechanisms are triggered, resulting in release of prolactin and oxytocin with their subsequent effect on the mammary gland (for more details see Breast Innervation, Chapter II, Section A, 4). The nipples are located in the center of the darker pigmented areola mammae and are elevated a few millimeters above the skin level. Only the tip of the nipple is not pigmented. Within the subcutaneous tissues of the nipple, numerous sebaceous and apocrine sweat glands are located; hair is absent. The corium of the nipple skin lacks fat; its texture is loose. The cylindric to bluntly conic nipples are located slightly below the center of each breast and about 12 cm from the thoracic midline; they point laterally and cranially. Each nipple contains 15–25 lactiferous ducts surrounded by fibromuscular tissue and covered by wrinkled skin containing large corial papillae. These ducts end as small orifices near the tip of the nipple. Within the nipple the lactiferous ducts may merge, and, therefore, the ductular orifices are often fewer in number than the respective breast lobi. The milk ducts within the nipple dilate at the nipple base in cone-shaped ampullae of the milk sinuses. Containing epithelial debris in nonlactating women, the ampullae function as temporary milk containers during lactation. Cranially, beyond the ampullae, the lactiferous ducts resume a normal width, and then the infundibular part of the duct ends, funnel-shaped, in the slightly indented surface of the nipple. Thus, the lining of the infundibular and ampullar part of the lactiferous ducts consists of an 8- to 10-cell-layered squamous epithelium. The bulk of the nipple is composed of smooth musculature, which represents a closing mechanism for milk ducts and milk sinuses of the nipple. This elastic-muscular system of the areola mammae and nipple forms an interdigitating meshwork and impinges on the overlying cutis (Fig. 4b). Because the milk ducts situated in the nipple are embedded in stretchable and mobile connective tissue (collagenous and elastic fibers), the inner more longitudinal muscular arrangement (fibers parallel to lactiferous ducts and interlaced in various directions) and the outer more circular and radial one will not greatly obstruct the milk ducts; albeit the fibromuscular arrangement functions as some sort of a "nipple sphincter." Tangential fibers also branch off from the more circular muscle fibers of the nipple basis to the outer more circular muscular ring of the areola mammae. The function of the muscular fibroelastic system of areola and nipple consists of decreasing the area of the areola mammae during nursing, nipple erection, and emptying of the lactiferous sinuses and ducts. Through nipple erection, which causes the mamilla to become smaller and firmer, the milk sinuses are emptied. Partly within the nipple and immediately below it, at its base, the lactiferous ducts dilate to form the lactiferous sinuses. The latter are the continuation of the mammary ducts coming from 15 to 25 breast lobi (Fig. 10). These lobi are surrounded by connective tissue, and each lobus is composed of many lobuli. Again, each lobulus is subdivided into 10–100 alveoli (acini), which are enveloped by a collagen sheath forming the basement membrane, the prolongation of the latter invests the collecting duct. A lobule containing several acini is enclosed by a somewhat thicker collagen envelope. Mammary alveoli are of tridimensional, circular to ovoid form and represent evaginations of terminal milk ducts. The alveolar walls of the virgin mammae lie close to each other and consist of a basement membrane with a sparse outer connective tissue layer; the inner surface of the basement membrane is covered with starlike myoepithelial cells and mammary epithelium.

The upper and central portion of the breast is predominantly composed of glandular tissue, which probably accounts for the higher incidence of breast cancer in the upper, outer quadrant. From a tonguelike mammary glandular tissue projection, the axillary tail of Spence, remarkably, passes through an opening in the axillary fascia into the axillary area; here, the mammary tissue is in direct contact with axillary lymph nodes lying beneath the axillary fascia. While the glandular tissue of the tail of Spence may visibly enlarge premenstrually, it enlarges even more during pregnancy and lactation. The mammary tissues are enveloped by the superficial pectoral fascia, and the breast is fixed by fibrous bands to the overlying skin and the underlying pectoral fascia (Cooper's ligaments); when invaded by cancer cells, the Cooper's ligaments become shortened and thickened, thus causing dimpling of the overlying skin. The deep layer of the superficial pectoral fascia at the base of the breast is separated from the deep pectoral fascia by a cleft, which allows considerable mobility of the breast. The deep surface of the pectoralis major muscle is covered by a strong clavipectoral fascia, which envelops the pectoralis minor muscle; the anterior thoracic vessels and nerves, as well as the cephalic vein, pass through the latter. The clavipectoral fascia also envelops the axillary vessels by forming a vascular sheath. The glandular part of the breast is surrounded by a fat layer belonging to the panniculus adiposus, and it seldom extends beyond the lower borderline of the pectoralis major. The surrounding mammary fat is framed by a septal system of connective tissue, which is derived from the connective tissue portion of the retinacula cutis. Also the mammary glandular connective tissue septa are in close contact with the overlying corium, serving together with the superficial pectoral fascia as support for the breast (Fig. 10 and Table 3) (Kopsch, 1955; Dabelow, 1957; Netter, 1965; Zilliacus, 1967; Leis, 1970).

2. Mammary Blood Supply

The major blood supply of the breast is provided by the internal mammary artery and the lateral thoracic artery (Fig. 12). The intercostal arteries play a subordinate role in this respect, and the participation of arterial branches of the axillary and subclavian arteries in mammary blood supply is minimal (Table 4).

About 60% of the total breast tissue (medial parts mainly) receives blood from the internal mammary artery (internal thoracic artery). The thoracic portion of this artery lies behind the cartilage of the six upper ribs just outside the parietal layer of the pleura. Its anterior rami mammarii pass through the second to fifth intercostal space 1–2 cm parasternally to reach the medial aspects of the breast. Usually either the first and fourth or the second and third branch (most often) of the internal mammary artery are thickest. All mammary branches lead transversely to the nipple and anastomose extensively with branches coming from the lateral thoracic artery. Anastomoses with the intercostal arteries occur less frequently or may even be absent. Some branches of the internal mammary artery may reach to the opposite breast. The lateral thoracic artery comes from the subclavian artery and descends along the lateral margin of the breast (lower border of the minor pectoral muscle); from it the external mammary artery diverges and turns around the edge of the pectoralis major. In rare instances the lateral thoracic artery may branch off from the thoracoacromial artery. Through its medial and lateral rami, the lateral thoracic artery provides blood to the lateral quadrants of the breast but chiefly to the upper outer quadrant. The outer lower quadrant is additionally supplied by anterior and lateral rami of the third, fourth, and fifth posterior intercostal arteries, derived from the descending thoracic aorta. The thoracoacromial, suprascapular, upper thoracic, and thoracodorsal arteries contribute only insignificantly to mammary blood supply (Pernkopf, 1964; Weitzel and Bässler, 1971).

Most arterial mammary vessels run in transverse and cranial directions; therefore, the upper half of the breast is vascularized best. The main arteries pass into the breast at its border course near the surface and then run in the plane of the outer limit of the fat lobules. From here on they undergo extensive branching and form anastomoses with skin and the glandular body. Thus, most parts of the mammary gland are supplied by two

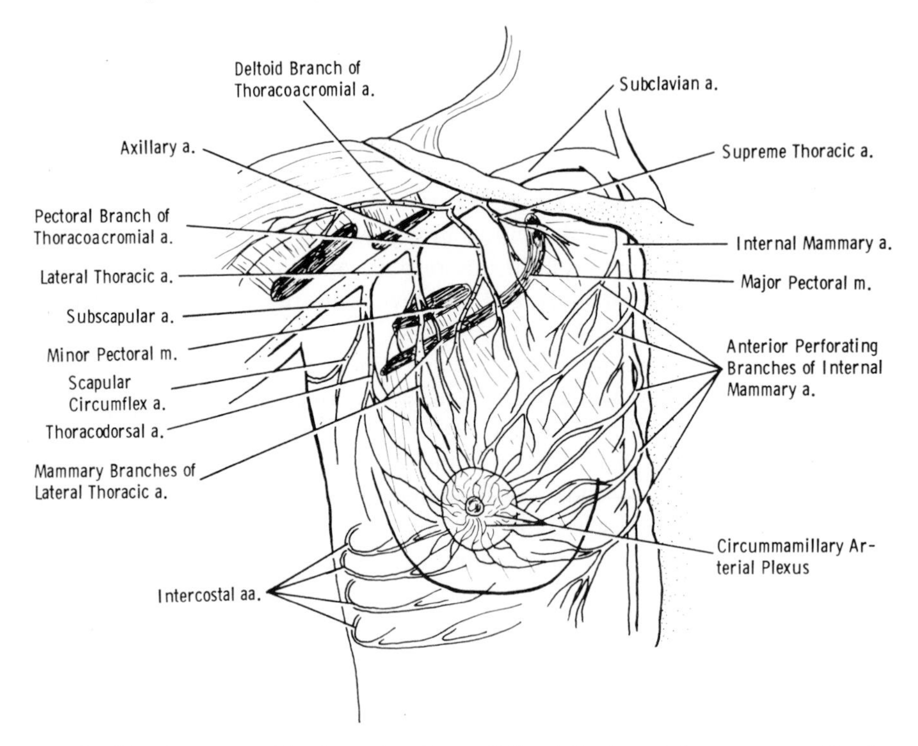

Fig. 12. Mammary blood supply. About 60% of the breast tissues receive blood from the anterior perforating branches of the internal mammary artery (internal thoracic artery) which enters the breast medially. By anastomoses of branches of the internal mammary artery with those of the lateral thoracic artery the circummamillary arterial plexus is formed. A similar arterial plexus is formed by these vessels in the deeper region of the corpus mammae; thus, many areas of the breast are supplied by two or even three different arterial sources. Approximately 30% of the mammary blood supply is provided by the lateral thoracic artery, which runs downward along the anterolateral thoracic wall, not far from the border of the pectoralis minor muscle. The mammary branches of the thoracic artery approach the breast tissues in the region of the upper outer quadrant laterally and from behind. Anterior and lateral branches from the third to fifth posterior intercostal arteries, which enter the breast laterally, also contribute to mammary blood supply (mainly lower outer quadrant of the breast). The anterior rami of the posterior intercostal arteries anastomose with the anterior perforating branches of the internal mammary artery. Additional minor sources of mammary blood supply are thoracodorsal (upper medial branches), thoracoacromial (pectoral branch), and supreme thoracic (upper peripheral breast tissues) arteries.

TABLE 4 Mammary Blood Supply

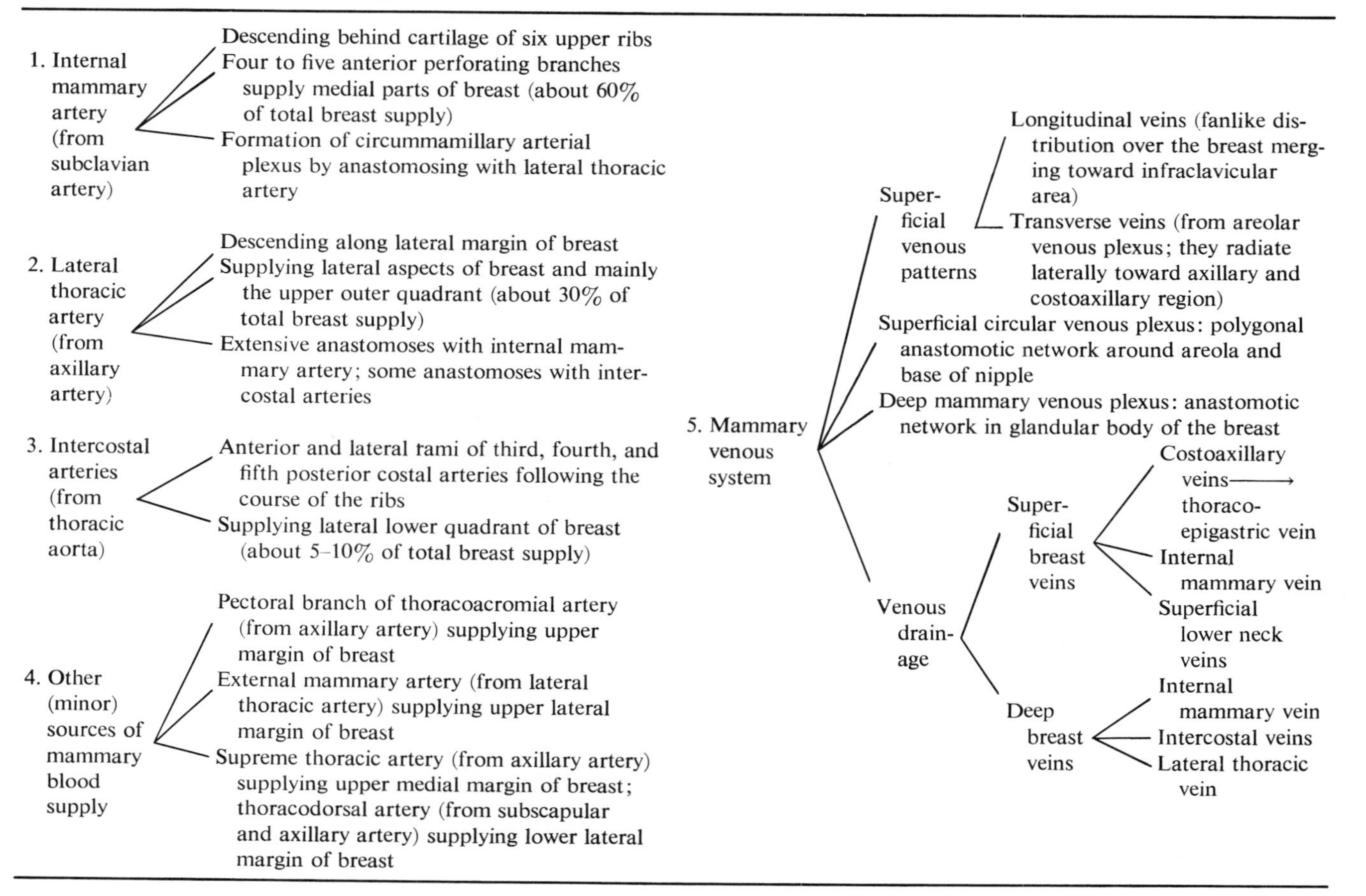

1. Internal mammary artery (from subclavian artery)
   - Descending behind cartilage of six upper ribs
   - Four to five anterior perforating branches supply medial parts of breast (about 60% of total breast supply)
   - Formation of circummamillary arterial plexus by anastomosing with lateral thoracic artery
2. Lateral thoracic artery (from axillary artery)
   - Descending along lateral margin of breast
   - Supplying lateral aspects of breast and mainly the upper outer quadrant (about 30% of total breast supply)
   - Extensive anastomoses with internal mammary artery; some anastomoses with intercostal arteries
3. Intercostal arteries (from thoracic aorta)
   - Anterior and lateral rami of third, fourth, and fifth posterior costal arteries following the course of the ribs
   - Supplying lateral lower quadrant of breast (about 5–10% of total breast supply)
4. Other (minor) sources of mammary blood supply
   - Pectoral branch of thoracoacromial artery (from axillary artery) supplying upper margin of breast
   - External mammary artery (from lateral thoracic artery) supplying upper lateral margin of breast
   - Supreme thoracic artery (from axillary artery) supplying upper medial margin of breast; thoracodorsal artery (from subscapular and axillary artery) supplying lower lateral margin of breast
5. Mammary venous system
   - Superficial venous patterns
     - Longitudinal veins (fanlike distribution over the breast merging toward infraclavicular area)
     - Transverse veins (from areolar venous plexus; they radiate laterally toward axillary and costoaxillary region)
   - Superficial circular venous plexus: polygonal anastomotic network around areola and base of nipple
   - Deep mammary venous plexus: anastomotic network in glandular body of the breast
   - Venous drainage
     - Superficial breast veins
       - Costoaxillary veins ⟶ thoraco-epigastric vein
       - Internal mammary vein
       - Superficial lower neck veins
     - Deep breast veins
       - Internal mammary vein
       - Intercostal veins
       - Lateral thoracic vein

to three different vascular sources. The arterial mammary network can be divided into a superficial circular plexus located around the areola mammae and the upper half of the breast and into a deep plexus lying within the corpus mammae. Whenever possible, breast incisions should be placed below the nipple in the lower quadrants to preserve blood supply and to make scars less visible. By and large, mammary venous drainage follows the arterial distribution pattern. The superficial subcutaneous veins form an extensive polygonal anastomotic network around the areola mammae and the base of the nipple; during lactation this plexus and the fanlike distributing superficial veins of the upper half of the mammary gland are enlarged and visible through the skin. These subcutaneous veins return the blood to deeper veins that run along the respective arteries. Superficial breast veins drain into branches of internal mammary veins, into superficial veins of the lower part of the neck, and some may even reach the cephalic vein. The deep breast veins return the blood into internal mammary, lateral thoracic, and intercostal veins; the latter are connected with the paravertebral venous plexuses (Table 4).

3. Mammary Lymphatic Drainage

The lymphatics of the breast originate in the lymph capillaries of the mammary connective tissue grid, which surrounds the parenchymal mammary structures. Lymph capillaries are similar to blood capillaries and are abundant in the breast. Several lymph capillaries unite to form larger lymphatic vessels, which resemble veins in structure but have thinner walls. Lymphatics are provided with valves to assure lymph flow in the direction of the venous system, i.e., away from tissues. Lymphatic vessels empty into lymph nodes, which represent collections of lymphocytes and their precursors held together by connective tissue; lymph nodes possess afferent and efferent lymphatics. When carcinomatous cells invade the lymphatic system, they reach the lymph nodes which act as a filter and retain the malignant cells. Within a lymph node the filtered cells grow at the expense of the node and gradually destroy it. Because, for a while, malignant cells do not migrate farther to other lymph nodes or into the venous system, cancer can often be eradicated entirely by removal of the primary lesion and the involved lymph nodes. Because 5 to 6% of all women suffer from breast cancer during their life cycle, the lymphatic drainage of the breast is of particular clinical importance. Some of the lymphatic mammary connections have been traced by clinical pathological studies rather than by anatomical dissections.

The lymph drainage of the breast consists of three main parts: (1) cutaneous, or superficial; (2) areolar; and (3) glandular or deep tissues (Table 5 and Fig. 13) (Gardner *et al.*, 1969).

a. SUPERFICIAL (CUTANEOUS) LYMPH DRAINAGE

Lymph from the skin of the breast, with the exception of that of the areola and nipple, collects in vessels that lead into the axillary glands of the ipsilateral side. Some cutaneous lymphatics of the upper part of the breast run along the thoracoacromial artery to reach the deeper region of the axilla. Some superficial lymphatics of the medial aspects of the breast may cross the midline and reach the opposite breast and axilla (cross- or trans-mammary pathway). From the lower border of the breast, cutaneous lymph vessels may pass to the lymphatic plexus of the epigastric region on the sheath of the rectus abdominis (Gerota's paramammary pathway); from here, lymphatics cross the abdominal wall, and the liver or abdominal lymph plexuses may be reached via subdiaphragmatic nodes (Fig. 13).

b. AREOLAR LYMPH DRAINAGE

The lymph formed in tissues of the areola mammae and nipple passes into the subareolar plexus of Sappey, which is drained by two major lymph vessels. The cutaneous inner (medial) and outer (lateral) mammary lymph channels unite laterally in the breast, and lymph flows through the newly formed duct into the pectoral nodes (anterior axillary nodes) (Fig. 13).

c. DEEP (GLANDULAR) LYMPH DRAINAGE

Most of the lymph produced in the corpus mammae flows into the anterior axillary (pectoral) nodes, which are located under the anterior axillary fold along the course of the lateral thoracic vein (region of the third rib). From the anterior axillary nodes, lymphatics drain into the central axillary nodes, which are situated in the fat of the upper part of the axilla along the border of the axillary vein. Enlargement of the central axillary lymph glands may exert pressure on the intercostohumeral nerve and thus cause pains in the axilla and upper inner arm. From the central axillary nodes, some lymphatics run to the deep axillary lymph glands (apical, subclavicular, or intraclavicular glands), which lie partly behind the costocoracoid membrane. These deep lymph nodes become continuous with the deep cervical lymph glands, which are located in the supraclavicular area. Lymphatics also drain from the central axillary nodes into the subscapular

TABLE 5

Mammary Lymphatic Drainage

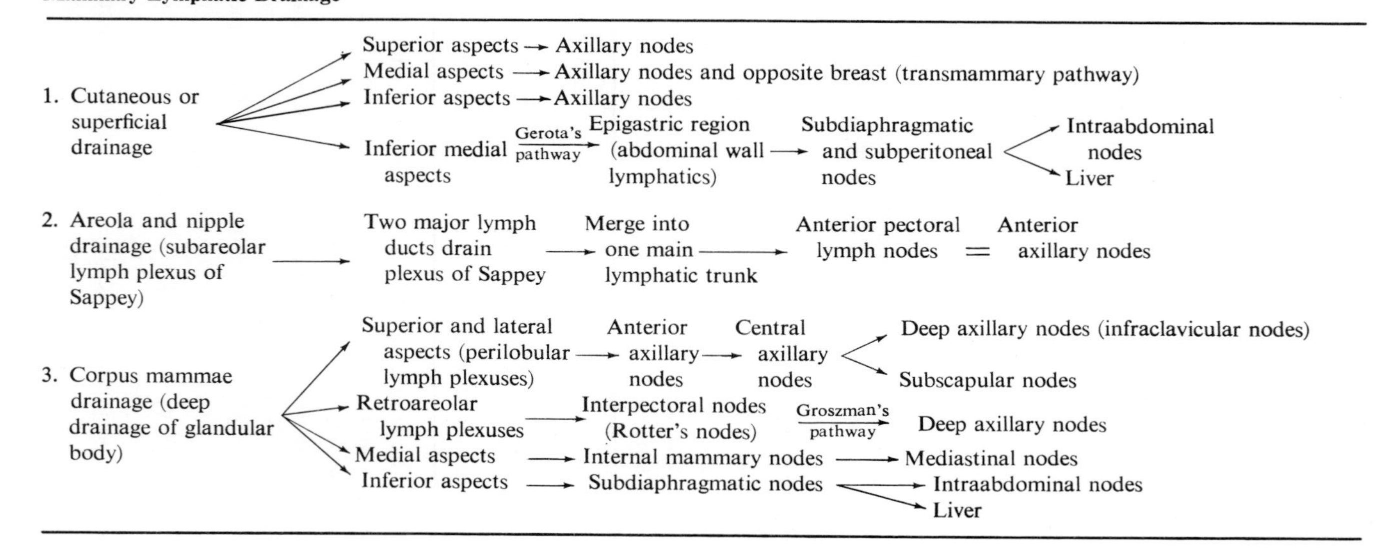

1. Cutaneous or superficial drainage	Superior aspects →	Axillary nodes		
	Medial aspects →	Axillary nodes and opposite breast (transmammary pathway)		
	Inferior aspects →	Axillary nodes		
	Inferior medial aspects (Gerota's pathway) →	Epigastric region (abdominal wall lymphatics) →	Subdiaphragmatic and subperitoneal nodes →	Intraabdominal nodes; Liver
2. Areola and nipple drainage (subareolar lymph plexus of Sappey) →	Two major lymph ducts drain plexus of Sappey →	Merge into one main lymphatic trunk →	Anterior pectoral lymph nodes = Anterior axillary nodes	
3. Corpus mammae drainage (deep drainage of glandular body)	Superior and lateral aspects (perilobular lymph plexuses) →	Anterior axillary nodes →	Central axillary nodes →	Deep axillary nodes (infraclavicular nodes); Subscapular nodes
	Retroareolar lymph plexuses →	Interpectoral nodes (Rotter's nodes) (Groszman's pathway) →	Deep axillary nodes	
	Medial aspects →	Internal mammary nodes →	Mediastinal nodes	
	Inferior aspects →	Subdiaphragmatic nodes →	Intraabdominal nodes; Liver	

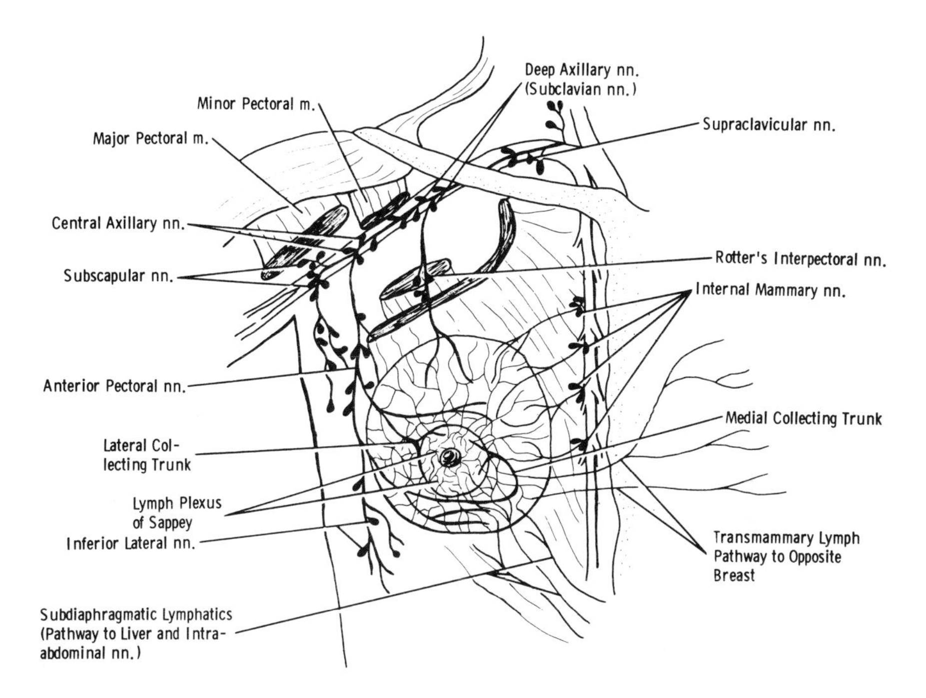

Fig. 13. Mammary lymphatic drainage. Most of the cutaneous lymphatics of the superior, medial, and inferior aspects of the breast, including the subareolar lymph plexus of Sappey, and a major portion of the deep mammary glandular body drain laterally toward the axilla. The first lymphatic filter station is represented by the anterior pectoral (anterior axillary) nodes, four to six of which are situated along the border of the pectoral muscles adjacent to the lateral thoracic artery and vein. From here the lymph flow proceeds to the central axillary nodes, situated along the axillary vein. These lymphatics continue toward the apex of the axilla where they form the deep axillary nodes (subclavian nodes) located at the borderline between axillary and subclavian vein. Some deep and superficial (inferior-medial aspects of breast) lymphatics also drain to the rectus sheath, and lymph flow continues to the subdiaphragmatic and subperitoneal lymph plexuses (Gerota's pathway); from there lymphatics of the liver and intraabdominal nodes are reached. Lymph vessels of the superficial medial aspects of the breast may lead to the opposite breast (transmammary pathway). Deep lymph plexuses of superior and retroalveolar parts of the glandular body extend to the interpectoral nodes (Rotter's nodes), and the lymph flow continues from here to the deep axillary nodes (Groszman's pathway). Lymphatics of the medial aspects of the corpus mammae drain along the intercostal branches of the internal mammary vessels into the internal mammary and mediastinal nodes. Lymph from some medial superficial and deep vessels may pass beneath the sternum to the anterior mediastinal nodes.

lymph glands, which lie in the subscapular fascia between the long thoracic nerve (serratus anterior muscle) and the thoracodorsal nerve (latissimus dorsi muscle).

Lymph from the retroareolar mammary area may pass into the interpectoral lymph nodes (Rotter's nodes), and from here lymph vessels run directly into the deep axillary nodes (Groszman's pathway). Lymph may flow from medial glandular tissues into the internal mammary nodes and from there into the mediastinal lymph nodes, which are situated in front of the thoracic aorta. Here the lymphatics follow the perforating rami of the internal mammary artery to enter the parasternal lymph nodes, which are located behind the inner intercostal muscles and in front of the thoracic fascia. The three to five parasternal nodes on each side communicate with pleural lymphatics. The lower aspects of the glandular breast may drain into subdiaphragmatic and subperitoneal nodes and, from there, to the liver and other intraabdominal nodes (Fig. 13, Table 5). Deep lymphatics of the breast anastomose freely with cutaneous lymph plexuses. Because breast lymph drains predominatly into the axilla, infections or cancer metastases are commonly transmitted into the axillary lymph nodes. Nevertheless, the secondary lymphatic channels of the breast (Groszman's pathway, lymphatics running along the perforating rami of the internal mammary artery, paramammary route of Gerota, superficial cross-mammary pathway) play an equally important role regarding the spread of breast cancer cells, and early diagnosis of lymphatic metastases migrating via these routes is often very difficult.

4. Mammary Innervation

Innervation of the breast is provided by somatic sensory and autonomous motor nerves. Somatic sensory innervation of the nipple and areola is abundant (close connection between autonomic and sensory nerves); whereas that of the corpus mammae (autonomic supply) is only sparse. Postganglionic, unmyelinated sympathetic fibers (gray fibers) that come from the thoracic paravertebral sympathetic ganglia accompany the mammary blood vessels; parasympathetic fibers are lacking in the breast. No ganglia are found in the breast, and only rarely do nerves impinge via knobs on mammary glandular cells. No innervation of mammary myoepithelial cells is observed. This indicates that secretory activities of the acinar epithelium are probably independent of nervous stimuli, and milk secretion is most likely due exclusively to the effects of prolactin and other supportive metabolic hormones. Stimulation of sensory free nerve fibers or of sensory receptors (tactile corpuscles of Vater-Pacini and Meissner, end bulbs of Krause, Ruffini-like endings) induces release of adenohypophysial prolactin and neurohypophysial oxytocin, via an afferent sensory reflex pathway, whereby stimuli reach the hypothalamus (see Chapter IV, Section B). Sympathetic mammary stimulation (postganglionic norepinephrine release) brings about contraction of the smooth musculature of the areola mammae and nipple (nipple erection) as well as of the mammary blood vessels (decreased blood flow). Simultaneously, under sympathetic excitation, the locally released norepinephrine induces stimulation of myoepithelial β-adrenergic receptors, causing muscular relaxation (Fig. 14). Because no parasympathetic fibers are present in the mammary gland, neither vascular nor myoepithelial smooth musculature stimulation, i.e., vasodilation and myoepithelial contraction, respectively, due to post-

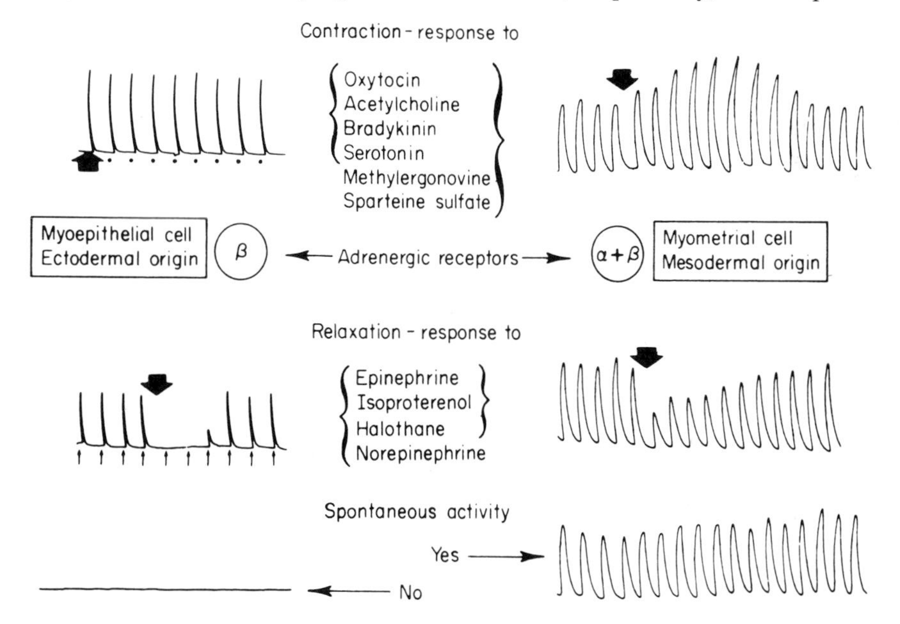

Fig. 14. Physiologic and pharmacologic characteristics of mammary myoepithelial and myometrial cells (from Vorherr, 1971, by kind permission of the publisher). Several endogenous substances can stimulate mammary myoepithelial cells to induce milk ejection and can produce contraction of myometrial cells (upper tracing). Exogenous compounds such as methylergonovine, prostaglandins of type E_1, E_2, and $F_{2\alpha}$, and sparteine sulfate can stimulate only myometrial cells but not the myoepithelium. Mammary myoepithelial cells are of ectodermal origin and contain β-adrenergic receptors only; whereas, the mesodermal myometrial cells are provided with adrenergic α- and β-receptors. Oxytocin-induced milk ejection is antagonized by catecholamines and halothane, as in the case of spontaneous myometrial contractions (middle tracing), excepting that norepinephrine is inhibitory to myoepithelial and stimulating to myometrial cells. In contrast to the myometrium, mammary myoepithelial cells display no spontaneous activity (lower tracing) and are 10–20 times more sensitive to oxytocin than myometrial cells.

ganglionic acetylcholine release is observed. Vascular α-adrenergic receptor stimulation, resulting in vascular smooth muscle contraction, is counteracted and balanced by β-adrenergic receptors; β-adrenergic receptor stimulation causes relaxation of vascular smooth musculature. In the absence of parasympathetic activity, a minor physiologic catecholamine inhibitory effect on the mammary myoepithelium may exist which is overcome by oxytocin release during suckling, inducing myoepithelial contraction. Therefore, it appears that no parasympathetic mammary nerves are necessary.

TYPE OF SENSORY INNERVATION OF THE BREAST SKIN

Generally, innervation of skin depends on both the degree of corial papillae development and the presence of hair follicles and other specialized dermal structures. Therefore, the nerve supply of breast skin has to be viewed distinctly because the innervation of areola mammae, nipple, and of the peripheral breast skin varies from that of other glabrous and hairy skin (Table 6). Differences in the type of innervation are most likely due to specific functions of the breast skin sensory receptors during lactation (see below).

Single and multibranched myelinated sensory and autonomic fibers supply the cutis of the breast. Sensory and motor fibers run parallel to, and partly envelop, the smooth musculature of the nipple and areola; some of these spiral nerve fibers may terminate as knoblike endings on muscle fibers. The breast skin peripheral to the areola mammae and nipple is hairy and possesses the same nerve endings as other hairy skin; also nerve elements pertinent to glabrous skin are found. The peripheral breast skin contains free circular nerve fibers and palisade nerve fibers that terminate in the form of expanded tips (Merkel's disks) on hair follicles. Because no encapsulated nerve endings are present within the shallow papillae of the hairy breast skin, sensory mechanisms must depend largely on the free nerve fibers and on the innervating elements associated with hair follicles. These free dermal nerve fibers extend from the corial papillae into the epidermis; there they run a short distance between the epithelial cells. Some expanded-tip nerve fibers are attached to the undersurface of the epithelium, but they are neither as abundant nor as regular as on other hairy skin. Dermal Ruffini corpuscles and Krause end bulbs are observed more regularly in peripheral breast skin than in other hairy skin (Table 6) (Cummins, 1966).

Abundant multibranched nerves are observed in the dermal parts of the areola mammae. The lactiferous ducts in the dermis of the nipple are in-

TABLE 6

Innervation of Breast Skin and of Other Glabrous and Hairy Skin

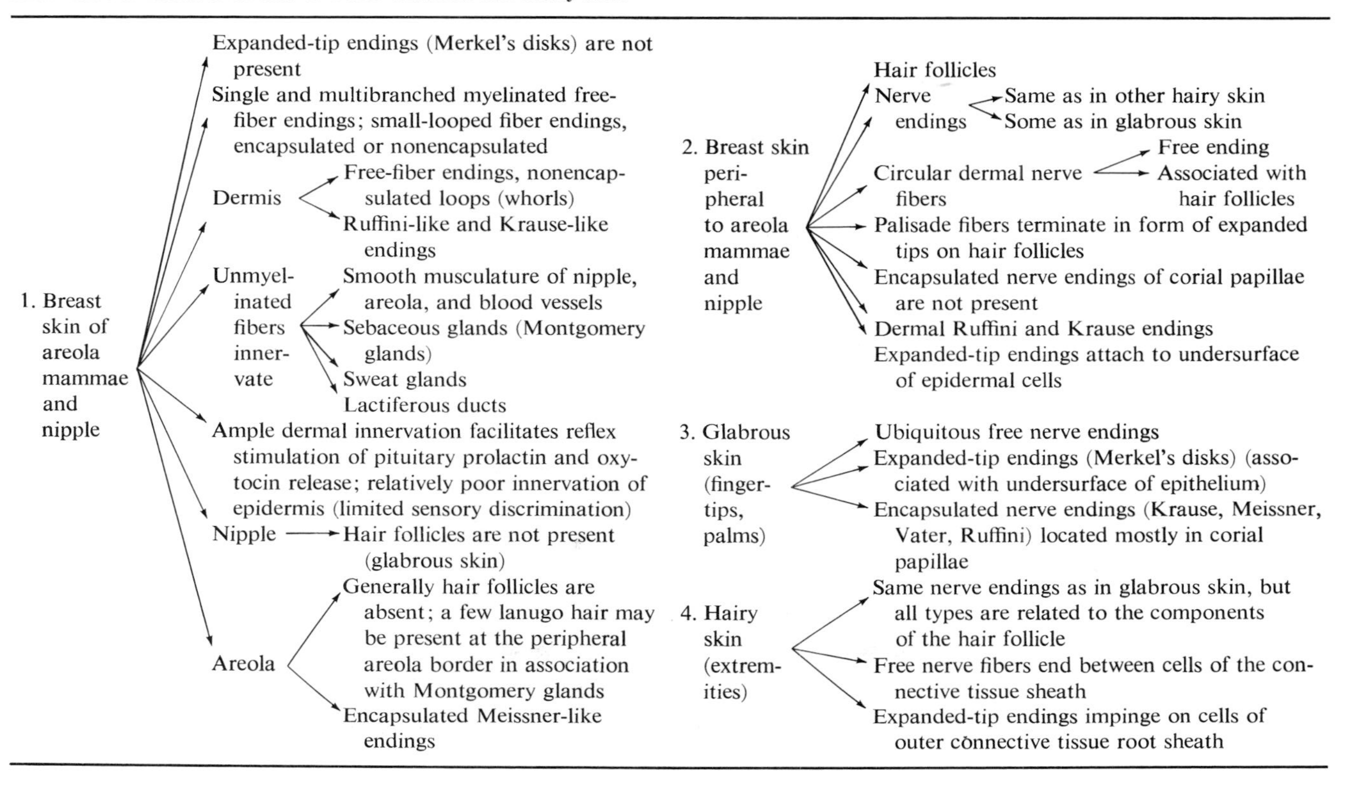

Skin	Structure	Innervation	Detail
1. Breast skin of areola mammae and nipple		Expanded-tip endings (Merkel's disks) are not present	
		Single and multibranched myelinated free-fiber endings; small-looped fiber endings, encapsulated or nonencapsulated	
	Dermis	Free-fiber endings, nonencapsulated loops (whorls)	
	Dermis	Ruffini-like and Krause-like endings	
	Unmyelinated fibers innervate	Smooth musculature of nipple, areola, and blood vessels	
	Unmyelinated fibers innervate	Sebaceous glands (Montgomery glands)	
	Unmyelinated fibers innervate	Sweat glands	
	Unmyelinated fibers innervate	Lactiferous ducts	
		Ample dermal innervation facilitates reflex stimulation of pituitary prolactin and oxytocin release; relatively poor innervation of epidermis (limited sensory discrimination)	
	Nipple	Hair follicles are not present (glabrous skin)	
	Areola	Generally hair follicles are absent; a few lanugo hair may be present at the peripheral areola border in association with Montgomery glands	
	Areola	Encapsulated Meissner-like endings	
2. Breast skin peripheral to areola mammae and nipple		Hair follicles	
		Nerve endings	Same as in other hairy skin
		Nerve endings	Some as in glabrous skin
		Circular dermal nerve fibers	Free ending
		Circular dermal nerve fibers	Associated with hair follicles
		Palisade fibers terminate in form of expanded tips on hair follicles	
		Encapsulated nerve endings of corial papillae are not present	
		Dermal Ruffini and Krause endings	
		Expanded-tip endings attach to undersurface of epidermal cells	
3. Glabrous skin (fingertips, palms)		Ubiquitous free nerve endings	
		Expanded-tip endings (Merkel's disks) (associated with undersurface of epithelium)	
		Encapsulated nerve endings (Krause, Meissner, Vater, Ruffini) located mostly in corial papillae	
4. Hairy skin (extremities)		Same nerve endings as in glabrous skin, but all types are related to the components of the hair follicle	
		Free nerve fibers end between cells of the connective tissue sheath	
		Expanded-tip endings impinge on cells of outer connective tissue root sheath	

nervated by unmyelinated nerve fibers. Poorly encapsulated Meissner-like endings are located in the small corial papillae of the areola. Generally, hair follicles are absent in the areola, although a few hairs (lanugo hair) may be found at the peripheral border of the areola in association with specialized sebaceous glands (Montgomery glands). The innervation of these areolar hair follicles is similar to that of other hairy skin, indicating an affinity of areola mammae skin to other hairy skin. The nipple skin definitely lacks hair follicles. In particular, the openings of the lactiferous ducts within the nipple are amply innervated by sensory fibers. The epidermis of the areola mammae, nipple, and the bordering dermal parts are poorly innervated, whereas the deeper portions of the dermis and the elements contained in it (smooth musculature, lactiferous ducts, sweat glands) are amply supplied with nerves. The relatively restricted innervation of the epidermal parts of the nipple and areolar skin is evidenced by a limited sensory discrimination: light touch is not well perceived, the characteristics of objects cannot be determined, and two-point discrimination is poor. The relatively large number of dermal nerve endings provides a high mammary responsiveness toward stimuli for elicitation of the suckling reflex. Thereby, the neural afferent reflex component induces adequate release of both adenohypophysial prolactin (milk synthesis and milk secretion into alveoli) and neurohypophysial oxytocin (milk ejection) to be humorally effective (efferent reflex component). Furthermore, the dermal smooth musculature of the areola and nipple can easily be stimulated to bring about nipple erection for the process of suckling. In particular, Krause and Ruffini-like endings located in the dermis of the areola and nipple are thought to contribute greatly to this increased skin sensitivity toward mechanical stimuli. It is not well understood as to what extent all these sensory and sympathetic elements influence mammary blood or milk flow. However, it appears that in addition to the mammary hormonal actions of prolactin and oxytocin, breast nerves also can influence the mammary blood supply and milk secretion. Dysfunction of sensory and autonomic nerve elements located in the skin of the areola and nipple may impair adequate lactation and cause nursing difficulties. Thus, nursing problems may be attributed not only to sufficient secretion or action of prolactin and oxytocin but also to local mammary nerval dysfunction.

i. SOMATIC SENSORY CUTANEOUS NERVE SUPPLY OF THE BREAST. Sensory fibers are provided by the supraclavicular nerves, which derive from the third and fourth branch of the cervical plexus. These twigs of supraclavicular nerves descend over the clavicle in the superficial fascia to innervate the upper cutaneous parts of the breast (Fig. 15 and Table 7). The major sensory cutaneous innervation of the breast, supplying the are-

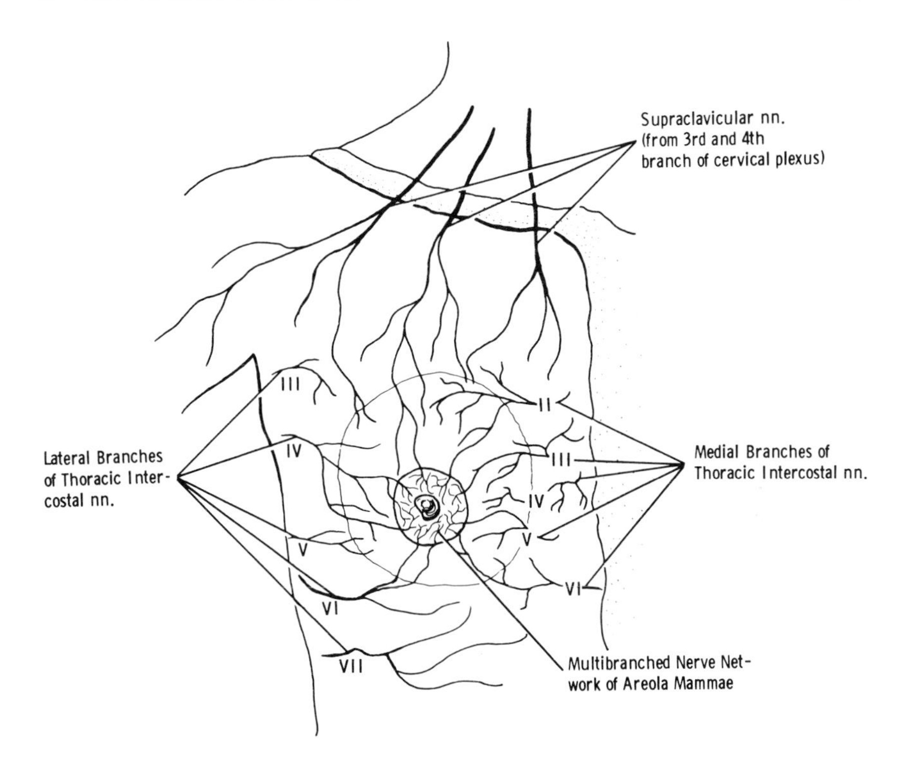

Fig. 15. Mammary innervation. The supraclavicular nerves supply sensory fibers for innervation of the upper cutaneous parts of the breast. The lateral branches (IV–VI) and the medial branches (II–VI) of the intercostal nerves provide the major sensory innervation of the mammary gland. Sympathetic sensory and motor fibers are derived from supracervical and intercostal nerves, respectively. Sympathetic fibers only run along the mammary gland supplying arteries to innervate the glandular body. The postganglionic sympathetic fibers stem from the ganglia of the paravertebral upper thoracic sympathetic chain.

ola mammae and nipple amply, is derived from the intercostal nerves, which also convey autonomic motor fibers to the smooth musculature of the areola mammae and nipple. Coming from the same nerves, sympathetic fibers run along the interlobular connective tissue septa and follow blood vessels and lactiferous ducts. The outer cutaneous parts of the breast receive nerves from the lateral mammary offshoots, which originate in the anterior rami of the lateral cutaneous branches of the fourth to the sixth intercostal nerves. These lateral cutaneous branches divide into anterior and posterior subbranches that run in the superficial fascia; the anterior subbranches send forth the lateral rami mammarii. The inner cutaneous aspects of the mammary gland obtain sensory nerves from the medial mammary rami of the anterior cutaneous branches of the second to

TABLE 7

Mammary Innervation

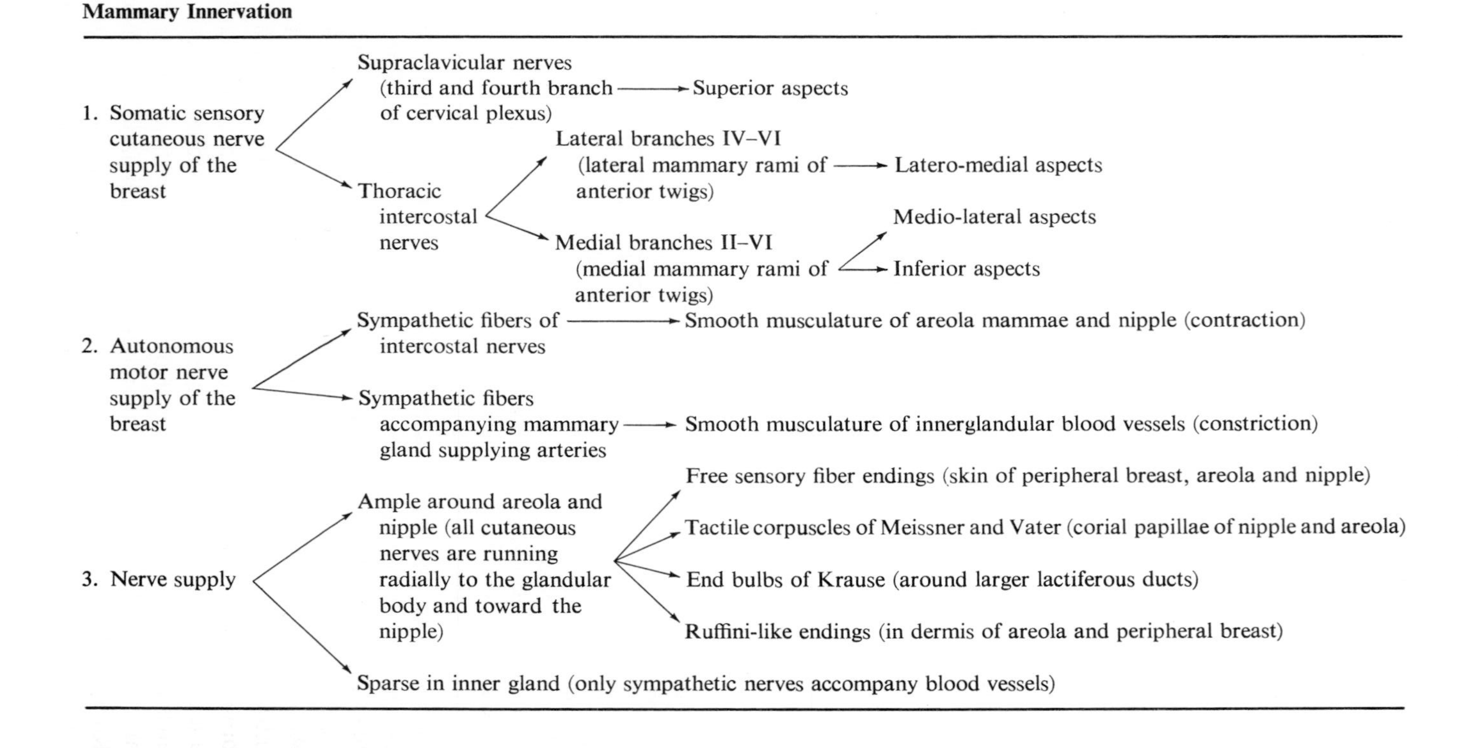

sixth intercostal nerves. They are small twigs penetrating the pectoralis major muscle in the thoracic region to reach the skin. These medial rami supply mainly the medial and partly the lateral aspects of the breast (Table 7, Fig. 15). All the sensory nerves perceiving warmth, cold, touch, and pain run in the skin of the breast radially to the glandular body toward the areola mammae and nipple. In addition to the free, nonencapsulated sensory endings, which are distributed in the epidermis, corium, and subcutaneous tissue, encapsulated afferent endings are observed in the breast. Within the epidermis nonencapsulated tactile corpuscles of Merkel are found; whereas encapsulated tactile corpuscles of Meissner are located in the corial papillae (nipple mainly). Encapsulated corpuscles of Vater-Pacini are observed in subcutaneous tissue and are supposed to sense pressure. End bulbs of Krause are situated in the corium; they are encapsulated sensory endings, probably perceiving cold and touch, and are assembled around lactiferous ducts. In the dermis of the areola mammae and the peripheral breast, Ruffini-like encapsulated endings, sensitive to touch, stretch, and warmth, are located.

In breast disease, pain sensation may be referred to the side of the chest or to the back (along the intercostal nerve trunks). Pain may also be felt above the scapula, on the medial side of the arm (along the intercostobranchial nerve), and in the neck (along supracervical nerves).

ii. AUTONOMOUS MOTOR NERVE SUPPLY OF THE BREAST. Sympathetic postganglionic fibers (gray rami) coming from the second to sixth intercostal nerves and from the supraclavicular nerves, and those alongside the arteries that supply the mammary gland (postganglionic fibers from paravertebral thoracic sympathetic ganglia), innervate the smooth musculature of the areola mammae, nipple, and mammary blood vessels. Although autonomic, intraglandular innervation by the sympathetic nerves of the vasculature is sparse, it influences mammary blood flow and milk formation (Table 7). In anxious postpartum women with increased sympathetic-adrenal tone, lactation may be impaired by pronounced catecholamine-induced relaxation of the mammary epithelium, antagonizing the milk-ejection effect of physiologic amounts of oxytocin released during suckling. As a consequence, lactation is impaired. Because in most of these cases an intramuscular injection of pharmacologic doses (5–10 IU) of oxytocin brings about milk ejection by overcoming the catecholamine antagonism and thus allowing successful nursing, it appears that increased sympathetic-adrenal tone does not interfere with the pituitary secretion and the mammary secretory effect of prolactin (Haagensen, 1971; Hollinshead, 1962; Kopsch, 1955; Lassmann, 1964; Miller and Kasahari, 1959; Netter, 1965).

B. MICROSCOPIC ANATOMY OF THE MAMMARY PARENCHYMAL TISSUE

During childhood the mammary ductular and alveolar lining consists of a 2-cell-layered basal cuboidal and low cylindrical surface epithelium. Ductular cells, in contrast to mammary alveolar cells, contain few mitochondria, sparsely endoplasmic reticulum, and some free ribosomes and tonofilaments; between ductular cells marked interdigitations are observed.

Under the influence of sex steroids, especially estrogens, the mammary glandular epithelium proliferates. It becomes multilayered, and it forms buds and papillae. The proliferating mammary epithelial cells differentiate into basal large clear cells and superficial basophilic dark cells (Fig. 16).

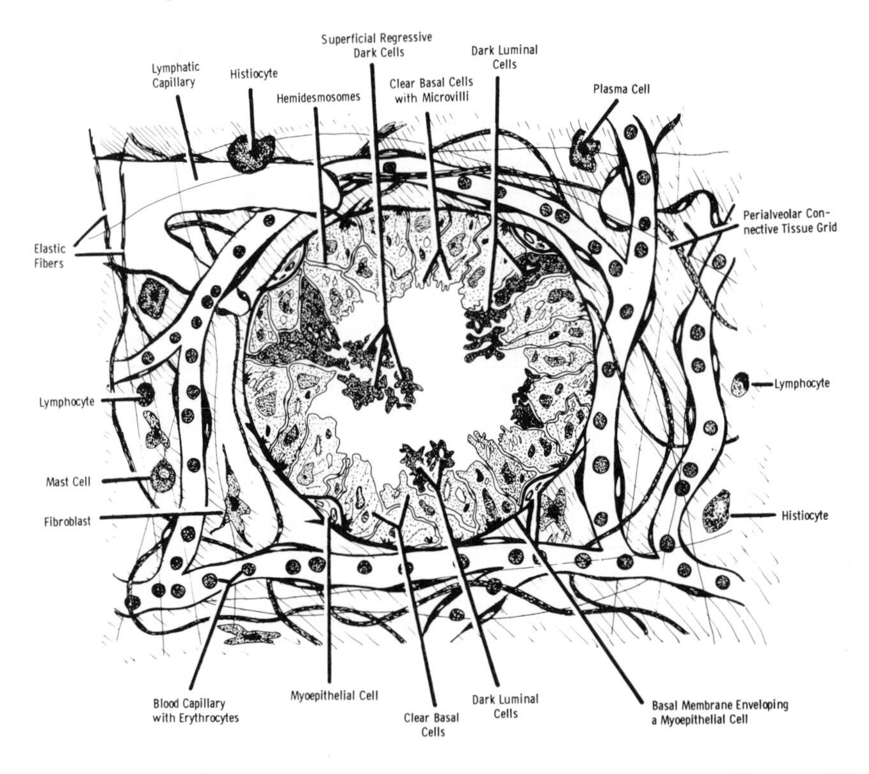

Fig. 16. Mammary alveolar morphology of the nonpregnant state. The essential structural elements in the acini of the mammary gland are the clear basal B-cells ("chief cells"), which are preponderant in number. Their cytoplasm is transparent; acidophylia is slight or lacking. The nuclei of clear basal B-cells are round or ovoid, and the smooth cell membranes are held together by hemidesmosomes. Some clear basal cells reach the alveolar lumen and develop apical microvilli. The cytoplasmic fibers of clear basal cells resemble those of mammary myoepithelial cells. It appears

Three different cell types have been described among the epithelial glandular structures of the breast of nonpregnant women (Bässler, 1970): (1) superficial (luminal) A-cells; (2) basal B-cells (chief cells); and (3) myoepithelial cells.

1. Superficial (Luminal) A-Cells

Luminal A-cells are rich in ribosomes and, therefore, appear dark. The cytoplasm of A-cells contains a large amount of RNA and ergastoplasm, indicating protein synthesis and accounting for the basophilic staining property. The proliferation of these superficial cells is induced by sex steroids, mainly estrogens. Some of these superficial cells may undergo intercellular dehiscence. Their ergastoplasm then displays wide fissures, and the mitochondria may be swollen, indicating regressive changes. Degenerating epithelial cell groups remain partially in contact with the epithelial surface and protrude into the lumen by forming buds and bridges.

2. Basal B-cells

These cells, also called "chief cells," preponderate numerically and seem to be the essential structural elements of the mammary epithelium. The cytoplasm of B-cells is clear and transparent and contains a round or

---

that clear basal cells function as stem cells for mammary myoepithelial and luminal dark A-cells. In contrast to cytoplasmic fibers of clear basal cells, those of the adjacent myoepithelial cells are denser and more contractile. Myoepithelial cells are opposed to the inner aspects of the basal membrane, which provides some sort of an envelope for them. The dark luminal A-cells are eosinophilic and contain numerous ribosomes causing the dark appearance of cytoplasm; their nuclei are rich in chromatin, and their nucleoli are prominent. Dark luminal cells display much more secretory activity than basal clear cells. They also show vacuolization, and it is thought that these vacuoles ("foam cells") may be derived from ascended and transformed clear basal cells developing metabolic activity and digesting extracellular and intracellular material. Some groups of dark luminal cells show intercellular dehiscenses and protrude more into the alveolar lumen; they have been considered as superficial regressive cells (Bässler, 1970). It appears that through stimulation of basal clear cells ('stem cells') by sex steroids (estrogen mainly) the basophilic luminal dark cells with a high ribosome content develop. The epithelium of the mammary ducts also consists of dark luminal cells and clear basal cells. At the beginning of menstruation, when luteal sex steroids greatly decline, pituitary prolactin is able to stimulate the alveolar epithelium to change temporarily into a more or less pronounced monolayered partially secretory lining as observed in the second half of pregnancy and exclusively during lactation (see also Fig. 18b). The cells of mammary alveoli are amply supplied with blood and lymphatic capillaries, which are embedded in the surrounding connective tissue grid with its typical fiber and cellular elements.

ovoid nucleus that is transparent, indicating an equal chromatin distribution. The cell membrane is smooth, and the cells are held together by desmosomes. Those cells reaching the luminal surface carry apical microvilli. Intracytoplasmic filaments and fibers are observed, indicating cellular differentiation processes toward myoepithelial cells. The intracytoplasmic fibers resemble myofilaments of myoepithelial cells, but the latter are denser and contractile only. The fibers of the B-cells appear similar to the stabilizing tonofilaments of blood capillary endothelium. The clear basophilic cells have also been considered as energy providers for the secretory luminal cells, and an endocrine activity of basal cells functioning as a "clear cell organ" has even been suggested (Dabelow, 1957).

3. Myoepithelial Cells

These ectodermal cellular units are located around alveoli and the small excretory milk ducts; they rest between the inner aspect of the basement membrane and the tunica propria and can be demonstrated clearly by Masson's trichrome stain. Myoepithelial cells display starlike branching, and their sarcoplasm contains myofilaments, 50–80 Å in diameter, which are inserted at the basal membrane of the cell via hemidesmosomes. The nucleus of mammary myoepithelial cells is ovoid, dense, relatively small, and its inferior part shows indentations. Nearby, and at the basal cell membrane of the myoepithelium, invaginated vesicles, 500–600 Å in diameter, are observed and are probably derived from pinocytotic processes. The myoepithelial basal membrane is in close contact with the basal membrane of the epithelial glandular cells. The cellular body of the myoepithelium contains dense filament bundles that display high alkaline phosphatase activity. It appears that the eosinophilic myoepithelial cells are free of sudanophilic lipids; they seem to represent a differentiated cell form derived from the clear basal glandular mammary epithelium. The myoepithelial cells proliferate during pregnancy and lactation, and their cytoplasmic branches form a dense fibrillar basketlike meshwork around alveoli and small milk ducts. The mammary myoepithelium is hormonally stimulated by sex steroids and prolactin to a similar extent as the other epithelial glandular cells; no innervation of myoepithelial cells is observed. Myoepithelial cells also participate in proliferative and neoplastic lesions of the breast. The morphology and topography of mammary myoepithelial cells is shown in Figs. 17 and 18a. As depicted in Fig. 17, ectodermal myoepithelial cells are irregularly shaped, show branching by elongated sarcoplasmic processes, and are oriented at right angles to the long axis of the underlying epithelial cells. They form a loose network around acini

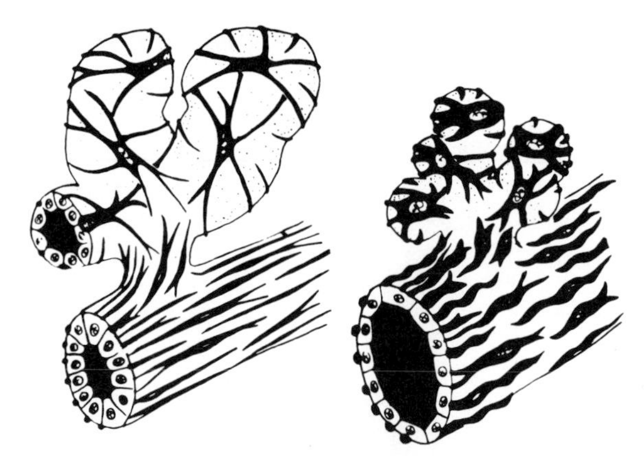

Fig. 17. Morphology and topographic anatomy of myoepithelial mammary cells (from Linzell, 1961; reproduced from "Veterinary Annual," W. A. Pool, ed., by kind permission of Wright, Bristol, 1961 and the author). The mammary alveoli, also called acini, are dilated and filled with milk (left side of the picture); the myoepithelial cells surrounding them are relaxed. Stimulated by the baby's suckling, endogenous oxytocin is released, and the myoepithelial cells located around the alveoli and smaller milk ducts contract as shown in the right part of the picture. Thereby, the milk is forcibly pushed from the alveoli and smaller milk ducts into the larger lactiferous ducts, which are dilated by the increased milk volume. Mammary myoepithelial cells are derived from the ectoderm and are probably a subtype of basal glandular mammary cells; they are 10–20 times more sensitive to oxytocin than uterine smooth muscle cells, which are derived from the mesoderm.

and are arranged parallel to the long axis of the mammary ducts. Myoepithelial cells lie within a widened (1000–2000 Å) basement membrane, on the outside of which collagen fibrils are located. Inside the myoepithelial cell, myofibrils are arranged in a parallel course to the basement membrane; these fibrils occupy mainly the basal part of the cell. The basal membrane functions as a tendinous plate, allowing the diminution of the alveolar diameter with ejection of its contents. No mesodermal smooth musculature has been observed around mammary alveoli or smaller ducts. Myoepithelial cellular size and number decrease with cessation of lactation, and they involute greatly during menopause (Dabelow, 1957; Langer and Huhn, 1958; Linzell, 1961; Schäfer and Bässler, 1969; Bässler, 1970).

4. Interrelationship of Mammary Glandular Cell Elements

It has been assumed that the basal glandular B-cells are the precursors of myoepithelial cells and luminal A-cells; the latter are believed to be actively involved in the process of milk synthesis and secretion during the

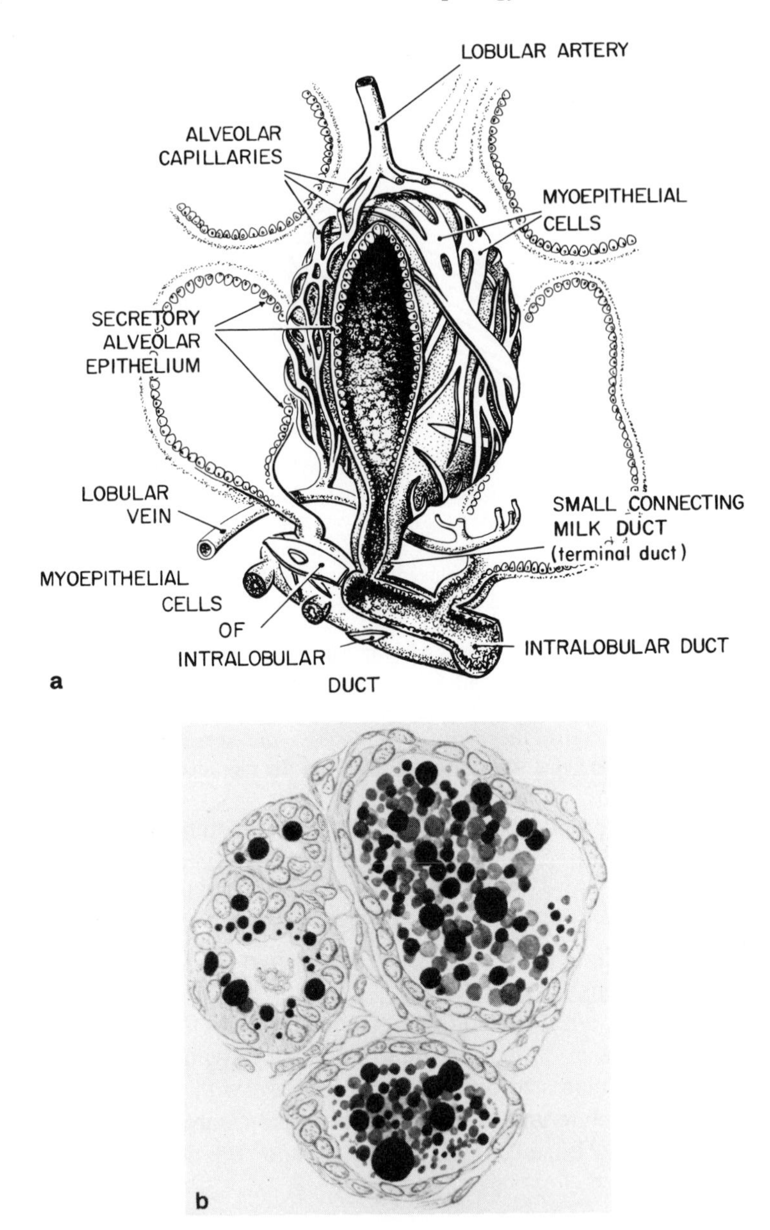

Fig. 18. Scheme of the functional mammary unit at lactation (part a illustrated by Norviel based on Turner, 1952; part b reproduced from Kopsch, 1955, by kind permission of the publisher). (a) Arrangement of secretory alveoli, myoepithelial cells, and vasculature. The secretory alveolar epithelium is monolayered and the epithelial lining of the milk ducts consists of two layers. Between the basis of the glan-

process of lactation. It is, however, also possible that both A- and B-cells represent morphologic structures derived from different types of epithelial stem cells. Recently, two different glandular cell types were found on mammary tissue of a woman at 20 weeks of gestation. A superficial eosinophilic cell layer abutting the alveolar lumen could be distinguished from a basal glandular layer of clear cytoplasmic cells resting upon the basement membrane. Both the dark eosinophilic superficial cells and the clear basal cells seemed to be involved in secretory processes, but the clear basal cells are believed to be the main source of glandular secretion at this stage of pregnancy. Again, it is not known whether the dark eosinophilic cells and the clear basal cells are derived from common or independent epithelial precursors (Toker, 1967). It seems, however, that the clear secretory cells are related to the monolayered secretory basophilic cells described by Dabelow for the second half of pregnancy. These basophilic cells contain sudanophilic lipids and show only low alkaline phosphatase activity.

The basal epithelial cells may also contain small cytoplasmic fibers that show a certain relationship to myofibrils of the adjacent myoepithelial cells. Thus, a transition from basal secretory glandular cells to myoepithelial cells seems possible. The superficial eosinophilic dark cells display large membrane-bound cytoplasmic lipid vesicles called "foam cells." It has been suggested that the latter are phagocytic histiocytes that migrate through the luminal parts of the epithelial lining (Toker, 1967). According to many investigators, the mammary alveoli are made up of a monolayered secretory epithelium from midpregnancy until cessation of lactation (Dabelow, 1957).

Although the specific function of the different cell types of the mammary glandular epithelium in nonpregnant women and in gravidas of

---

dular epithelial cells and the tunica propria, starlike myoepithelial mammary cells surround the alveolus in a basketlike fashion. The mammary arteries approach the alveoli at their tip, and the venous blood drains basally and laterally from an alveolus. *Note:* This scheme, as designed for the cow's udder, is very similar to the lactating mammary gland of man. The difference being, the terminal mammary ducts in humans are wider than in the cow. (b) Mammary alveoli of a lactating woman 14 days postpartum. Characteristic for the lactating mammary gland is the form-variety of alveolar cells dependent on the respective secretory state and the extent of alveolar milk filling. The two alveoli in the left part of the picture show narrow lumina and thick cells containing intracellular fat droplets, a status encountered shortly before milk secretion into the acinar lumen. The two alveoli on the right part of the picture display a thin epithelium, with larger lumina, a state observed after milk secretion into the alveolar lumen.

early pregnancy has not yet been clarified, the secretory function of the glandular monolayered epithelium during the last trimester of pregnancy and during the period of lactation is well known. The physiology of the myoepithelial cells is also well understood; they function as contractile ectodermal smooth muscle cells (Figs. 17 and 18a) that closely resemble the mesodermal smooth musculature in their cellular structure and organelles. Myoepithelical cells are 10–20 times more sensitive to oxytocin than myometrial cells. During the stimulus of suckling, oxytocin is released from the posterior pituitary gland inducing contraction of mammary myoepithelial cells; as a result ejection of milk from the alveoli and smaller milk ducts into the lactiferous ducts and lactiferous sinuses takes place. Hence, the milk can be removed by the suckling baby; this subject will be discussed in more detail later.

C. CYCLIC CHANGES OF THE ADULT MAMMARY GLAND

1. Proliferative Breast Changes

These changes are presented in Table 8. In the proliferative phase of the menstrual cycle, ovarian estrogens induce parenchymal proliferation with formation of epithelial sprouts. This state of hyperplasia with increased cellular mitoses extends into the secretory phase of the menstrual

TABLE 8

Cyclic Mammary Changes of Menstruating Women

1. Premenstrual sex steroid-induced increase in breast volume due to
   a. Enhanced water retention (lobular edema)
   b. Thickening of epithelial basal membrane
   c. Enlargement of alveolar luminal diameters and appearance of intraalveolar secretory material
   d. Infiltration of perilobular mammary stroma with fluids, lymphoid and plasma cells
2. Moderate intraalveolar mammary secretion during menstruation as a consequence of luteal sex steroid withdrawal permitting limited milk-secretory prolactin action
3. Postmenstrual regressional mammary changes involving
   a. Degeneration and necrosis of some glandular cells with loss in cellular order; subsequent hyalinization with homogenization of basal membrane and surrounding connective tissue
   b. Reduction in lobular-alveolar size with narrowing of alveolar lumina
   c. Disappearance of tissue edemization
   d. Tissue reparation by phagocytic activities of lymphoid and plasma cells
   e. Size of breast smallest at days 4–7 of cycle
4. Incomplete postmenstrual mammary "involution"; with each ovarian cycle glandular-alveolar growth continues slightly up to the age of 30–35 years

cycle. Estrogens stimulate cellular RNA synthesis whereby the density of the nucleus increases, the nucleolus enlarges, and other cellular organelles such as Golgi complex, ribosomes, and mitochondria increase in number and size; this is found not only in the mammary glandular epithelium but also in the endometrium and myometrium. With increasing amounts of luteal progesterone (this applies for an ovulatory cycle only) mammary ducts become somewhat dilated, and the mammary alveolar cells differentiate into secretory cells that may assume a partly monolayered alveolar arrangement. Under the influence of progesterone and (in addition to estrogens) metabolic hormones, some lipid droplets appear in the alveolar cells, which may show signs of secretion of cellular contents into the alveolar lumen.

2. Premenstrual Breast Changes

In the luteal phase of the menstrual cycle, the mammary blood flow is increased. From 3 to 4 days before the onset of menstruation, enhanced mammary turgescence, tension, fullness, tightness, heaviness, and pains can be observed in some women, and on the average the breast volume is increased by 15–30 cm^3 (Dabelow, 1957). The premenstrual engorgement with augmented density and nodularity of the breast is due to increased water retention in the connective tissue (interlobular edema) and enhanced ductular-acinar sprouting, growth, and probably new formation of mammary alveoli; all of this is induced by the increased amounts of estrogens and progesterone produced by the corpus luteum. In Table 9 values

TABLE 9

Female Sex Steroid Secretion, Plasma Levels, and Urinary Excretion during the Menstrual Cycle [a]

Female sex steroids	Menstrual cycle		
	Midfollicular phase	Ovulation	Midluteal phase
Daily estrogen (mainly 17β-estradiol) secretion rate (μg)	50	350	200
Plasma estrogens (pg/ml)	140	300	270
Daily estrogen urinary excretion rate (μg)	15	80	60
Daily progesterone secretion rate (mg)	5	12	30
Plasma progesterone (ng/ml)	2	3	13
Daily pregnanediol urinary excretion rate (mg)	1.5	2	5

[a] Data from Ganong (1971) and other literature sources.

for female sex steroid secretion, blood levels, and excretion are presented indicating increased sex steroid secretion during the luteal phase of the cycle (Ganong, 1971).

The mammary edema formation may be attributed to the estrogen-induced histamine effect on the mammary microcirculation with leakage of fluid (Zeppa, 1969). Premenstrually, the alveolar lining consists of one or two epithelial cell layers with basal small, chromatin-rich nuclei that show signs of secretory activity; some acinar lumina may contain exfoliated cells and/or colostrum. It appears that premenstrually the luminal layer of dark cells (Fig. 16) becomes gradually desquamated, whereas the basal clear cell layer undergoes slight secretory changes, as observed during pregnancy. The premenstrual mammary changes that are more marked in nulliparous and premenopausal women can be summarized as follows: (a) interlobular edema; (b) thickening of basal membrane; (c) increase of alveolar luminal size and appearance of colostrum; (d) vacuole formation (fat droplets) of basal glandular alveolar cell layer; and (e) slight infiltration of perilobular mammary stroma by lymphoid and plasma cells.

3. Menstruation

With the onset of menstruation, secretory activity is apparent, and toward the end of menstruation regressive tissue processes take place. At this time, the acini display signs of cellular regression, and their borders are less clearly defined. The basal membrane of some epithelial cells, including parts of the surrounding connective tissue, show hyalinization and homogenization. The lobuli become smaller, and the alveolar lumina narrow considerably.

4. Postmenstrual Regressive Changes

Postmenstrually, the perilobular swelling and tissue edematization wanes rapidly, and the mammary stroma again appears more condensed. The cellular regressive changes cease early in the new menstrual cycle because increasing ovarian estrogen production begins. Thus, the mammary proliferation induced by ovarian estrogens during a menstrual cycle never fully returns to the starting point of the preceding cycle. Accordingly, each ovulatory cycle slightly fosters mammary development, which continues until the age of about 35 years. In the early postmenstrum the infiltration of perilobular stroma with lymphoid and plasma cells is most pronounced, indicating the phase of intensive repair of epithelial-glandular degeneration processes. A minimum of breast volume is observed at 5–7 days postmenstruation. With increased estrogen secretion from newly maturing ovarian follicles, glandular and mammary connective tissue elements are stimulated anew for progressive changes, which are most apparent near the end of the second half of the cycle due to the effect of luteal estrogens and progesterone (Dabelow, 1957).

D. BREAST CHANGES DURING PREGNANCY

A remarkable ductular-lobular-alveolar growth occurs during pregnancy; it is most apparent in the peripheral mammary areas. This intensified mammary stimulation is evoked by luteal and placental sex steroids, placental lactogen, prolactin, and, probably, by chorionic gonadotropin (Fig. 19). In studies on rats, estrogens were found to stimulate prolactin synthesis and to facilitate pituitary prolactin release by diminishing the hypothalamic synthesis of the prolactin-inhibiting factor (MacLeod *et al.,* 1969). Treatment of adult nonpregnant rats with pharmacologic doses of estradiol and progesterone over a period of 8 days results in mammary development resembling that observed during pregnancy. Discontinuation of sex steroid treatment, allowing secretory prolactin action on mammary glandular epithelium, leads to a lactation-like state, enabling the nonpregnant, sex-steroid-pretreated rat to "nurse" offspring taken from another lactating mother rat, albeit under greater difficulties and remarkable growth retardation of the "nursed" pups (H. Vorherr, unpublished observations, 1971). In Fig. 20 the increasing ductular-alveolar proliferation is demonstrated in rats during early, middle, and late pregnancy. According to most recent reports (Tyson *et al.,* 1972 a,b; Tyson and Blizzard, 1972; Berle and Apostolakis, 1971), prolactin seems to be increasingly released during human pregnancy and thus may be involved with stimulating growth and secretory activities of mammary tissues. Application of specific prolactin radioimmunoassays, clearly allows discrimination between the activities of pituitary growth hormone, placental lactogen, and pituitary prolactin. During pregnancy, plasma levels of FSH, and probably also of LH, are greatly reduced (Jaffe *et al.,* 1969; Hanson *et al.,* 1970), and prolactin is increasingly released as pregnancy advances (Fig. 26) (Friesen, 1972). A decrease in anterior pituitary gonadotropin secretion is usually accompanied by an increase in prolactin secretion, as observed in nonpuerperal patients with abnormal lactation (galactorrhea) and amenorrhea.

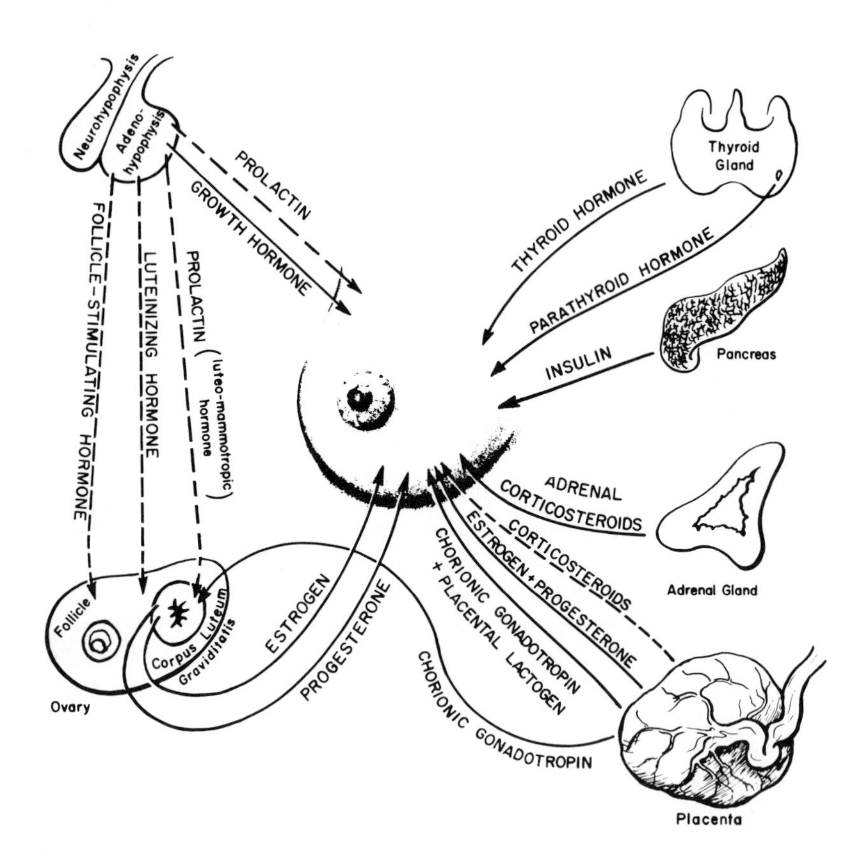

Fig. 19. Hormonal stimulation and preparation of the breast in pregnancy (from Vorherr, 1972d, by kind permission of the publisher). During the period of gestation, luteal and placental sex steroids induce enhanced ductular-lobular-alveolar development of the breast. The adenohypophysis may contain up to 50% actively prolactin-synthesizing cells (prolactin cells = "pregnancy cells"), and prolactin is increasingly secreted into the circulation as pregnancy advances. The number and activity of basophilic pituitary gonadotropin-secreting cells are diminished, resulting in low plasma levels of gonadotropins (FSH and LH). Besides prolactin, placental lactogen, growth hormone, and probably HCG, support the lobular-alveolar tissue growth of the breasts and contribute, together with insulin and cortisol, to the differentiation of glandular epithelial stem cells into presecretory mammary cells. Adrenal corticosteroids (mainly cortisol) and placental corticosteroids derived either from trophoblastic sex steroid metabolism or possibly from active placental synthesis, and thyroid-parathyroid hormone, also contribute to mammary cellular differentiation for the process of galactopoesis. The interrupted line for prolactin indicates that prolactin cannot induce lactation during gestation; also in humans, in contrast to rodents, prolactin probably cannot exert a luteotrophic effect.

The extent of mammary changes during pregnancy depends on the individual's constitution in regard to preformed size and number of mammary lobuli as well as on age and parity of the woman (Kon and Cowie, 1961, pp. 292–294).

Within the first 3 to 4 weeks of gestation, an intensified ductular sprouting, branching, and lobular formation is observed, exceeding the respective premenstrual glandular changes (Table 10, Fig. 8). At 5 to 8 weeks of pregnancy a definite enlargement of the breast with dilation of superficial veins and a feeling of heaviness are noticed. At the same time pigmentation of areola and nipple is intensified, and the area of the areola mammae is enlarged. In early gestation dichotomic ductular sprouting is most pronounced (mainly estrogen-induced), whereas from the third gestational month on, the lobular formation exceeds the number of primitive dichotomic ductular sprouts (beginning progesterone influence). The increase in mammary lobular tissue is due to new formation of lobuli and hypertrophy of preexisting lobular-alveolar structures (Fig. 21). At the end of the third month the alveoli begin to show some secretory material (colostrum), which rarely contains any fat. The alveolar cells become single layered at midpregnancy, and a grouping of duplicated milk ducts and of alveolar structures with formation of the prominent mammary breast lobules is noted. At the same time, the alveolar lumina begin to dilate (Fig. 7), and, from midpregnancy on, small amounts of colostrum may be found in the acini. This colostral secretion is most likely due to glandular cell stimulation by prolactin, which is increasingly released from the anterior pituitary gland as gestation advances; increasing secretion of placental lactogen may also play a part. Concomitantly, the mammary blood supply is enhanced by increasing vascular luminal diameters and by new formation of capillaries around the mammary lobuli. Toward midpregnancy the rate of mammary epithelial proliferation (increase in size and number of smaller ducts and alveoli) gradually declines; whereas the alveolar epithelium differentiates, assuming a secretory function that rapidly increases due to the effect of prolactin, progesterone, placental lactogen, and other metabolic hormones (Fig. 22). The continued increase in breast size during the second half of pregnancy is attributable mainly to progressive alveolar dilation, filling with colostrum, and enhanced mammary vascularization; fat and connective tissue decrease relatively.

The hormonal stimulation of the mammary tissues is already rather effective during the first half of gestation, as proven by the fact that lactation may occur in cases of interrupted pregnancy of 16 weeks and beyond. Toward the end of the first half of pregnancy the definite development of ductular-lobular-alveolar structures is accomplished, and in the second half

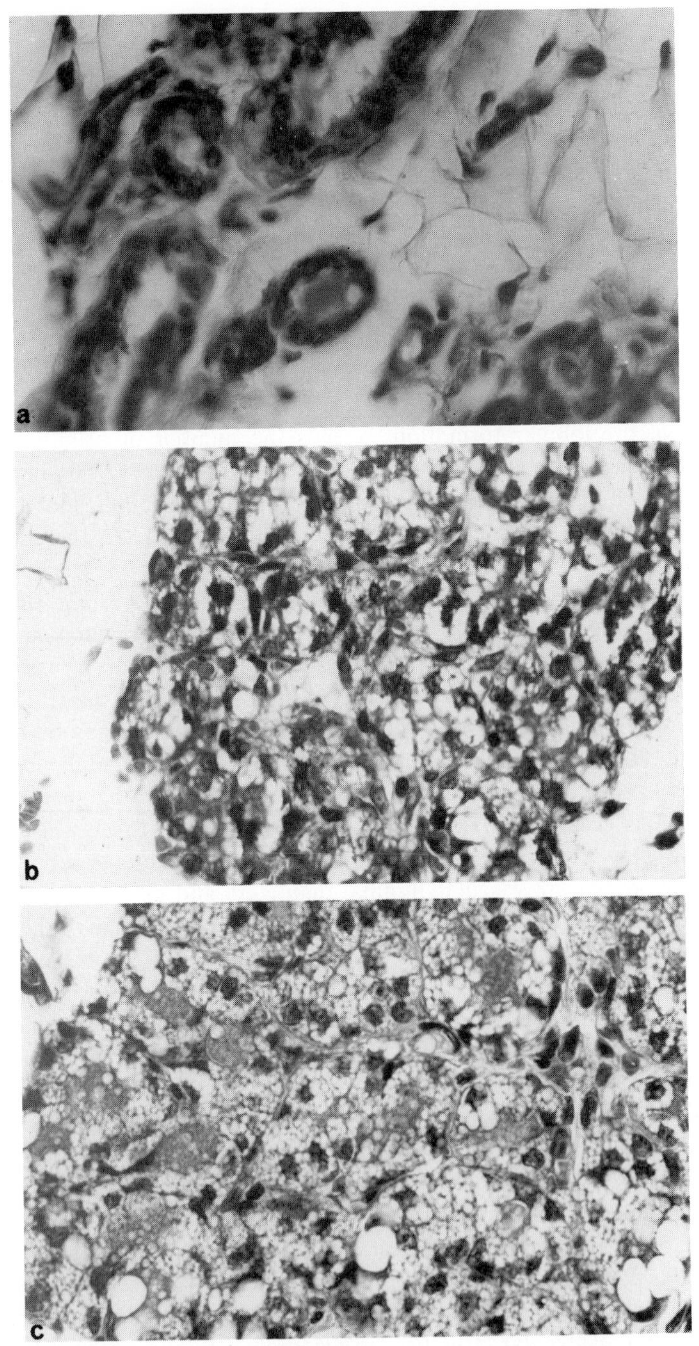

Fig. 20. Mammary gland: nonpregnant, pregnant, and lactating rat (× 160) (from H. Vorherr, unpublished observations, 1971).

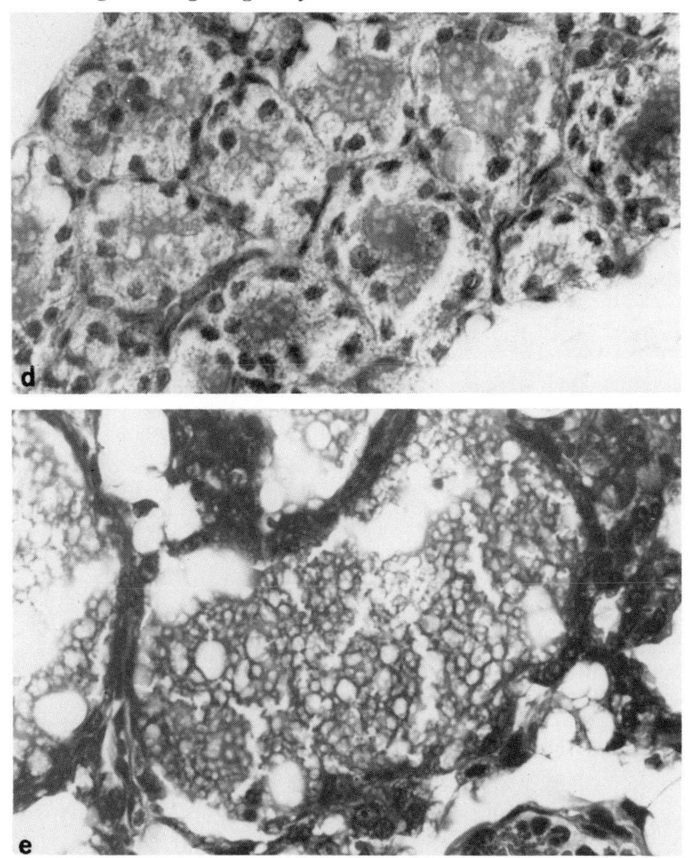

Fig. 20. (*continued*) (a) The parenchymal tissue of the mammary gland of a nulliparous nonpregnant adult female rat in diestrus consists of a few ducts and sparse alveoli. Both ducts and alveoli are lined mainly with a bilayered epithelium and are embedded into mammary connective tissue and fat; only a few alveoli display some secretory activity and possess a monolayered epithelium. (b) At day 7 of gestation a typical mammary "adenosis of pregnancy," i.e., alveolar mammary proliferation and growth, is discernable. The epithelium of mammary alveoli consists of a monolayered cylindrical lining, and the alveolar epithelium shows intracellular secretory activity, but no milk is released into the alveolar lumen. (c) At 14 days of gestation the mammary adenosis has advanced far, and major areas of former mammary fat are replaced by proliferating parenchymal tissues; at this stage intracellular secretory activities of the glandular epithelium are most evident by formation of clusters of fat droplets. At this stage of pregnancy some colostrum secretion into the alveolar lumina is visible. (d) At 19 days of gestation (two days before delivery) the mammary alveolar development and secretory activity has progressed to a stage where the conditions are optimal for the process of upcoming lactation. The alveolar epithelium displays maximal intracellular secretory activity, and its cells are filled with secretory products to be released into the mammary alveolar lumen upon onset of lactation. (e) Postpartum, mammary alveolar lumina are filled with milk, leading to distention and dilation of alveoli; most of the alveolar cells are small and empty.

TABLE 10

Histologic Mammary Changes during Pregnancy [a]

1. First trimester
   a. Increased terminal ductular sprouting and beginning of lobular-alveolar formation at 3–4 weeks of gestation; glandular epithelial buds invade connective tissue and mammary fat and grow by replacing the fat matrix tissue
   b. Appearance of migratory round cells and fibroblasts in the connective tissue neighbored to the sprouting glandular elements
   c. Change of the 2-cell layer alveolar epithelium into a monolayer secretory unit beginning at the end of the third month of pregnancy

2. Second trimester
   a. Intensified dichotomic ductular proliferation and formation of lobular-alveolar structures
   b. Increased appearance of lymphocytes in surrounding connective tissue
   c. Enhanced activity of secretory epithelial cells leading to accumulation of colostrum within the mammary alveoli

3. Last trimester
   a. Pronounced accumulation of fat droplets in secretory alveolar cells
   b. Increased filling of mammary alveoli with colostrum
   c. Relative diminution of interlobular connective tissue and fat by extensive lobular-alveolar proliferation ("adenosis of pregnancy")

[a] From Vorherr (1972b).

of gestation only a minor additional sprouting of mammary tissues occurs. Proliferative effects encountered during the follicular and luteal phase of the menstrual cycle are much more pronounced during the first trimester of gestation. As in the last stage of the biphasic mentrual cycle, so it is around midpregnancy, mammary proliferative processes are replaced by mechanisms of differentiation and secretory function. This secretory mammary function reaches its culmination point at the time of postpartum lactation. Thus, the mammary changes appear to follow a principle of nature, whereby proliferating mammary tissues cannot exert specialized functions, and, conversely, differentiating mammary epithelium cannot effectively proliferate. Until term, each breast gains approximately ¾ lb in weight and becomes firm and full and ready for lactation. The mammary subcutaneous veins are greatly dilated and prominent; the mammary blood flow is almost doubled compared to the nonpregnant state. Augmentation and dilation of milk ducts and alveoli, which are partially filled with colostrum, contribute to the enhanced size of the breasts; hypertrophy of mammary myoepithelial cells and connective tissues and an increased deposition of fat are other factors responsible for the augmentation of mammary

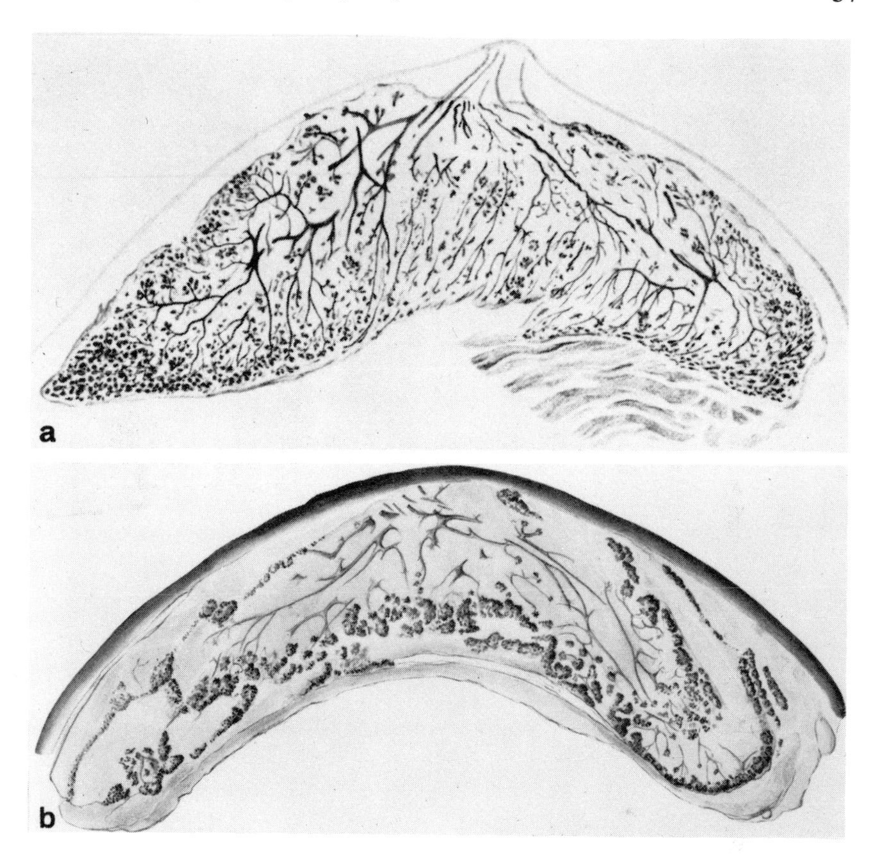

Fig. 21. Parenchymal mammary structure of a nonpregnant and a pregnant woman (from Dabelow, 1957). (a) Section through the breast of a nulliparous normally cycling woman (thick slice, Alum Carmine staining) showing a well-differentiated ductular and peripheral lobular-alveolar system. (b) Similar section through the breast of a pregnant woman during the third gestational month demonstrating ductular sprouting and intensive peripheral lobular-alveolar development which is proceeding peripherally and centrally.

size (Zilliacus, 1967; Netter, 1965; Bässler, 1970; Schäfer and Bässler, 1969). The epithelial mammary lining of ducts and alveoli in pregnant women prior to the third and fourth month of gestation and in nonpregnant women consists of two cell layers. Toward the end of the third month of pregnancy, the mammary alveoli begin to lose their superficial epithelial layer; whereas the 2-cell layer of the milk ducts is maintained throughout pregnancy and lactation. This alveolar transformation into a monolayered epithelium in the second half of human pregnancy continues to exist during lactation and is also observed in the pregnant and lactating

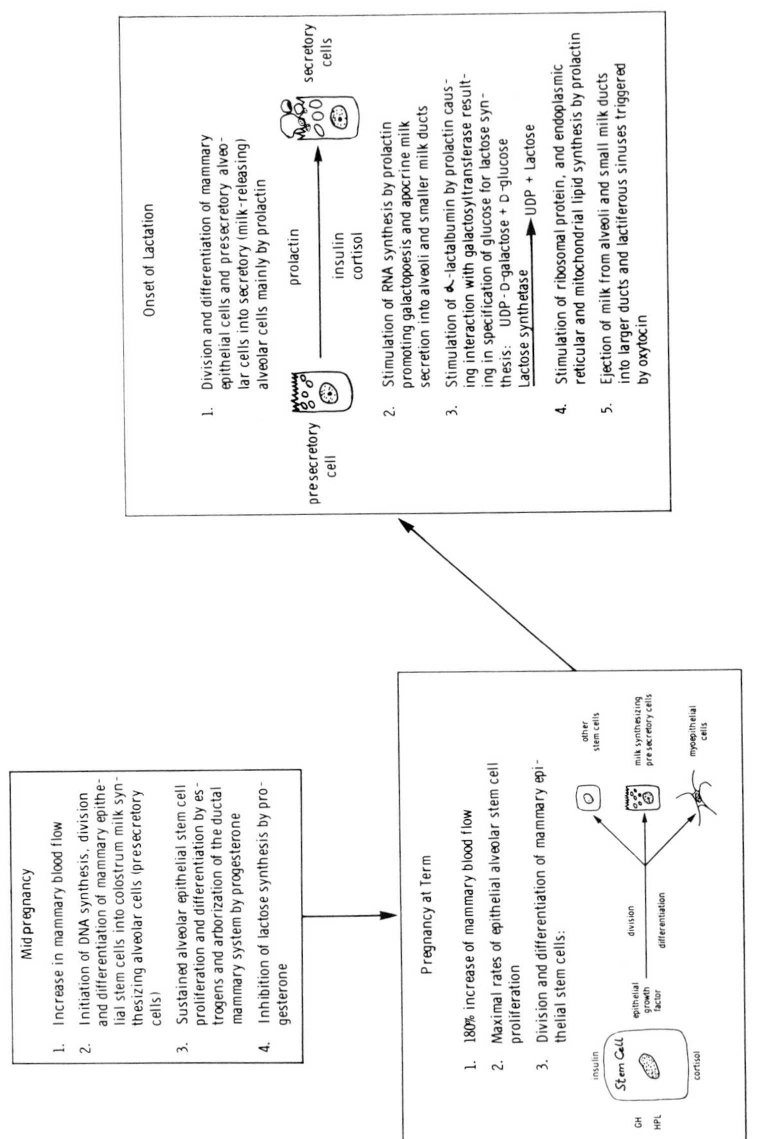

See legend on opposite page.

rat (Fig. 20). During the second and third trimester of pregnancy, the remaining basal epithelial layer of the mammary alveoli differentiates and matures into secretory cells and is called the "colostrum cell layer" (Dabelow, 1957); it has lost its ability to proliferate further in a dichotomic fashion. Accordingly, in the last trimester these secretory alveolar cells rarely show mitoses, which occur most frequently in proliferating mammary glandular tissue of midpregnancy. The alveolar lumina contain secretory material (colostrum) from the fourth month of pregnancy on. The colostrum is composed of desquamated glandular and phagocytic cells ("foam cells"), which are believed to be degenerated epithelial cells or migratory lymphoid phagocytes. Closely assembled around the mammary alveoli are eosinophilic cells, plasma cells, and mononuclear and polymorphonuclear leukocytes. As pregnancy advances, colostrum is increasingly secreted and thus fills up and dilates the mammary alveoli. This dilation and the proliferation of mammary glandular, vascular, and connective tissue structures contribute to the increasing size of the breast. Because during gestation the mammary glandular-alveolar development is predominant, the amount and distribution of connective mammary tissue appears to be relatively decreased (Dabelow, 1957).

The sprouting of glandular mammary elements during pregnancy is so pronounced that the term "adenosis of pregnancy" (Dabelow, 1957) has been created. The most extensive proliferation of parenchymal tissues is observed around fat lobuli, adjacent to the directory channels of connective tissue. Epithelial mammary cells seem to invade a fat lobule at its per-

---

Fig. 22. Pregnancy-induced mammary changes effecting milk synthesis, colostral milk-secretion, and lactation postpartum (from Vorherr, 1972b). Already at midpregnancy, increased DNA synthesis of mammary epithelial tissues is observed, followed by division and differentiation of the glandular epithelium. Whereas the synthesis of proteins increases at midpregnancy, that of lactose is greatly suppressed; probably by progesterone. As pregnancy advances toward term, the mammary metabolic processes as well as proliferation and differentiation of glandular cells become intensified. It is thought that glandular stem cells undergo division and differentiation into milk-synthesizing, presecretory and myoepithelial cells due to the influence of prolactin, growth hormone, human placental lactogen, insulin, cortisol, and an epithelial growth factor. Although glandular luminal cells are actively synthesizing milk fat and protein, only small amounts of them are released into the alveolar lumen. With postpartum withdrawal of luteal and placental sex steroids and lactogen, pituitary prolactin is able to stimulate these presecretory glandular cells to actively synthesize and release milk into the alveoli and smaller milk ducts by apocrine and merocrine secretion. During suckling, in addition to prolactin, oxytocin is also released, causing ejection of milk from the alveoli and smaller ducts into the larger ducts and lactiferous sinuses.

iphery, grow through it toward its center, and in this manner fat is replaced by glandular tissue. In summary, the following subsequent mammary gestational changes are observed.

1. Mammary changes in the first trimester of pregnancy
   a. Sprouting of terminal ductular tubules
   b. Appearance of migratory cells and fibroblasts in connective tissues adjacent to the sprouting elements
   c. Invasion of mammary fat tissue by glandular epithelial buds, which grow by replacing the fat tissue
2. Mammary changes in the second trimester of pregnancy
   a. Dichotomic ductular formation of endtubuli and endvesicles (alveoli); change of 2-cell-layered alveolar epithelium into a monolayered secretory unit
   b. Accumulation of secretional products in the alveolar lumina
   c. Appearance of lymphocytes in the surrounding connective tissue
3. Mammary changes in the last trimester of pregnancy:
   a. Increasing accumulation of intraplasmic fat droplets in alveolar epithelial cells
   b. Progressing dilation of mammary alveoli by enhanced secretion of colostrum
   c. Diminution of interlobular connective tissue carrying a large number of dilated and blood filled capillaries

E. MILK SYNTHESIS AND MILK SECRETION DURING PREGNANCY

Pituitary prolactin secretion increases as pregnancy advances (Friesen, 1972; Fig. 26), and it appears that luteal and placental sex steroids antagonize the full secretory prolactin effect on the mammary epithelium. In studies on pregnant and lactating rats it was observed that sex steroids antagonize the galactopoetic prolactin effect on the mammary epithelium (H. Vorherr, unpublished observations, 1971). In the second trimester of pregnancy, however, the mammary glandular cells show signs of beginning protein synthesis. The DNA and RNA content of the cellular nuclei increases during pregnancy and is highest at lactation. In rat experiments it has been found that maternal hormones are the primary stimulant of mammary development (DNA content) and metabolism (RNA content) during the first half of pregnancy, whereas during the second half of gestation substances from the fetal placenta effectively stimulate mammary development

(Desjardins *et al.*, 1968). On the basis of experiments with rodents (Lyons, 1958; Turkington, 1968 a,b), it is thought that during pregnancy glandular epithelial stem cells are stimulated by sex steroids, placental lactogen, insulin, growth hormone, cortisol, and a specific epithelial growth factor to initiate DNA synthesis and cellular mitosis (scheme of Fig. 22). The placenta provides a major part of sex steroids and other lactogenic substances. The injection of rat placental extracts to virgin rats can induce lobular-alveolar mammary development and lactation (Lyons, 1958). Milk synthesizing presecretory glandular cells, myoepithelial cells, and other stem cells seem to be formed through stem cell division and differentiation. It is likely that basal alveolar clear B-cells function as such stem cells. These glandular proliferation processes increase as pregnancy advances and hormonal stimulation grows. Although from midpregnancy on, alveolar mammary cells are actively synthesizing milk fat and proteins, only small amounts of them are released into the acinar lumen. It seems, therefore, that only postpartum, due to sex steroid withdrawal, can prolactin fully stimulate presecretory alveolar cells to synthesize at a maximal pace and to secrete milk into the alveoli and thus to become typical secretory cells. From midpregnancy on, however, some colostrum is secreted into the alveoli and can be expelled from the nipples. The synthesis of colostral milk during pregnancy is most likely due to the effects of prolactin, human placental lactogen, and metabolic hormones under the presence of luteal and placental sex steroids. It appears that during pregnancy, however, synthesis and secretion of milk into the alveolar lumen are largely inhibited by the antagonistic effects of estrogens and progesterone toward prolactin acting directly on the mammary epithelium. Formerly, it was believed that estrogens and progesterone stimulate the hypothalamic prolactin-inhibiting factor (PIF) into continued function resulting in restriction of transmembranal prolactin release from pituitary lactotrophs. However, since high serum levels of prolactin have recently been observed (by specific radioimmunoassay) during the last trimester of gestation, it seems that mammary synthesis and alveolar secretion of milk is most likely suppressed by a direct inhibitory effect of estrogens and progesterone on the alveolar epithelium rather than by stimulation of PIF; it has been suggested that progesterone inhibits the specifier protein α-lactalbumin, a component of lactose synthetase. In the last trimester of pregnancy two types of highly restricted milk secretion seem to exist (Dabelow, 1957); these are apocrine (cellular apical decapitation) and merocrine (no change in cellular morphology) secretional processes. Also, in the last trimester cytoplasmic protrusions appear at the cellular apex of the alveolar epithelium. In this evagination of cytoplasm, fat and Golgi substances (proteins and lactose) are present. Occasionally, the protruding part is constricted away from the cell body into the alveolar lumen (apocrine secretion).

In addition, albeit to a lesser extent, fat and proteins may escape via apical cellular pores. During the process of milk secretion, the rather high and cylindrical alveolar cells may release proteins by merocrine secretion; these proteins seem to adhere from the outside to the apically bulging membrane to form protein "caps." Inside the cell, beneath the bulging apical membrane, fat droplets usually appear. It is also believed that the fat droplets may be released by protruding through membranal openings (pores) into the alveolar lumen, carrying with them the protein cap attached to their outside. Through this secretory mechanism the cell remains largely intact. This process of milk synthesis and alveolar milk secretion will be discussed in detail in Chapter III.

F. PREGNANCY INDUCED CHANGES OF THE ANTERIOR PITUITARY GLAND

It was known as early as 60 years ago that the anterior pituitary increases in size during pregnancy and that a specific cell type termed "pregnancy cells" ("Schwangerschaftszellen") appears (Erdheim and Stumme, 1909). In the pituitary of a nonpregnant woman or in the male, both chromophobe and chromophile cells are observed in equal amounts. The chromophilic cells are subdivided into acidophilic cells (70%) and basophilic cells (30%) (Table 11). In Table 12 the weight, composition, blood supply, and cell types of the human hypophysis are presented.

The acidophilic-erythrosinophilic pregnancy cells (prolactin cells) are increased in number as well as in hormone-synthesizing activity during pregnancy and lactation (Tables 11, 12, 13, and 14; Figs. 23 and 24). A transitional stimulation of these prolactin cells is also found in the pituitary of newborns; this is due to maternal transplacental sex steroids. During childhood practically no prolactin cells can be seen under the light microscope. Even in young women, pituitary prolactin cells (lactotrophs) are poorly developed; they are few, small, and contain hardly any secretory granules (Pasteels, 1970). Pituitaries of adult males and females contain equal amounts of prolactin (1.0–1.5 μg per mg wet weight) (Sinha *et al.*, 1973c); in gravidas and lactating women the pituitary prolactin content is perhaps 10- to 20-fold higher than in nonpregnant women.

The cells that were formerly called pituitary "pregnancy cells" are now

TABLE 11

Adenohypophysial Epithelial Cells in Nonpregnant and Pregnant Women [a]

	Nonpregnant state [b]				Pregnant state [b]		
Pituitary cell types	Cellular size (diameter)	Size of nucleus	Distribution of epithelial pituitary cells	Main localization in pituitary	Cellular size (diameter)	Size of nucleus	Distribution of epithelial pituitary cells
Chromophobes	7 μm	5.7 μm	49%	Anterior glandular parts (forming cellular cords 50–70 μm thick)	7 μm	5.8 μm	51%
Acidophils							
Somatotrophs	11 μm	5.4 μm	35%	Posterior glandular parts	10 μm	5.1 μm	22%
Lactotrophs	12 μm	5.5 μm	0.5%	Anterior and lateral glandular parts (forming cellular cords 250 μm thick)	14 μm	8.3 μm	18%
Basophils	12.5 μm	5.6 μm	15.5%	Anterior glandular parts	12.5 μm	5.6 μm	9%

[a] Data derived from various sources as indicated in the text.
[b] Mean values given.

TABLE 12

Human Pituitary Gland

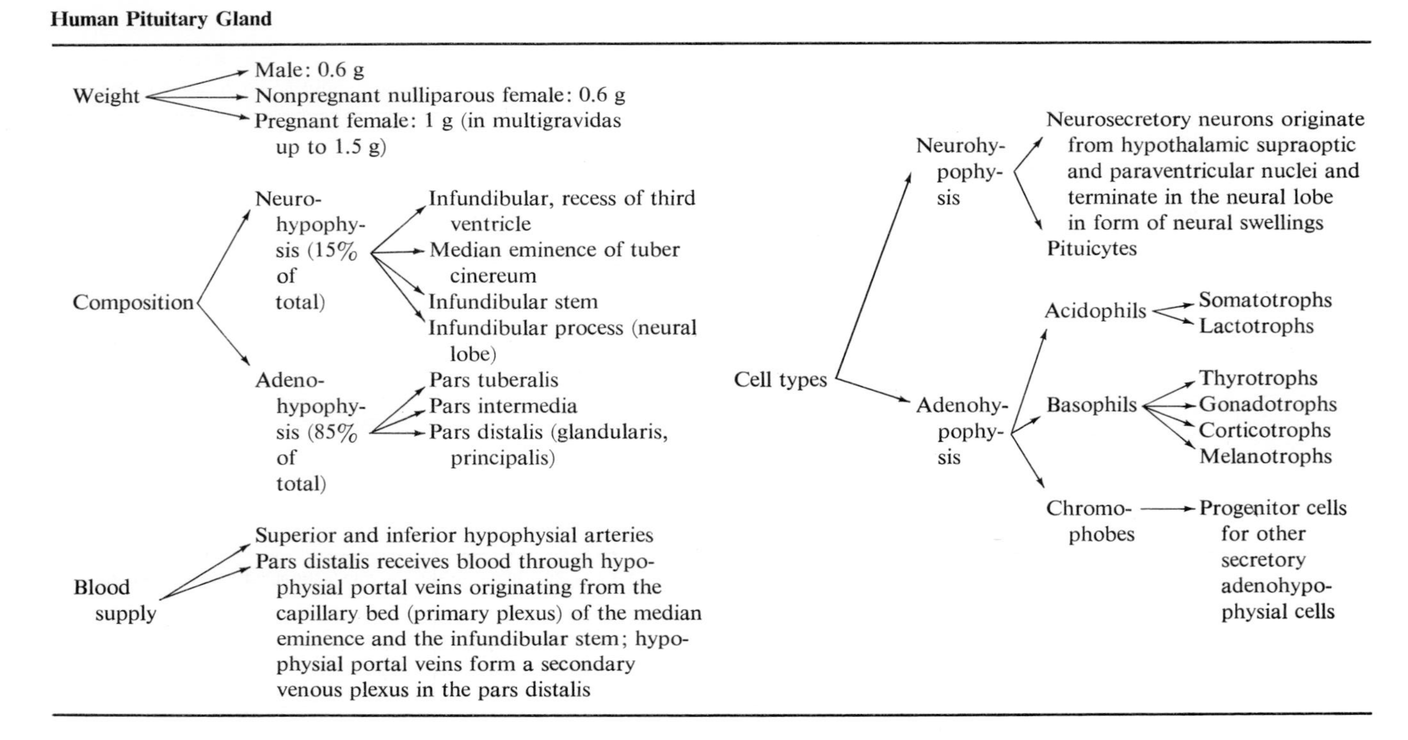

TABLE 13

Morphologic and Physiologic Characteristics of Prolactin Cells (Lactotrophs)

1. Large, ovoid, granulating erythrosinophilic cells, well established around the third month of gestation (in nonpregnant women the adenohypophysial content of lactotrophs is negligible)
2. Arranged as cellular cords in the center of pituitary alveoli and surrounded by somatotrophs; with advancing pregnancy, prolactin cell cords invade the whole pars distalis of the pituitary gland
3. Pituitary chromophobes are considered as progenitor cells for the generation of prolactin cells; toward the end of gestation lactotrophs represent 30–60% of all acidophilic pituitary cells
4. During the last trimester of gestation prolactin cell hypertrophy (nucleus, cytoplasm) reaches a maximum; also an increase in size and number of cellular organelles (mitochondria, Golgi apparatus, rough-surfaced endoplasmic reticulum) is observed; a variable degree of degranulation of lactotrophs indicates excretory function, i.e., prolactin secretion
5. In nonlactators prolactin cell involution takes place within 4–8 weeks postpartum; first an accumulation of secretory material in prolactin cells occurs, followed by autolytic regressive cellular processes

recognized as a transitional phase of chromophobe-chromophile-chromophobe changes of pituitary chromophobe cells (Floderus, 1949). These cells are actively synthesizing prolactin. Their plasma contains large granules, and their nucleocytoplasmic ratio is high. Accordingly, the nuclei are large, and their nucleoli are also rather voluminous. These "pregnancy cells" or prolactin cells (lactotrophs) stain selectively bright red with carmoisine L and with erythrosine and thus can be distinguished from other acidophilic anterior pituitary structures, which stain preferentially yellowish with Orange G (Table 14, Fig. 23). "Pregnancy cells" can also be distinguished morphologically from the chromophobe pituitary cells by their larger nuclei and cytoplasmic fraction containing large granules (chromophobe cells show no cytoplasmic granules under the light microscope); in some lactotrophs cytoplasmic granules may be absent. Whereas according to an earlier publication (Erdheim and Stumme, 1909) an accumulation of lactotrophs in the peripheral parts of the adenohypophysis was observed (see also Table 11), no regional differences in the distribution of lactotrophs were reported by Pasteels *et al.* (1972).

A relationship seems to exist between chromophobe-acidophilic-orangeophilic (growth hormone producing), chromophobe-acidophilic-erythrosinophilic (prolactin secretion), and chromophobe-basophilic (gonadotropin, corticotropin, and thyrotropin secretion) cells, which may all derive from the common chromophobe progenitor cell upon specific hypothalamic

TABLE 14

Morphologic Characteristics of Pituitary Prolactin and Growth Hormone Cells

Prolactin cells (lactotrophs) (acidophilic–erythrosinophilic)	Growth hormone cells (somatotrophs) (acidophilic–orangeophilic)
1. Cytoplasmic granules stain with erythrosine in a brilliant rose color; sequential staining with Orange-G phosphotungstick acid cannot displace the erythrosine	1. Cytoplasmic granules stain with erythrosine in a brilliant rose color; sequential staining with Orange-G phosphotungstick acid displaces erythrosine leaving yellow-orange cytoplasmic granules
2. Polyhedral cellular shape, larger than somatotrophs; cytoplasmic granules are coarse and sparsely distributed; high nucleocytoplasmic ratio (large nuclei and nucleoli)	2. Ovoid cells, smaller than lactotrophs with dense, fine cytoplasmic granules, small, dark, round nucleus
3. Electron microscopy: irregularly shaped nucleus with multiple indentations and dense chromatin; homogenously dense polymorphic secretory granules (400–700 $m\mu$ in size) dispersed throughout cytoplasm; enlarged mitochondria; dilated rough endoplasmic reticulum; enlarged Golgi apparatus; vaguely defined plasma membrane	3. Electron microscopy: round or elongated nucleus with visible nucleolus and fine chromatin; circular closely packed secretory granules (300–500 $m\mu$ in size) dispersed along plasma membrane or throughout cytoplasm; small to normal-sized mitochondria; prominent, lamellar oriented rough endoplasmic reticulum; moderately developed Golgi apparatus; thin, electron dense plasma membrane with desmosomes and pinocytotic vesicles
4. Antibodies against growth hormone do not localize in prolactin cells (no discernible fluorescence)	4. Reduced number during pregnancy and lactation (somatotrophs are probably suppressed to a chromophobic state through feedback inhibition by the growth hormonelike effect of placental lactogen)

Acidophilic (eosinophilic) pituitary adenomas may consist of:
- Somatotrophs → Acromegaly
- Lactotrophs → Galactorrhea

(chromophobe adenomas may contain degranulated secretory active lactotrophs or somatotrophs)

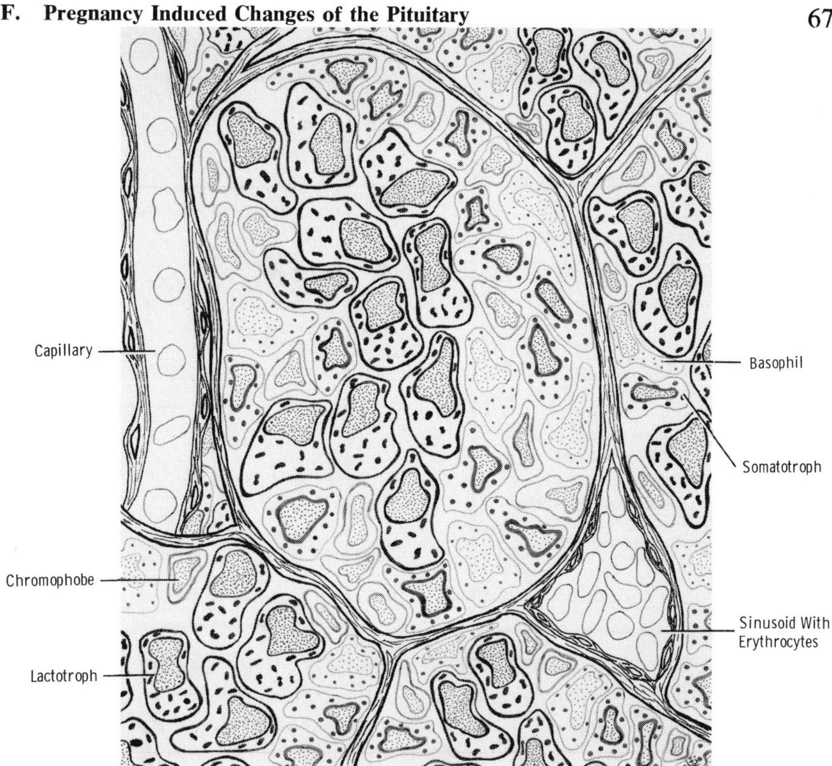

Fig. 23. Distribution of epithelial cells in the adenohypophysis of pregnant women. (1) Lactotrophs (prolactin cells): Lactotrophs are large ovoid or irregularly shaped cells that are rarely found in the hypophysis of infants, men, or nonpregnant women. Prolactin cells ("pregnancy cells") increase in number and size during gestation. Already at the third month of pregnancy many lactotrophs are present, and at midpregnancy these cells contain large secretory granules of about 600 mμ in diameter. In some gravidas near term, up to 50% of all epithelial cells of the pars distalis of the pituitary gland may consist of lactotrophs, which are arranged as cellular cords of 100–250 μm thickness in the center of the adenohypophysial alveoli. (2) Somatotrophs (growth hormone cells): Somatotrophs are smaller than lactotrophs and contain fine, dense granules of about 300 mμ in diameter. In pituitaries of gravidas, somatotrophs are more scattered due to dispersion by the increasing number of hypertrophied lactotrophs. It is thought that during pregnancy the number of somatotrophs is reduced by their suppression to a chromophobic state by placental lactogen; this suppression persists into the phase of lactation. (3) Chromophobes (progenitors of pituitary epithelial cells): Chromophobes possess no intraplasmic granules. Because they are progenitor cells of lactotrophs, and since somatotrophs may be suppressed into a chromophobe-like state by human placental lactogen, a slight increase in the number of chromophobes is observed during gestation. (4) Basophils (thyrotrophs, gonadotrophs, corticotrophs, melanotrophs): Basophils are almost as large as lactotrophs, with a rather round nucleus and intraplasmic granules of about 150 mμ in diameter. A reticular meshwork supports the epithelial elements arranged in alveoli, which are surrounded by large vascular sinusoids (for further information on staining and morphology of lactotrophs and somatotrophs, see Table 14).

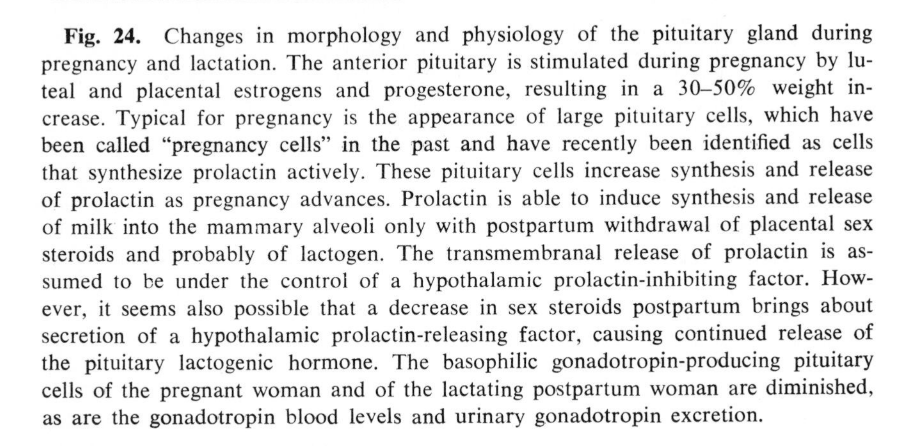

Fig. 24. Changes in morphology and physiology of the pituitary gland during pregnancy and lactation. The anterior pituitary is stimulated during pregnancy by luteal and placental estrogens and progesterone, resulting in a 30–50% weight increase. Typical for pregnancy is the appearance of large pituitary cells, which have been called "pregnancy cells" in the past and have recently been identified as cells that synthesize prolactin actively. These pituitary cells increase synthesis and release of prolactin as pregnancy advances. Prolactin is able to induce synthesis and release of milk into the mammary alveoli only with postpartum withdrawal of placental sex steroids and probably of lactogen. The transmembranal release of prolactin is assumed to be under the control of a hypothalamic prolactin-inhibiting factor. However, it seems also possible that a decrease in sex steroids postpartum brings about secretion of a hypothalamic prolactin-releasing factor, causing continued release of the pituitary lactogenic hormone. The basophilic gonadotropin-producing pituitary cells of the pregnant woman and of the lactating postpartum woman are diminished, as are the gonadotropin blood levels and urinary gonadotropin excretion.

stimulation. It appears also probable that the chromophobes derive from regressed and secretory degranulated acidophilic and basophilic pituitary cells. It appears rather unlikely that the acidophilic pituitary cells represent one cell type capable of producing both growth hormone and prolactin. Antibodies produced in rabbits against greatly purified human growth hormone and placental lactogen did not localize in pregnancy cells, indicating that pregnancy cells do not produce a hormone similar to growth-hormone or placental lactogen; this speaks in favor of the specific function of the pregnancy cell regarding prolactin synthesis (Friesen and Guyda,

1971). Conversely, antibodies produced against ovine prolactin, yield immunofluorescence only in pituitary lactotrophs but not in somatotrophs. This pituitary prolactin cell hypertrophy (Fig. 23) is induced by luteal and placental sex steroids and leads to a 30–50% increase in pituitary size during gestation. In nonpregnant, nulliparous women, in primigravidas and in multigravidas near term, the average pituitary weight is 0.62 g, 0.85 g, and 1.1 g, respectively. In some multiparous women the anterior pituitary grows during pregnancy to such an extent that visual disturbances through pressures on the optic nerve develop temporarily; in these gravidas pituitary weights of up to 1.5 g may be observed. This increase in pituitary size and weight during gestation involves only the adenohypophysis and is mainly due to hyperplasia and hypertrophy of lactotrophs. Lactotroph and somatotroph counts in human hypophyses clearly show the increase in the number of pituitary lactotrophic cells during gestation and lactation (Goluboff and Ezrin, 1969; Table 15).

Recently, increased prolactin synthesis and release into the incubation medium was observed in pituitary glands of pregnant rhesus monkeys (Friesen and Guyda, 1971). In pregnant or lactating rhesus monkeys the number of pituitary gonadotropin-producing basophilic cells and of growth hormone-synthesizing acidophilic somatotrophs is reduced, whereas the lactotrophs greatly increase in number. The data in Table 16 represent the distribution of lactotrophs, somatotrophs, and the prolactin and growth hormone concentration in the pituitary of the rhesus monkey (Pasteels *et al.*, 1972).

TABLE 15

Lactotroph and Somatotroph Count in Human Hypophyses [a]

Subjects	Cell count: mean numbers of cells (acidophils) per visual field		Lactotrophs as percentage of total acidophils
	Lactotrophs	Somatotrophs	
Children, nonpregnant women, and men	0–0.6	21—35	0–2
Pregnant women			
1st Trimester	6	7	15–60
2nd Trimester	5	22	5–40
3rd Trimester	17	7	50–90
Lactating women	14	6	40–90

[a] Data from Goluboff and Ezrin (1969).

TABLE 16

Lactotroph and Somatotroph Count, Pituitary Prolactin, and Growth Hormone Concentration in Rhesus Monkey Hypophyses [a]

	Cell count in percent of area of visual field		Hormone concentration (ng/mg wet weight)	
Subjects	Lactotrophs	Somatotrophs	Prolactin	Growth hormone
Infant	0	32	4	16
Adult male	8	20	4	8
Adult female	8	29	6	9
Lactating female	30	27	11	9

[a] Data from Pasteels *et al.* (1972).

It is possible that during pregnancy the increased secretion of placental lactogenic substances (HPL) somewhat restrains the function of pituitary growth hormone cells; it appears unlikely that HPL inhibits prolactin secretion through a hypothalamic mechanism or by a direct effect on the adenohypophysis. Pituitary growth hormone secretion is decreased in late pregnancy, and its output remains reduced during the first 6 weeks postpartum. During gestation, increased plasma cortisol and placental lactogen (negative feedback at hypothalamo-pituitary level) appear to counteract synthesis, release, and metabolism of pituitary growth hormone (Spellacy and Buhi, 1969). Moreover, the adenohypophysial growth hormone content is slightly decreased in pregnant and lactating rhesus monkeys (Pasteels *et al.,* 1972).

Increased production of luteal and placental sex steroids also produces a negative hypothalamic feedback mechanism, leading to diminution of pituitary gonadotropin secretion during pregnancy (see also Chapter VI) (Herlant, 1964; Amoroso and Porter, 1966; Herlant and Pasteels, 1967; Goluboff and Ezrin, 1969; Jaffe *et al.,* 1969; Schelin and Lundin, 1971).

CHAPTER III

Galactopoiesis, Galactosecretion, and Onset of Lactation

A. ANTENATAL PREPARATION OF THE BREASTS FOR NURSING

For a gravida who intends to nurse her baby, the antenatal preparation of the breasts is important for the process of lactation. The volume of puerperal milk secretion is closely related to the extent of mammary glandular development during pregnancy, which in turn depends on proper antenatal and immediate postnatal breast stimulation for lactation. The prepregnant size of the breasts shows little correlation with the amount of milk produced postpartum. Good physical health and positive psychologic attitude toward nursing are prime factors for successful breast feeding.

During the last 6 to 8 weeks of pregnancy, turgescence of breast and nipples is enhanced, and the skin of areolae and nipples is strengthened by massaging breast and nipple areas twice daily with a wet washcloth. In gravidas with dry and parched skin, nonirritant creams (Nivea, Polysorb Hydrate, Nutraderm Cream) may be massaged into the skin of the breast and areolar area. The application of lotions or solutions containing alcohol or other volatile substances, which may dry and harden the skin of the breast, should be avoided. Inverted nipples or poor nipple protraction may greatly interfere with the nursing process. Such nipples should be drawn outward each day; if necessary, a milk-suction pump may be used, and nipple shields with a well-fitting brassiere should be worn. During the last 4 to 6 weeks of gestation, the gravida should be advised to express colostrum from the nipples twice daily, in preparation for the upcoming process of lactation. Proper nutrition during pregnancy is necessary to meet the

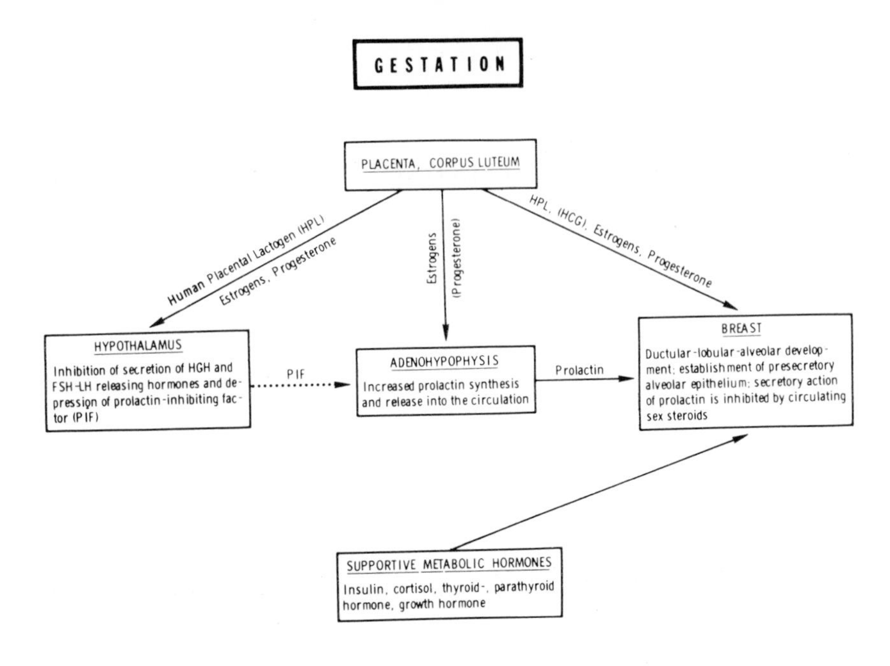

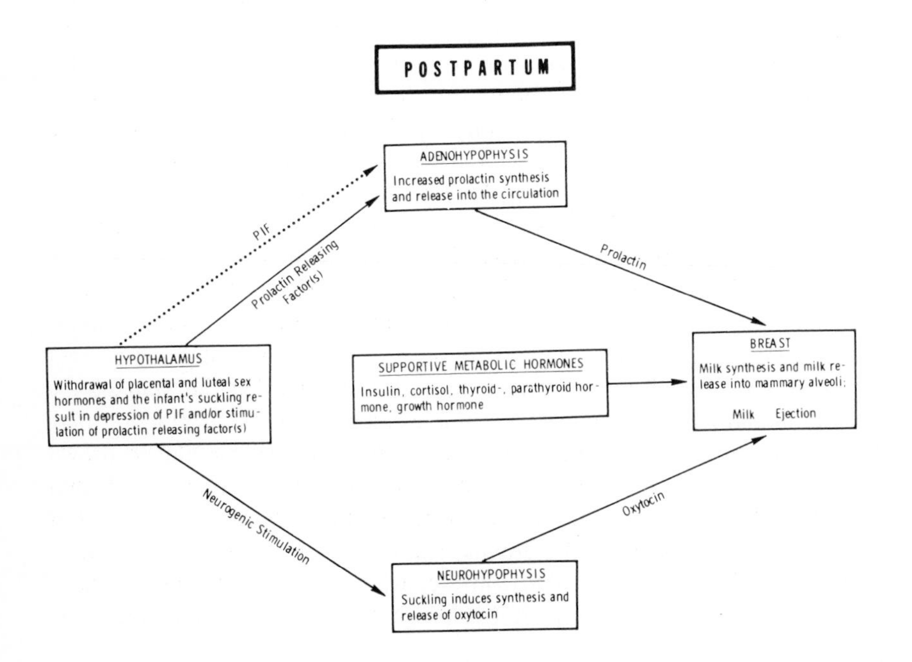

See legend on opposite page.

many metabolic requirements of mother and fetus and to allow adequate preparation of the organism for lactation (Table 27) Zilliacus, 1967; Benson, 1968; Millar, 1969).

B. MILK SYNTHESIS AND COLOSTRUM FORMATION DURING PREGNANCY

Pituitary prolactin secretion slowly increases during the first half of pregnancy, and, in the second trimester of pregnancy when blood prolactin levels are 3- to 5-fold increased, some mammary glandular cells begin to show signs of protein synthesis. The DNA and RNA content of the cellular nuclei increases during pregnancy and is highest at the time of lactation. Experiments with rodents (Turkington, 1968 a, b) indicate that during pregnancy glandular epithelial stem cells are stimulated by sex steroids, placental lactogen, insulin, growth hormone, cortisol, and a specific epithelial growth factor to initiate DNA synthesis and cellular mitosis (Figs. 19, 22, and 25). Milk synthesizing presecretory glandular cells, myoepithelial cells, and other stem cells seem to be formed through stem cell division and differentiation; as mentioned before, it is likely that basal alveolar B-cells function as stem cells from which presecretory, secretory, and myoepithelial cells derive. These glandular proliferation processes increase as pregnancy advances and hormonal stimulation grows. Although from midpregnancy on, alveolar mammary cells are actively synthesizing milk fat and proteins, only small amounts of these are released into the acinar lumen; only with postpartum withdrawal of luteal and placental sex

---

Fig. 25. Hormonal preparation of the breast for lactation. Increased amounts of sex hormones stimulate prolactin synthesis in "pregnancy cells" (prolactin cells) of the adenohypophysis, and prolactin is increasingly released into the bloodstream as gestation advances; prolactin-inhibiting factor (PIF) function is decreased (dotted line). Placental hormones and maternal metabolic hormones bring about ductular-lobular-alveolar breast development in preparation for lactation. Postpartum withdrawal of placental and luteal sex steroids permits full secretory effect of prolactin on the mammary alveolar epithelium. During the infant's suckling, adenohypophysial prolactin and neurohypophysial oxytocin are released into the circulation stimulating further milk synthesis and milk ejection, respectively. Metabolic hormones such as insulin, cortisol, thyroid–parathyroid hormone, and growth hormone are important cofactors in the process of postpartum milk synthesis. Effective breast stimulation by the suckling newborn, adequate removal of milk, intact hypothalamo-hypophysial function, normal sympathetic-adrenal tone, and appropriate nutrition are key factors of successful nursing.

steroids, and probably of placental lactogen, is prolactin fully capable of stimulating synthesis and secretion of milk into the alveoli and thus changing the mammary alveolar epithelium into typical secretory cells. From midpregnancy on, however, some colostrum is secreted into the mammary alveoli and can be expelled from the nipples. Besides water and minerals, colostrum is composed of fat droplets, round cells (lymphocytes, monocytes, histiocytes), and Donné bodies, which are a coagulation of lymphocytes, round cells, and desquamated phagocytic alveolar cells (foam cells).

C. ONSET OF LACTATION

After postpartum withdrawal of placental lactogen and sex steroids, prolactin-induced lactation sets in. It appears that during pregnancy luteal and placental sex steroids prevent prolactin-induced milk secretion considerably by antagonizing the prolactin effect on the mammary secretory epithelium (Fig. 26). After placental delivery a rapid decline of sex steroids in the plasma occurs; also the luteal sex steroid production, which is maintained during pregnancy until term, ceases; and around the fourth to the fifth postpartum day only low systemic levels of estrogens and progesterone comparable to those in the follicular phase of the cycle are present (LeMaire *et al.,* 1971). In this manner, a conversion of mammary glandular cells from presecretory into actively milk-synthesizing and milk-releasing cells is achieved (Fig. 22). It is also possible that the postnatal decrease of sex steroids and placental lactogen may induce hypothalamic secretion of a prolactin-releasing factor, which may be discharged into the hypophysial portal system for further stimulation of transmembranal secretion of prolactin from pituitary lactotrophs (Fig. 25) (Sulman, 1970).

It is assumed that the synthesis of prolactin is hardly increased by small amounts of sex steroids as encountered postpartum but that low sex hormone levels promote transmembranous prolactin release. Nevertheless, postpartum mothers having ovulatory cycles and thus higher concentrations of circulating ovarian sex steroids (Table 9) can nurse successfully; this indicates that physiologic amounts of luteal estrogens and progesterone cause no measurable inhibition of prolactin-induced milk secretion postpartum. On the other hand, estrogens and progesterone are not necessary for the process of lactation because ovariectomized women and animals can nurse properly. Because of the effect of prolactin and other supportive metabolic hormones, milk is synthesized and released into the mammary alveoli and smaller milk ducts; this process is fully established around

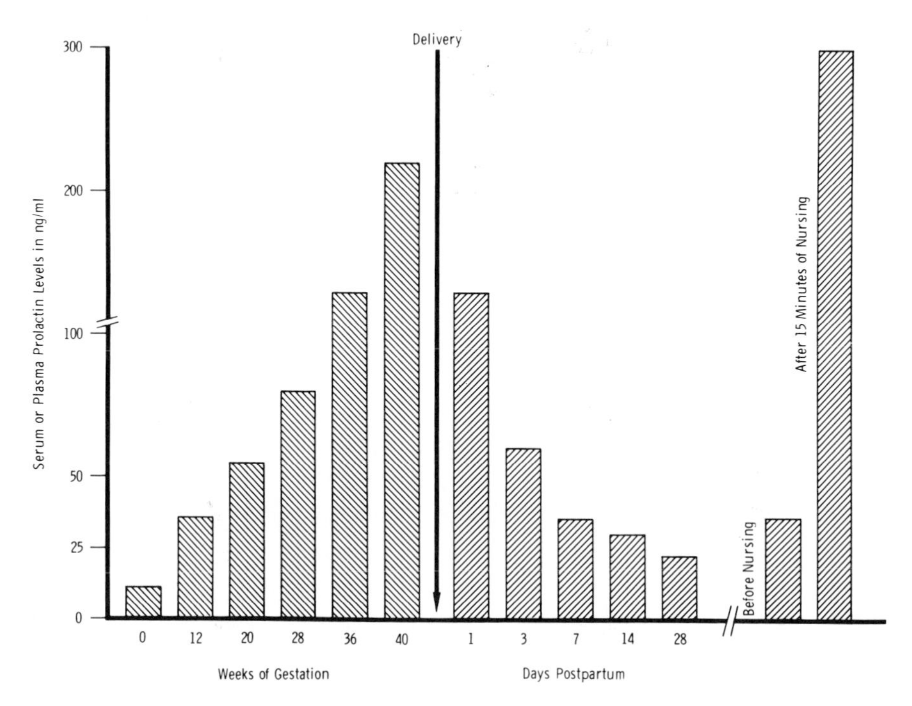

Fig. 26. Human prolactin blood levels during gestation and lactation (data mainly from Friesen, 1972, see also Table 28). The serum or plasma prolactin levels in nonpregnant females are about 10 ng/ml. During gestation serum prolactin concentrations increase gradually and reach peak values near term. After delivery, prolactin serum levels drop rather sharply; 4 weeks postpartum the serum prolactin levels are 20–30 ng/ml in lactating women and about 10 ng/ml in nonlactating puerperas. Stimulated by the infant's suckling, a prolactin spurt from the pituitary into the circulation occurs, temporarily increasing the prolactin blood levels about 10 times over the basal presuckling levels (right side of picture).

2 to 5 days postpartum, i.e., when sex steroid blood levels are low, resulting in dilated alveoli filled with milk. The breasts become engorged and tender due to this process of milk "coming in" or "shooting in" at the onset of lactation. The apparent tightness, swelling, and redness of the breasts are due to alveolar accumulation of milk and dilation of mammary blood vessels, as well as to venous and lymphatic stasis and mammary tissue edema. One third of the breast volume can be attributed to milk secretion and its storage in alveoli and smaller milk ducts (Fig. 27). Mammary tissue elasticity allows milk storage for up to 48 hours before the rate of milk synthesis and milk secretion rapidly begins to decrease (Dabelow, 1957; Zilliacus, 1967). Prolactin is the requisite hormone for milk syn-

Fig. 27. Section through an engorged mammary gland of a lactating woman 3 weeks postpartum, 48 hours after cessation of nursing (from Dabelow, 1957; thick slice Alum Carmine staining). Apparent is the dichotomic ductular-alveolar branching with maximally dilated milk-filled mammary alveoli. A similar picture may be observed postnatally at the onset of lactation ("milk coming in") when the breasts are engorged and the alveoli are maximally filled with milk. *Note:* Dashed line indicates two primitive parenchymal sprouts that did not proliferate during pregnancy.

thesis and milk release into the alveoli. Although mammary development for lactation is stimulated mainly by luteal and placental sex steroids, these hormones act fully on the breast only in the presence of placental lactogen, prolactin, ACTH (cortisol), insulin, thyroid, and parathyroid hormone. As mentioned, lactation is independent of ovarian function (estrogens and progesterone). In animals, hypophysectomy abolishes milk secretion, but it can be restored by administration of prolactin, growth hormone, and ACTH. Also, the amount of milk secreted is reduced by ablation of either the pancreas, thyroid, parathyroid, or adrenal glands, but the milk yield is restored to normal by administration of the respective hormone. More so, in cows the milk yield could be increased by 30% through administration of thyroxine or thyroglobulin. During lactation, the secretion of thyrotropic and thus of thyroid hormone is enhanced. Although lactation may occur in thyroidectomized animals, the amount of milk secreted is substantially reduced. Provided milk is removed regularly from the mammary gland, the alveolar cells continue to secrete milk almost indefinitely. Thus rats with a life span of about 2 years have been able to lactate for one year; women were found to lactate continuously for 4 years and cows for 14 years (Newton, 1961; Linzell, 1971; Linzell and Peaker, 1971).

D. BIOCHEMISTRY OF MILK SYNTHESIS AND CELLULAR MECHANISMS OF MILK RELEASE

The superficial luminal glandular cells of the acini and smaller milk ducts are involved in active milk synthesis and secretion into the alveoli and smaller milk ducts. Pituitary prolactin represents the main stimulus causing glandular cells to synthesize and secrete milk; the prolactin effect on mammary secretory cells is supported by metabolic hormones such as cortisol, insulin, thyroid hormone, parathyroid hormone, and growth hormone. Because the prolactin release is highest during the stimulus of suckling, most of the milk is synthesized during the process of nursing or immediately thereafter. Also pituitary release of ACTH is increased during suckling, resulting in enhanced cortisol plasma levels, which support the process of alveolar milk synthesis (Sar and Meites, 1969). The mammary secretory cells are of cuboidal or cylindrical shape with nuclei at their bases or at their tips depending on the stage of the mammary milk producing cycle (Fig. 28 and 18b). Mammary glandular cells are uninuclear, and, due to accumulation of secretory products, they assume a cylindrical shape shortly before milk secretion. Owing to apocrine and merocrine

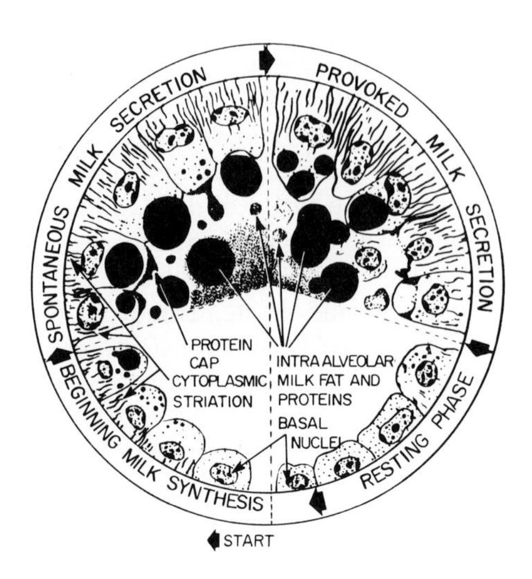

Fig. 28. Scheme of the secretory mammary cycle (from Grynfeltt, "Étude du Processus Cytologique de la Sécrétion Mammaire," Archives d'Anatomie Microscopique, 1937, by kind permission of Masson and Cie, Paris). In the resting or recovery phase of milk secretion the alveolar cells are flat; their cytoplasm is finely granulated and contains only a few fat vacuoles. In this stage the nucleus is located at the cellular basis, and a moderate secretory activity is maintained. In the process of renewed milk synthesis the glandular cells become thicker, especially when the mammary alveoli are emptied of milk. Also, cellular secretory activity increases, and the nucleus is moving toward the apical part of the cell. In addition, cellular water uptake is enhanced, leading to a longitudinal cytoplasmic striation. This phase of milk synthesis is followed by the process of spontaneous milk secretion into the glandular alveoli. At this stage the thick, secretory, clublike epithelial cells contain apical membranal caps, which are then constricted away with parts of the cytoplasm, leaving the apical cell membrane intact. The secretory activity of alveolar cells seems to be even more intensified during the process of nursing (provoked milk secretion).

milk secretion into the mammary alveoli, the amount of cytoplasm decreases, and the cells become greatly flattened. When milk is removed by suckling, the alveoli partially collapse, and the flattened glandular cells with nuclei at their base protrude into the alveolar lumen. At this stage, milk synthesis and secretion is continued to a small degree. The flattened glandular cells reveal a finely granulated cytoplasm with vacuoles and droplets indicating continued cellular activity; their nucleus rests near the basal membrane (Dabelow, 1957).

In the process of morphologic rebuilding, the cellular volume becomes at first enlarged below the basally located nucleus. Through the enhanced uptake of water, the cellular turgor increases, and the cytoplasm appears longitudinally striated from the cell basis to its apex. The glandular secretory cells are mononuclear with round or ovoid nuclei; the axis of the ovoid nuclei corresponds to the longitudinal axis of the cell. The nuclear diameter of the alveolar cell of the nonlactating breast is smaller (5 μm in diameter) than that of the milk synthesizing glandular alveolar cell (6.3 μm in diameter).

During the process of secretory activities, the nuclear chromatin content and distribution change depending on the secretory activity of the respective glandular cells. Thus, it may be assumed that these nuclear chromatin changes are related to the activities of nuclear DNA and messenger RNA, inducing cellular RNA, protein, lactose, and fat synthesis.

During the process of lactation, mitosis of secretory glandular cells is rarely observed. Mitochondria are abundant in the transparent cytoplasm, providing energy. The Golgi apparatus of secretory glandular cells is mainly located around the nucleus and is enlarged during lactation. This hypertrophy of the Golgi apparatus is most pronounced at the height of cellular milk synthesis, shortly before milk is released into the alveolar lumen. At the peak of cellular milk synthesis, the secretory cell becomes rather large and extends as a cylindrical unit into the alveolar lumen; at this stage its apical cell membrane is very thin (Dabelow, 1957).

While the glandular cell is recovering from its excretory losses, it becomes more and more cuboidal with its nucleus located near the basis (Fig. 28). The glandular cells are surrounded by starlike myometrial cells at their base, and both glandular and myometrial cells rest on a narrow basal band of connective tissue containing thin-walled capillaries. The mammary lobular arteries approach the alveoli at their tip, from where they branch into arterioles and capillaries running along each alveolus to its ductal origin. Venous drainage occurs at the acinar base near its junction with the secondary small milk duct (Turner, 1952; Dabelow, 1957) (Fig. 18a).

During lactation the secretory, monolayered cells of alveoli and smaller milk ducts are involved in active milk synthesis and passage of the formed products into their respective lumina (Table 17). Approximately 1–2 ml of milk per gram mammary tissue per diem are produced. The milk's content of fat, proteins, and electrolytes may vary, but the lactose concentration remains rather constant. It has been suggested that lactose is a controlling factor for the volume of secreted milk.

As mentioned before, milk is secreted by apocrine and merocrine mechanisms. It seems unlikely that holocrine milk secretion, characterized by total cell disintegration, takes place, since a high mammary glandular mitosis rate is not observed in the lactating breast. Fragments of epithelial

TABLE 17

Synthesis and Transmembranal Secretion of Milk into the Alveolar Lumen [a]

1. Proteins. *De novo* synthesis of casein, α-lactalbumin, and β-lactoglobulin from plasma amino acids on ribosomes of endoplasmic reticulum (ER); aggregation of protein "vesicles" in the Golgi apparatus

2. Fat. *De novo* synthesis of fat from plasma long-chain fatty acids on ER (malonyl-CoA pathway); mammary glandular synthesis *de novo* of short-chain fatty acids. Esterification of fatty acids and lipid accumulation at ER cisternae. Coalescence of smaller fat droplets and localization of larger fat globules at the cellular apex

3. Lactose. Synthesis from glucose; catalyzation of the reaction UDP-galactose + glucose→ lactose + UDP by lactose synthetase in the Golgi vesicles. Swelling of Golgi apparatus through osmotic effect of synthesized lactose; localization of enlarged Golgi apparatus near apical cell membrane

4. Secretory mechanisms for milk constituents to be released into the mammary alveolar lumina
   a. Fat, mainly by apocrine secretion
   b. Proteins, by apocrine and merocrine secretion
   c. Lactose, mainly by merocrine secretion
   d. Ions, by diffusion and active transport

5. Alveolar epithelial membrane permeability
   a. Cell membrane is permeable in both directions (cell inside to alveolar lumen and conversely) to glucose, water, sodium, potassium, calcium, chlorine, iodine, phosphate, and sulfate
   b. Cell membrane is impermeable from alveolar lumen to inside of secretory cell for lactose, sucrose, citrate, proteins, and fat ("Milk–Blood Barrier")

6. Primary alveolar milk is diluted secondarily by extracellular fluid to reach plasma isotonicity

[a] From Vorherr (1972b).

cells, however, are found in colostral milk of humans, indicating the possibility of a small amount of holocrine milk secretion. The secreting mammary cell (Fig. 28 and 29) displays a spherelike nucleus with increased amounts of ribosomes, rough endoplasmic reticulum, and an enlarged Golgi apparatus lying between nucleus and cellular apex. Alveolar cells contain a large amount of fat globules, protein granules, and microvilli at their apex. Their basal membrane is folded; this can be attributed to transport processes for amino acids and sugar from the surrounding capillaries. The glandular cells are connected laterally by tight desmosomal junctions that barely allow passage of fluid between cells. The alveolar epithelium changes form and size (Fig. 18b and 28) depending on the stage of the secretory cycle. Milk is a suspension of proteins and fat in an aqueous lactose–mineral solution. In the following section the cellular synthesis of the various milk constituents will be discussed (Linzell and Peaker, 1971; Kon and Cowie, 1961, pp. 371–479).

1. Protein Synthesis

The postpartum synthesis of milk proteins is initiated by prolactin and supported by insulin and cortisol. Protein synthesis *de novo* produces casein, lactalbumin, and lactoglobulin, but essential and nonessential plasma amino acids are also utilized in this process. It is assumed that prolactin stimulates nuclear RNA polymerase for synthesis of messenger and transfer RNA from the DNA template. Subsequently, due to enzymatic catalytic processes, the genetic information contained in messenger RNA is conveyed to the protein-synthesizing centers of the cells. In these cells specific milk proteins are formed with the help of ribosomes and transfer RNA, the latter transports active amino acids to messenger RNA which are attached to ribosomes. Prolactin may also affect the nuclear transcription for α-lactalbumin synthesis, the "specifier" milk protein. Through these prolactin-induced mechanisms, synthesis of milk proteins (lactalbumin, lactoglobulin, casein) is achieved at ribosomes, which are attached to the endoplasmic reticulum (Fig. 29, Table 17). This rough endoplasmic reticulum is extensively augmented in mammary epithelial cells, and its cavities are increasingly dilated. The synthesized proteins are in turn discharged into the cavities of the rough endoplasmic reticulum, and, from there, they are transported to the Golgi apparatus. Here, the proteins become condensed and, thus, appear as visible secretory granules moving toward the cellular apex. Proteins are discharged mainly through apocrine secretion, which may be compared with an inverse process of pinocytosis (Fig. 30). The swollen Golgi vesicles seem to fuse with the apical plasma membrane before lactose and proteins can be discharged into the alveolar lumen. Also, protein-rich Golgi vesicles, together with fat globules and other cytoplasmic constituents, may be constricted away from the cell body and released into the alveolar lumen. Here the apical membrane envelope undergoes lysis, and all secreted milk particles become mixed in the acinar lumen. Merocrine protein secretion connotes the cellular discharge of proteins and other cellular constituents through apical membranous pores leaving the cellular structures intact. Because neither proteins nor lactose can reenter the alveolar cell, they most likely cannot leave it with-

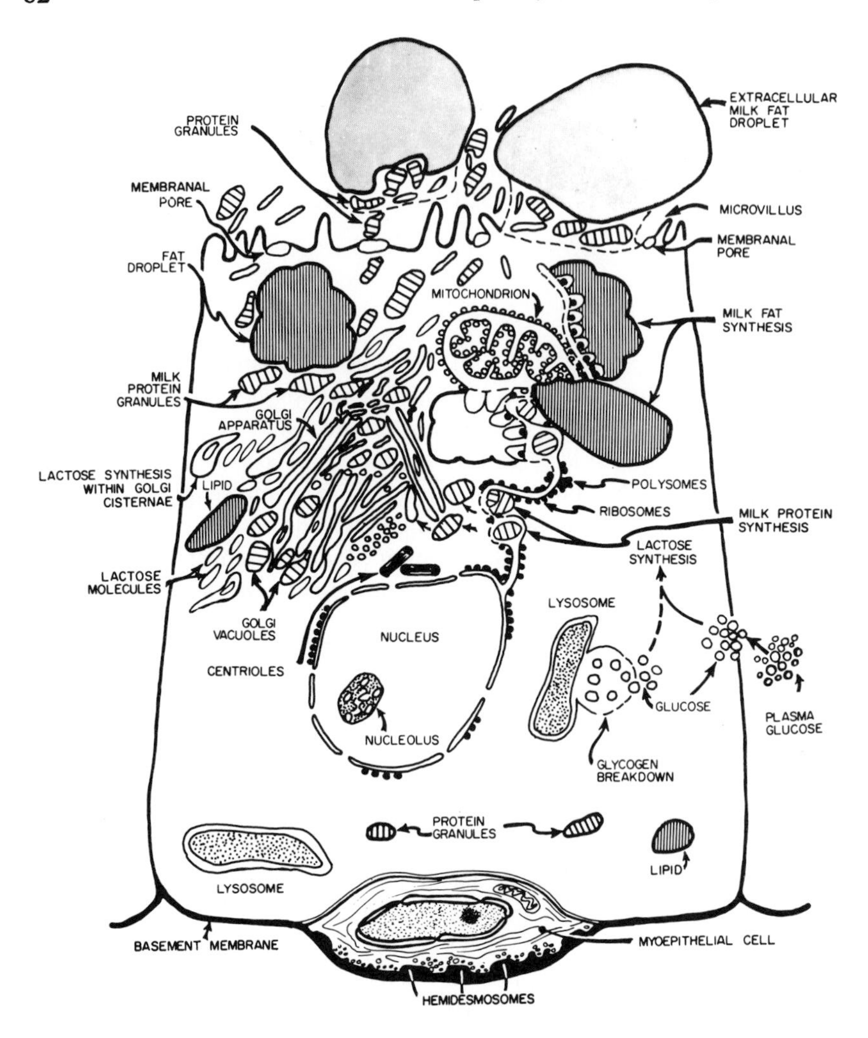

Fig. 29. Mammary synthesis and secretion of protein, fat, and lactose (from Vorherr, 1972b). Postpartum, mammary milk synthesis and its alveolar secretion are induced by prolactin. Milk-synthesizing mammary glandular cells are cuboidal with scattered microvilli on the luminal surface, and their lateral cell borders are held together by junctional interdigitating complexes. The nucleus is located toward the cellular base and contains a distinct nucleolus. A large Golgi apparatus is situated above the nucleus, and accumulation of fat droplets is largest near the apex of the cell. The other cytoplasmic organelles consist of large mitochondria with closely packed lamellar cristae and lysosomes. Protein synthesis is initiated on ribosomes, which are attached to the endoplasmic reticulum; this rough-surfaced endoplasmic reticulum is most prominent at the base of the cell. The synthesized proteins (casein, lactalbumin, lactoglobulin) are discharged into the tubules of the rough endoplasmic

out involving special secretory mechanisms, as mentioned above. It is also possible that proteins released into the alveolar lumen form "protein caps" or "signets" on the outside of the apical membrane and thus protrude into the alveolar lumen. It is believed that these signets are condensed cytoplasmic organelles such as endoplasmic reticulum, mitochondria, and Golgi protein vesicles (Kurosumi *et al.,* 1968; Hollmann, 1968; Linzell, 1971).

2. Milk Fat Synthesis

The agranular and granular endoplasmic reticulum is thought to be the locus of synthesis of milk lipids. It has also been suggested that fat synthesis in the mammary secretory cell may occur in the ergastoplasm on mitochondria (Fig. 29, Table 17). Mammary alveolar cells are capable of *de novo* synthesis of short-chain fatty acids via the malonyl-CoA pathway requiring NADPH, which is largely supplied by the pentose cycle. Thus a large proportion of cellular glucose is oxidized by the pentose pathway. The cow's secretory mammary epithelium is able to utilize carbon from plasma acetate and β-hydroxybutyrate for fat synthesis (Linzell, 1971). This is the reason why cow's milk contains a high amount of short-chain fatty acids of C_4–C_{14}. As in the case of protein synthesis, fatty acids (long-chain fatty acids) from blood plasma are also utilized for milk fat synthesis; in addition, plasma supplies the triglyceride glycerol and glucose. Most of the glycerol contained in milk fat derives from intracellular glucose. In human milk the short-chain fatty acids are synthesized from acetate; whereas the long-chain fatty acids of milk fat are derived from

---

reticulum and are transported to the Golgi apparatus where the proteins condensate as visible granules. Fat synthesis may take place on the rough endoplasmic reticulum and in the ergastoplasm on mitochondria. Synthesis of lactose probably occurs within the cisternae of the Golgi apparatus whereby α-lactalbumin functions as a specifier protein to induce lactose synthesis from glucose as substrate; the energy for this process is provided by the mitochondria. Lipid droplets and parts of the osmotically enlarged (swollen) Golgi complex, containing proteins, lactose, and water, may be secreted into the alveolar lumen by an apocrine mechanism whereby fat droplets and portions of the apically located Golgi complex are pinched off from the cytoplasm (reversed pinocytosis; see Fig. 30). Also, proteins and lactose may leave the secretory cell via apical membranous pores, i.e., merocrine secretion. In addition, diffusional, osmotic, and active transport mechanisms for intraalveolar secretion of water, ions, lactose, vitamins, and drugs have to be considered. Inserting at the basal cell membrane are myoepithelial cells which show hypertrophy during pregnancy and lactation.

blood plasma. The respiratory quotient of actively secreting glandular cells is usually over 1.0 due to synthesis of fat from carbohydrates. The esterification of fatty acids takes place in the rough endoplasmic reticulum, as

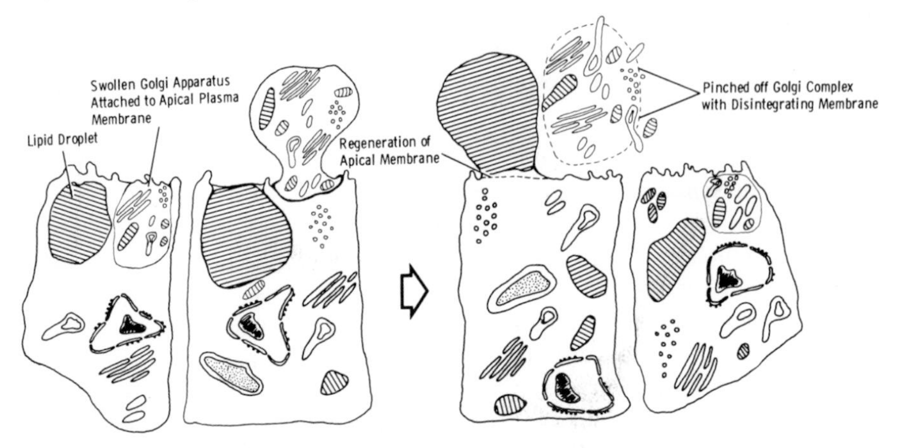

Fig. 30. Apocrine secretory mechanisms for milk lipids, proteins, and lactose. Intracellular lipid droplets and the metabolically active Golgi apparatus containing proteins, lactose, minerals, and water may be secreted into the alveolar lumen by a process of reversed pinocytosis (reversed membrane vesiculation). At the beginning of this apocrine secretory action the membrane around a larger lipid droplet or the enlarged apically located Golgi complex starts to fuse with the apical plasma membrane. This process is facilitated by the development of protoplasmic husks between the secretory granules and the apical membrane. As shown in the alveolar cell on the left side of the figure, the swollen Golgi apparatus attaches to the apical membrane by means of protoplasmatic husks; the membrane of the lipid droplet has not yet reached the plasma membrane. After fusion of the Golgi membrane with the apical plasma membrane, evagination occurs along the area of membranal attachment, and the evaginating secretory granule remains in contact with the rest of the apical membrane. This is demonstrated in the adjacent alveolar cell (right alveolar cell, left part of figure); here the fat droplet has become attached to the apical cell membrane. In the next step of the cellular secretory process (alveolar cell next to tip of arrow, right side of figure) the membrane of the swollen Golgi complex is constricted away from the alveolar cell, and the Golgi membrane begins to disintegrate, allowing admixing of the Golgi contents with the alveolar milk. After its complete apical fusion the lipid droplet evaginates and is then pinched off from the apical plasma membrane, which is already in the stage of regeneration as indicated by the dashed line in the figure. In the adjacent outer cell (right part of figure) the process of renewed milk synthesis with apical ascension of lipid droplets and Golgi apparatus is depicted; the apocrine secretory process has not yet started. It appears that milk secretion by reversed pinocytosis is only possible in actively secretory alveolar cells, and cellular energy producing processes seem essential. Therefore, active transport mechanisms seem to exist for apocrine milk secretion by reversed pinocytosis.

demonstrated with ^{3}H-labeled oleic acid. Lipids then accumulate inside and between the cisternae of the rough endoplasmic reticulum; several cisternae seem to produce and harbor a fat droplet. The smaller lipid droplets sit mainly on the cellular base, and, following subsequent coalescence, they ascend as larger globules toward the cellular apex. Synthesized fat droplets are discharged by apocrine secretion (Fig. 30). Thereby, apical parts containing fat globules, proteins, and small amounts of cytoplasm are lying beneath the plasma membrane; they first bulge and eventually become detached from the glandular cell body and are pinched off into the alveolar lumen.

3. Lactose Synthesis

The synthesis of lactose from intracellular glucose is stimulated exclusively by prolactin. In some experimental systems the amount of mammary lactose synthesis has been used as an indicator for the amount of prolactin released from the anterior pituitary. Most of the intracellular glucose is derived from blood glucose, and only a small amount seems to be provided by mammary intracellular glycolysis. The synthesis of lactose is catalyzed by the specifier protein, α-lactalbumin, and probably takes place on ribosomes and mitochondria according to the reaction

$$\text{UDP-galactose}^{*} + \text{glucose} \xrightarrow{\text{lactose synthetase}} \text{lactose} + \text{UDP}$$

(Figs. 22 and 29, Table 17). The rate-limiting enzyme in the biosynthetic pathway of lactose is lactose synthetase, consisting of galactosyltransferase and α-lactalbumin; this enzyme is probably inhibited by progesterone during gestation. This would also account for the fact that practically no lactose synthesis occurs during pregnancy when prolactin is present. The whey protein, α-lactalbumin, also called "B-protein," interacts with the A-protein, the galactosyltransferase, to specify glucose as a substrate for transfer into lactose. In the absence of α-lactalbumin, lactose is found only in very low amounts; it seems that progesterone may directly inhibit the induction of α-lactalbumin synthesis during gestation (Turkington, 1968a, b; Armstrong, 1970). With the reduction and withdrawal of high plasma progesterone and estrogens after parturition and the release of prolactin, synthesis of α-lactalbumin is induced to a high degree, leading to the formation of lactose from glucose. The B-protein (α-lactalbumin) is synthe-

* Uridine diphosphogalactose.

sized in large quantity on ribosomes in association with endoplasmic reticulum and is then transported to the Golgi complex (Fig. 29, Table 17). Here the B-protein vesicles, consisting partially of a rough endoplasmic reticulum envelope, contain free soluble proteins or proteins attached to microsomes. The B-protein vesicles then associate with the A-protein, which is bound to a microsomal cell fraction of the inner surface of the Golgi apparatus, to create the condition (lactose synthetase) necessary for lactose formation. The Golgi vesicles, containing lactose and milk proteins, swell by drawing water in (osmotic effect of lactose and proteins) and then move toward the cellular apex for apocrine and merocrine release into the alveolar lumen. The existence of lipoproteins and UDP-galactosyltransferase in skim milk has been related to shedding of Golgi, apical plasma, and microvilli membranes. It is thought that during the process of mammary lactose and protein synthesis and alveolar secretion the swollen Golgi apparatus, containing lactose and proteins, becomes located near the apical cell membrane. Here the membranes of the Golgi apparatus fuse with the apical plasma membrane and, by means of apocrine secretion, parts of the apically located cell constitutents are constricted away (inverse pinocytosis) (Fig. 30). Thereby, an excess of plasma membrane and vesicles derived from Golgi apparatus and microvilli, which may be a product of fusion of the Golgi membrane with the apical membrane, appears in the alveolar milk. These apical membrane parts with attached vesicles (50–60 Å in diameter) have been identified in skim milk by electron microscopy (Stewart *et al.,* 1972). Once secreted, lactose cannot reenter the alveolar cells as shown in experiments with labeled lactose. About 3% of the lactose produced in the glandular cell does not enter the alveolar lumen but is released into clefts between cells and reaches the systemic circulation via lymphatics and capillaries. However, when the mammary alveoli are distended by milk, the lactose released into the circulation may reach values up to 50% of the amount synthesized; it seems, therefore, that Golgi vesicles can open into intercellular clefts for release of lactose. The intraalveolar release of lactose seems to be due partly to apocrine but mainly to merocrine apical porous secretion (Brew, 1969; Linzell, 1971). Plain transcellular diffusion may account for the extrusion of lactose as well as water and various electrolytes. Vitamins, or drugs, may also enter the secretory mammary cells from the bloodstream and reach, via the pericapillary space, the acinar lumen by way of apocrine and merocrine secretion from alveolar cells. Furthermore, drugs, vitamins, and water may mix with alveolar milk directly from the systemic circulation via mammary capillaries and intercellular fluid (Fig. 31).

4. Secretion of Ions and Water

Although lactose is the major mammary intracellular and intraalveolar milk osmole, ions such as sodium, potassium, chlorine, magnesium, calcium, phosphate, sulfate, and citrate also play a role. These and other ions, as well as water, can pass the secretory cell membrane in both directions, i.e., from inside the cell into the alveolar lumen and conversely. Water may be added to milk from alveolar cells and via the interstitial fluid (intercellular fluid), serving as an isotonic diluent. But, the rather tight desmosomal junctions between alveolar cells seem not to allow a major passage of interstitial water between alveolar cells into the acinar lumen. Therefore, it appears that most of the aqueous milk phase is derived from water of the secretory cells. Passage of plasma water into the alveolar cells is dependent on the availability of intracellular glucose to be converted into lactose. Lactose and proteins stored in Golgi vesicles draw water from plasma into the cell. It appears unlikely that other plasma constituents besides water can pass directly into the alveolar lumen without first going through the secretory cell. The aqueous phase of milk is isosmotic to plasma and represents a mixture of a lactose solution with a fluid containing mainly potassium, calcium, and chlorine but also small amounts of sodium, magnesium, and iron. Accordingly, the major osmole of the aqueous phase of milk is lactose; the concentration of sodium and chlorine in milk is much lower than in blood plasma (Table 18), and it seems that cellular fluid as a whole may pass into the alveolar lumen. The ionic content is higher in milk immediately after its secretion (primary milk) than it is in extracellular fluid or final alveolar milk, indicating that water is taken up from the blood plasma of the surrounding capillaries, thereby diluting the alveolar primary milk. Intracellular sodium, chlorine, and potassium are in an equilibrium with the ions of blood plasma and alveolar milk. Sodium, potassium, chlorine, and water may leave the cell apically by membrane diffusion. On its basal and lateral surface sites the mammary secretory cell possesses the conventional ionic pumping mechanisms as indicated by increased ATPase activity at these loci. An apical pumping mechanism has been postulated for chlorine extrusion, whereas sodium, potassium, and part of the intracellular chlorine can pass into the milk according to their electrochemical gradients. Because the final concentration of the above ions in acinar milk is much lower than in the alveolar cell (Fig. 31), albeit in a similar ratio, a secondary dilution of alveolar milk with water must be taken into consideration. In order to maintain the intracellular–extracellular ionic equilibrium, the basal and lateral cellular pump ac-

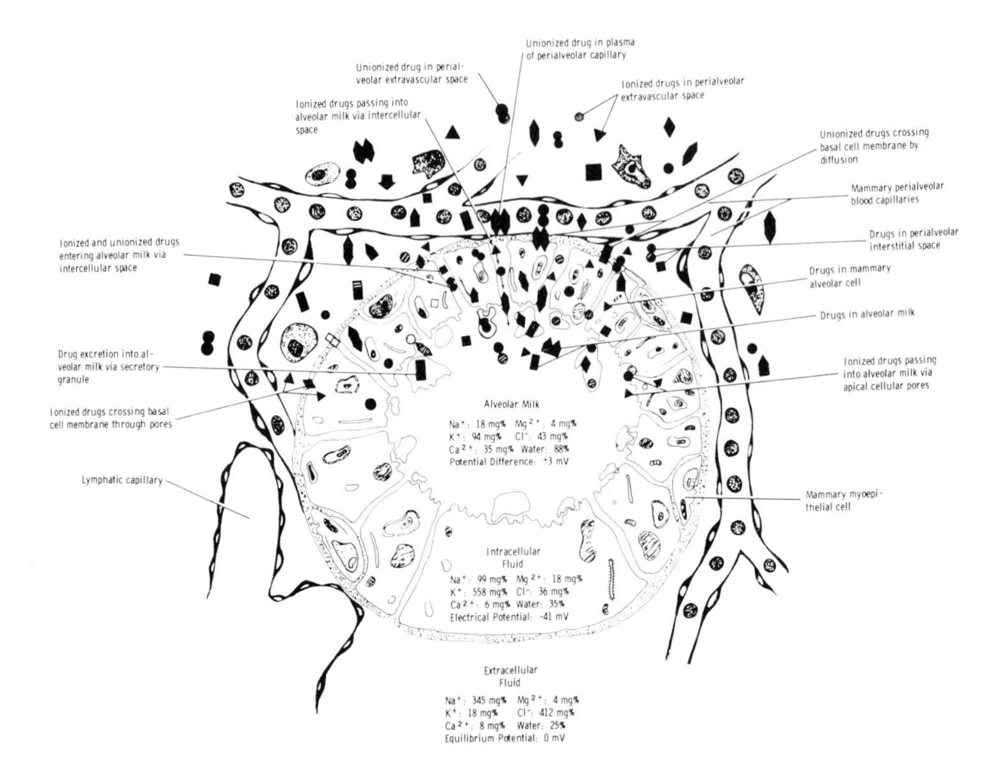

Fig. 31. Transport of drugs into milk—electrolyte concentration of alveolar milk and of intracellular and extracellular fluid. The mammary alveolar epithelium represents a lipid barrier with water-filled protein pores. Drug excretion into the milk depends on the drug's degree of ionization, its solubility in fat and water, and the pH difference between plasma and milk. Most drugs enter the alveolar cells in the unionized, nonprotein-bound form, either by diffusion or by active transport through the basement lamina and the basal plasma membrane. Drug molecules leave the alveolar cell via apical pores, but they may also be contained in constricted-away apical cellular parts and thus enter the mammary alveolar lumen. Pharmacologic agents may also pass into the alveolar milk via intercellular spaces of the alveolar epithelium. In general, not more than 1% of a given drug dosage appears in milk; therefore it is usually harmless to the suckling infant. The electrolyte composition of milk in the mammary alveoli has to be maintained during its storage. It appears that monovalent ions (Na^+, K^+), which are lower in milk than in intracellular fluid, may freely pass the semipermeable apical alveolar cell membrane in both directions; a concentration gradient exists from cell-inside to the alveolar milk. The sign of potential difference (+3 mV in alveolar milk and −41 mV cell-inside) is opposite to that of the concentration gradient. For chlorine, which is higher in milk than in the alveolar cell, an active apical transport that pumps Cl^- back into the alveolar cell may exist; also, Cl^- may be held back passively in alveolar milk. Changes of sodium and potassium by apical diffusion into milk are balanced by active basal cellu-

tively extrudes sodium and allows potassium to enter; chlorine is actively held back by a consistent cell inward pumping mechanism. Thus, the cellular pumping mechanisms maintain the ionic concentrations in the cell and outside the cell (extracellular fluid and alveolar milk). In animal experiments (rats, guinea pigs) it was observed that the electrical potential difference between mammary blood and milk is very low (Linzell and Peaker, 1971) Fig. 31), but the alveolar secretory cell inside is to an extent of −35 to −40 mV negative as compared to blood (Linzell, 1971). This indicates that active transport mechanisms occur in alveolar milk-synthesizing cells. The apical membrane seems to be penetrable to calcium and magnesium only by an active process of extrusion. The calcium and magnesium content of the milk is 35 mg% and 4 mg%, whereas their intracellular concentration is 6 mg% and 18 mg%, respectively. Because these divalent ions are largely bound to milk proteins (60–70%), their concentration is lower in the aqueous milk phase but still higher than in plasma (Table 18). Prosphate is bound mainly to milk proteins; some is also dissolved in the aqueous phase of milk. In contrast, citrate is mostly found in the aqueous milk phase; it is formed in the endoplasmic reticulum of the Golgi apparatus and discharged with Golgi vesicles into the alveolar lumen. The concentration of vitamins B_1, B_2, and C is higher in milk than in blood plasma, whereas niacin, vitamin B_{12} and vitamin D are less concentrated in milk than in blood. Therefore, active transport mechanisms for vitamins B_1, B_2, and C from secretory alveolar cells into the acinar lumen seem to exist. Vitamins or drugs may also leave the secretory

---

lar membrane transport mechanisms pumping potassium in and actively extruding sodium. Ion transfer from the secretory cell into the alveolar milk occurs also via extrusion of the swollen Golgi complex. Thus, the ion concentration in primary alveolar milk (bulk of milk removed during initial phase of milking) is higher than in secondary milk, which is diluted by attraction of water due to the osmotic effect of lactose, the major osmole of milk. The apical cell membrane is also permeable to calcium and magnesium. It is not clear whether there is an apical active transport process for calcium, but the higher concentration of calcium in alveolar milk suggests the existence of such a mechanism. It is also possible that the binding of calcium to milk proteins (60–70% calcium bound) causes the higher concentration of calcium in alveolar milk. Intracellular water is freely exchanged in both directions. Because the water content of milk is much higher than that in the alveolar cell, the intracellular equilibrium is maintained via water uptake through the basal membrane; also, luminal water rediffusion seems possible. A potential difference of the same order and sign is observed between inside the alveolar cell and alveolar milk as between the intracellular and the extracellular fluid, indicating active cellular transport mechanisms. Above data on electrolytes, water, and potential difference are based on a concept derived from animal experiments by Linzell and Peaker (1971).

TABLE 18

Difference in Composition of Human Milk and Blood Plasma [a]

	Specific gravity	Osmolarity	pH	Calories	Water	Carbohydrates	Fat	Protein: Albumin	Protein: Globulin	Iron	Na^+	K^+
Human mature milk	1031	295	7.3	65 kcal/100 ml	87.5 g%	7.0 g% (lactose)	3.7 g%	0.3 g%	0.2 g%	0.15 mg%	15 mg%	57 mg%
Blood plasma	1033	285	7.4	35 kcal/100 ml	92 g%	80 mg% (glucose)	200 mg%	4.5 g%	2.5 g%	125 μg%	320 mg%	18 mg%

	Ca^{2+}	Mg^{2+}	Cl^-	Phosphorus	Sulfur	Vitamins: Vit A [b]	Vitamins: Vit B_1	Vitamins: Vit B_2	Vitamins: Niacin	Vitamins: Vit C	Vitamins: Vit D
Human mature milk	35 mg%	4 mg%	43 mg%	15 mg%	14 mg%	280 IU/100 ml	20 μg%	50 μg%	172 μg%	5 mg%	5 IU/100 ml
Blood plasma	10 mg%	2.5 mg%	365 mg%	4 mg%	2 mg%	50 μg%	10 μg%	0.5 μg%	500 μg%	1 mg%	188 IU/100 ml

[a] Data derived from various sources as indicated in the text.

[b] One μg of vitamin A corresponds to the activity of 3 IU of vitamin A.

mammary cell via apical pores and by diffusion, or they may be present in cytoplasmic apical evaginations, which are then pinched off into the alveolar lumen (Figs. 30 and 31). Furthermore, water, electrolytes, vitamins, and drugs coming directly from the permeable mammary capillaries may mix with alveolar milk via interstitial fluid. The concentration of iodine is higher in milk than in blood plasma, which also suggests an active transport mechanism. Iodine is strongly bound to milk casein, which thus is capable of exerting a thyroxine-like action. The secretory epithelium of the mammary gland may effectively compete with the thyroid gland for iodine and can even limit thyroxine production.

5. Intrinsic Regulation of Milk Secretion

Mammary alveolar cells produce milk very efficiently. About 60% of the energy value of food exceeding the maternal metabolic requirements is used for milk synthesis. The amount of blood passing through the breast (500–700 ml/minute) is about 400–500 times the volume of milk secreted. The rate of mammary blood flow and oxygen consumption is similar to that of the brain. The mammary tissue of lactating animals may amount to as much as 7–10% of the total body weight; whereas in nursing women the breast tissue contributes only 3% to the body weight. The hypertrophy of mammary tissues during pregnancy and lactation (each breast gains about 400 g of weight) is accompanied by some hypertrophy and hyperfunction of the liver and gastrointestinal tract with increased blood flow in these organs. Variation of food intake has little effect on composition or volume of milk. However, if the diet is quantitatively inadequate, the milk yield is reduced, but the milk constituents remain equally balanced because the mother draws from the resources of her own tissue stores. If, for instace, the mother does not receive enough calcium, her organic bone matrix breaks down and thus the bones become "bendy." In such cases, when the maternal reduced calcium intake persists for a long period of time, permanent bone deformation may be encountered. Also maternal deficiency in fat-soluble vitamins may be reflected by decreased amounts of vitamins A, D, E, and K in the milk.

Intraalveolar milk is isosmotic to plasma. The initial aqueous phase of milk is also isotonic to plasma but different in its composition; it may become subsequently diluted by extracellular interstitial fluid that passes between acinar cells into the alveolar milk. The aqueous milk is mainly a lactose solution containing electrolytes. Because the electrolyte composition of intracellular fluid is similar to that of aqueous milk (high in potassium, low in sodium and chlorine), it appears that intracellular fluid may pass unchanged into the alveolar lumen. Extracellular markers such as sucrose and inulin cannot enter into the alveolar milk. It also appears that fluid exchange between alveolar cells is difficult as consequence of tight desmosomal intercellular junctions.

Plasma glucose enters the secretory epithelial mammary cells easily and is essential for the process of milk synthesis. Even in the absence of plasma fatty acids and amino acids, milk is produced as long as glucose is available. In the absence of glucose only a small amount of milk, rich in fat and proteins, is synthesized, and the secretion of the aqueous milk phase is only resumed when glucose is available in adequate amounts. No other sugar or metabolite can substitute for glucose in the process of milk secretion. In cows, goats, and pigs it was found that the lactating mammary glands use up 50–85% of glucose and amino acids entering circulation. In radioisotope studies it has been shown that the alveolar epithelium is permeable in both directions, i.e., from the inside of the cell into the lumen and, conversely, for water, sodium, potassium, calcium, chlorine, iodine, phosphate, and sulfate. However, the mammary secretory cells are impermeable from the alveolar lumen to the inside of the cell for lactose and citrate (Linzell, 1971; Linzell and Peaker, 1971).

Experiments with ouabain have disclosed that ATPase activity related to cellular ionic pumping mechanisms is localized mainly on the basal and lateral site of the secretory cell, whereas its apical luminal part is free of sodium pump activity. Accordingly, the basal and lateral ionic pump regulates intracellular composition and thus, indirectly, intraalveolar ionic composition. Lactose is the main regulator for intracellular and alveolar milk volume by drawing water osmotically into the Golgi apparatus; after lactose secretion into the mammary alveoli, additional water is osmotically attracted into the acinar lumen.

The mammary blood flow greatly influences the amount of milk secreted. Mammary blood flow decreases in response to stress (increased sympatheticoadrenal tone), to epinephrine and norepinephrine injections, and to fasting. Consequently, milk secretion declines with restricted blood flow because the mammary supply of oxygen, glucose, fatty acids, and amino acids is reduced. Reduced mammary blood flow and restricted milk secretion diminishes the lactose and potassium content of milk while enhancing its sodium and chlorine concentration.

Some sodium and water may be reabsorbed into the bloodstream from the acinar lumen, whereas potassium may be taken up from the secretory cell in exchange for lactose. Although milk is isotonic with plasma, its sodium and chlorine concentration is lower than that of plasma, and its potassium, calcium, and magnesium content is higher. When milk is removed from the alveoli by suckling or milking, the lactose content of milk removed toward the end of the nursing procedure rises rapidly, due to increased lactose secretion into the acinar lumen.

The apical membranal microvilli of the alveolar cell are not only involved in secretory processes; they also contribute to active reabsorption of ions from the alveolar contents. Nevertheless, compared to the microvilli of intestinal cells or tubular cells of the kidney, the microvilli of the mammary acinar epithelium are rather vestigial. The exact processes of secretion as well as of reabsorption of ions and water from the ductal mammary system are not clearly understood. It is possible that secretory and reabsorptive mechanisms apply to alveoli and smaller ducts only; the latter closely resemble the alveoli in respect to the morphology and physiology of their monolayered epithelial lining. In contrast, the medium-sized and larger milk ducts display a 2-cell layer of epithelial lining, which is impermeable to the constituents of milk; only a negligible transfer of radioactive sodium, potassium, and chlorine from ductular milk into the bloodstream takes place (Linzell, 1971). Because the mammary ducts are permeable to water only, ductular milk osmolarity is maintained at the same level as that of plasma solely by osmotic factors, and lactose, potassium, sodium, and chlorine are the main osmoles. The physical barrier of the 2-layered mammary duct epithelium is considered essential for the maintenance of intraductular milk composition. The apical membrane of ductular cells seems to represent this physical barrier for the various other milk constituents, resembling more the transitional epithelium of the urinary bladder, which is highly impermeable for ions and water. Thus, the ductal mammary system can be considered a "milk–blood barrier" comparable to the CSF–blood barrier of the brain; this also holds true to a great extent for the alveolar epithelium regarding the movement of milk constituents back into the secretory cell. The content of proteins and of some ions ($Na+$, $K+$, $Cl-$) is higher and that of lactose is lower in colostrum than in mature milk. The lower content of lactose in colostral milk may be due to increased pressures of the dilated mammary alveoli at the onset of lactation, preventing optimal lactose release from the alveolar cells. Thus milk obtained early in suckling or milking, as mentioned before, contains less lactose than milk removed toward the end of the nursing process.

Apart from mammary cellular proteins synthesized and released into the alveoli, plasma proteins may also enter directly into the alveolar milk. The amount of γ-globulins especially, is relatively high in milk due to the concentrating function of secretory alveolar cells. The difference in composition between milk and plasma with respect to proteins and fat is quite obvious (Table 18). Colostral milk contains the highest amounts of immunoglobulins, which are a major part of the γ-globulins. Whether these colostral immunoglobulins (in addition to transplacentally received maternal immune bodies) contribute to effective passive immunization of the fetus and newborn against bacterial and viral diseases cannot be determined yet. However, in regard to the phylogenetic development it is of interest to note that the placenta of the horse, pig, cow, sheep, goat, and dog, in contrast to the human, monkey, rabbit, guinea pig, and rat placenta, is impermeable for maternal immune bodies. Thus, it seems almost essential for survival of the newborn from species lacking transplacental immune body supply that immune substances supplied by colostrum and milk be able to substitute for the deficiency in placental transport. In late pregnancy the mammary alveolar cells of many species, including man, become highly permeable to plasma proteins, and the colostrum contains large quantities of γ-globulin antibodies directed against microorganisms that the mother has met and overcome. Whether local mammary production of immunoglobulins plays a major role for provision of passive immunity to the breast-fed baby cannot be answered yet. In cattle, the newborn's gut is very permeable during the first days postpartum, allowing absorption of the essential immunoglobulins without major losses through digestive inactivation (Linzell, 1971). Furthermore, in bovine colostrum a much higher trypsin inhibitor activity is observed than in human colostrum (Pedersen *et al.,* 1971). This colostral trypsin inhibitor is of low molecular weight (6,000–10,000) and is heat and acid stable. It appears that the function of the trypsin inhibitor is important for the absorption of colostral immune bodies from the intestines of newborn mammals, in the fetal life of which no antibodies could be supplied via the placenta. Although antibodies are present in human milk, no essential intestinal antibody absorption and thus no passive immunization in the breast-fed baby appears to take place. This is underlined by the fact that no transmission of rhesus antibodies from the lactating mother to the suckling rhesus incompatible baby has been observed. If this should occur, hemolytic disease in the rhesus positive newborn would be the consequence, which has not yet been reported. It is also possible that in humans as well as in cattle immunoglobulins are selectively secreted into colostral milk and partially into transitional and mature milk. While the main immunoglobulin in man is IgA, in the cow, pig, and sheep, the immunoglobulin types of IgG and IgM are predominant. It is thought that most of the immune bodies in milk are derived from local plasma cells located around the mammary secretory cells, but immune globulins may also pass from the dilated and highly permeable mammary capillaries into the alveolar milk (Linzell, 1971).

Regarding mammary lactose synthesis as the carbohydrate nutrient, questions that may be asked are, why is this particular disaccharide produced, and why is the monosaccharide glucose not utilized as a milk carbohydrate? The answers seem to be found in the following facts:

a. Mammary secretory cells are not permeable for lactose; therefore, once lactose is secreted it is trapped in the alveoli and cannot be reabsorbed. Thus, no additional energy is needed to retain lactose in the alveolar ductular lumen, and, in addition, through the osmotic effect of lactose, adequate water supply is provided for the suckling infant.

b. Lactose is not as easily fermented into other products in the gut as is glucose; for instance, glucose in the milk diet of lambs leads to inebriation within a few days of feeding due to the synthesis of ethanol from glucose by the gut's yeast content.

c. During the process of evolution mammary protein synthesis most likely preceded that of lactose, because a specifier protein for lactose synthesis from glucose has to be present first. Accordingly, with the availability of the specifier protein, α-lactalbumin, the pathway was created for lactose synthesis from glucose.

d. The disaccharide lactose (galactose-glucose) possesses bifidus factor activity that facilitates the growth of the bifidus flora in the neonate intestines for better resorption of nutrients.

In the nonlactating mammary gland one may find milky fluid, the composition of which is very similar to blood plasma. Here, the area of the apical cell membrane and the whole size of alveolar superficial cells are reduced. Under these circumstances the sodium and potassium activated cellular ATPase is located more apically, indicating a functioning apical ionic pump. Now chlorine may be pumped out of the alveolar cell instead of potassium. It is also possible that the existence of an apical ionic cellular pump may account for the change of milk composition in late lactation, when sodium and chlorine contents in milk rise and lactose and potassium concentrations fall, a reversal of the situation of normal active lactation (Fig. 31). It has been suggested that a decrease in prolactin secretion in late lactation may be largely responsible for these changes in cellular ATPase activity, with an apically located ionic pump leading to altered transport mechanisms for various ions, lactose, and fatty acids.

In addition, a morphological change of the semipermeable apical membrane in late lactation has to be considered. While in early lactation water, electrolytes, and lactose easily escape into the alveolar lumen, fat and larger protein molecules (casein) are retained to a large extent in the secretory cell; they can be most efficiently released by cellular decapitation (apocrine secretion) immediately after a major component of alveolar milk is removed. This situation varies greatly from that in late lactation when the alveoli are much less filled with milk. In active lactation a pronounced intraalveolar milk accumulation between nursing periods is observed due to diffusional, osmotic, and other active transport mechanisms, leading to a filling pressure and not allowing further cellular escape of fat and proteins. This explains why milk secreted into the alveoli before nursing or milking contains relatively higher concentrations of lactose and lactalbumin but lower amounts of fat and casein than milk removed toward the end of the nursing or milking process. When milk, which accumulates during the nursing-free interval, is removed (nursing, milking), the intraalveolar filling pressure decreases, allowing subsequent increased secretion of fat and larger proteins into the acinar lumen; this accounts for the higher fat and protein content of the "after-milk." During the stage of decreased milk secretion as observed in the late phase of lactation, with diminished prolactin stimulation of secretory cells, the intraalveolar filling pressure is low due to reduced secretory activities; therefore, the milk fat concentration is higher initially in the preformed milk than at the end of the nursing period (Linzell, 1971).

It is not understood to what extent adjoining secretory alveolar cells may be able to exchange water, electrolytes, lactose, and amino acids and to what degree fatty acids can be used interchangeably in those adjacent cells more actively engaged in the process of milk synthesis; nor is it known whether ductular cells may change their barrier capacity during lactation, allowing water and ions to penetrate more or less actively in either direction.

6. Nutritional Value of Milk for the Newborn

The postpartum mammary secretion of colostrum (1 week in duration) is followed by a 2–3 week period of transitional milk secretion gradually leading to production of mature milk. Colostrum and transitional and mature milk fulfill valuable immunological, biochemical, and physical functions for the baby (Tables 19 and 20) (Ling *et al.*, 1961; Lovell and Rees, 1961; Macy and Kelly, 1961; White *et al.*, 1964; Oser, 1965; Cantarow and Schepartz, 1967).

a. COLOSTRUM

In animals secretion of colostrum enhances the newborn's chances for survival by supplying nutrition and protection against infectious diseases. The secretion of colostrum and milk represents a specific phylogenetic

TABLE 19

Nutritional Value and Properties of Human Milk

1. Ideal food with optimal adaptation to nutritional needs of newborn; adequately supplying all nutrients during first 3–4 months of postnatal life with the exception of vitamin D and possibly vitamin C
2. Milk volume at the beginning of the second week postpartum: 300–500 ml/day (400 ml of mammary blood flow yield secretion of 1 ml milk)
3. Growth factor for infant's intestinal bifidus flora (*Lactobacillus bifidus*) through supply of lactose and active oligosaccharide compounds containing galactose, fucose, *N*-acetylglucosamine, and *N*-acetylneuraminic acid (sialiac acid)
4. Relatively large amount of nucleotides (much lower in cow's milk) facilitating nucleic acid and protein synthesis for the infant's growth
5. Ideal source for calcium, phosphorus, other electrolytes, and water
6. Outstanding source for vitamins, exceedingly rich in vitamin A and vitamin B_1
7. When human or cow's milk stand for a long time at room temperature, anaerobic glycolysis by *Streptococcus lactis* and related organisms takes place producing lactic acid and precipitation of casein (souring of milk)
8. Biologic value of human milk is slightly higher than that of cow's milk
9. Thermostabile "antistaphylococcal factor" present in phosphatide fraction; this "resistance factor" possesses no direct antibiotic properties

adaptation characteristic for mammals. Human colostrum is a yellowish sticky fluid (specific gravity: 1040–1060) secreted in amounts of 10–40 ml per diem during the first days postpartum; it contains less water-soluble vitamins, fat, and lactose than transitional or mature milk (Table 21). Proteins, fat-soluble vitamins, and mineral elements are present in colostrum in larger amounts than in mature milk. The high protein content of colostrum, reaching up to 10 g%, is half composed of true globulins, which seem to be identical with γ-globulins of blood plasma. During intrauterine life the fetus eventually has to cope with bacteria and viruses or their antigens, which come from the maternal organism, possibly penetrating the placental barrier. In man and in various rodents, maternal immune bodies (antitoxins, agglutinins), directed against bacterial and viral agents, are provided to the fetuses transplacentally. In some species (horse, cow, pig, and dog) the placental structure does not allow an effective transport of immune bodies to the fetus. These species therefore depend critically on postnatal transfer and passive immunization via colostrum, accomplished in the early postpartum days by intestinal absorption of undigested immune globulins. After passing the intestinal mucosa, the immune substances enter the newborn's lymphatics and veins of the portal circulation and from there reach the systemic bloodstream providing passive immunization against various infective agents. This temporary capacity of mucosal

TABLE 20

Nutritional Value and Properties of the Various Constituents of Human Milk

Constituent	Properties
1. Milk proteins (casein, albumins, globulins)	Essential for growth of muscular, neural, and glandular tissue
	Essential for growth and multiplication of cartilage cells for the formation of new bone
	Some serum albumins and globulins are transferred intact from blood into milk; casein and β-lactoglobulin are synthesized in mammary alveolar cells from plasma amino acids
	Breast milk provides about 2 g of protein/kg of infant's body weight
	Complete intestinal digestion
4. Water and electrolytes of milk	Water is primary solvent for milk and is essential for metabolic and temperature regulating mechanisms in the newborn
	Calcium and phosphorus for bone formation are abundantly provided
	Iron is an important element for the erythropoesis; sulfur, sodium, potassium, chlorine, magnesium, and the trace elements (copper, zinc, molybdenum, manganese, cobalt) are the osmotic intra- and extracellular components for tissue growth function and homeostasis of body fluids
	Maternal calcium stores account for 10–15% of milk calcium; iron and probably copper supply via breast milk is inadequate

2. Milk fat (fatty acids, phospholipids, cholesterol)
- Most complex food fat
- Synthesized chiefly in mammary gland
- Fine emulsion yielding easily digestible curd
- Primary source for energy and myelinization of human brain
- Milk is only product from which fat can be easily separated
- Provision of fat soluble vitamins (vit A, carotenoids, vit D, vit E, vit K)
- Almost completely digestible

3. Milk lactose
- Split on the surface of intestinal mucosal cells by enzyme lactase into glucose and galactose
- Facilitates intestinal uptake of calcium
- Lack of sweetness, less tendency to cloy the appetite
- Providing beneficial medium for intestinal resorptive activity
- Support of intestinal bifidus flora

5. Milk vitamins
- Essential for biological growth mechanisms and enzymatic systems involved in energy metabolism of protein fat and carbohydrates
- Initiation and catalysis of the functions of respiratory, digestive, and excretory systems
- Among the B-complex vitamins (vit B_1, vit B_2, niacin, vit B_6, pantothenic acid, biotin, folic acid, and vit B_{12}) only niacin and folic acid are higher in human milk than in cow's milk
- Prolonged deficiency of any of the B-complex vitamins will hamper the infant's growth and development
- Vitamin C is adequately supplied by milk only during the first 3–4 months postpartum

6. Milk enzymes
- Lipase, amylase, catalase, peroxidase, alkaline phosphatase (cow's milk contains additional xanthine oxidase and aldolase)
- Lipase liberates free fatty acids from triglycerides; free fatty acids are made available by milk lipase before reaching the digestive phase in the newborn's intestine

TABLE 21

Composition of Human Milk and Cow's Milk [a]

Milk elements	Colostrum (percent)	Mature human milk (percent)	Mature cow's milk (percent)
Water	87.0 g	87.5 g	86.0 g
Lactose	5.3 g	7.0 g	4.8 g
Fat	2.9 g	3.7 g	4.3 g
Total protein	5.8 g	1.2 g	3.3 g
Casein	1.2 g	0.4 g	2.8 g
Lactalbumin	1.1 g	0.3 g	0.4 g
Lactoglobulin	3.5 g	0.2 g	0.2 g
Ash	0.3 g	0.2 g	0.7 g
Iron	0.1 mg	0.15 mg	0.1 mg
Sodium	48 mg	15 mg	58 mg
Potassium	74 mg	57 mg	138 mg
Calcium	31 mg	35 mg	125 mg
Magnesium	4 mg	4 mg	12 mg
Chlorine	91 mg	43 mg	103 mg
Phosphorus	14 mg	15 mg	120 mg
Sulfur	22 mg	14 mg	30 mg
Vitamin A	470 IU/100 ml	280 IU/100 ml	180 IU/100 ml
Vitamin D	— [b]	5 IU/100 ml	2.5 IU/100 ml
Vitamin C	4.5 mg/100 ml	5 mg/100 ml	1.5 mg/100 ml
Specific gravity	1050	1031	1033
Milk pH	—	6.8–7.3	6.8
Bacteria	0	0	+
Calories	57 kcal/100 ml	65 kcal/100 ml	65 kcal/100 ml

[a] The values presented in this table are derived from various sources of literature data; they are expressed as grams or milligrams per hundred grams of milk, unless otherwise indicated.

[b] A dash indicates no information available.

cells for pinocytosis of undigested γ-globulins is probably aided by a colostral trypsin inhibitor, delaying the digestive hydrolysis of protein in the gastrointestinal tract. The newborn's passively acquired antibodies decline greatly within the first 3–6 months of postnatal life regardless of whether the antibodies were supplied by placental transfer during gestation or postpartum by colostral milk. Colostrum is also rich in fat-soluble vitamins and seems to contain one factor that promotes removal of intestinal meconium and one that facilitates the establishment of bifidus flora in the digestive tract of the newborn. Among the essential fatty acids of colostrum, octadecadienoic acid is most valuable for nutrition; the colostral content of phospholipids (constituent of plasma membranes) is high. Carotenoids contained in colostral lipids cause the yellow color of colostrum.

b. TRANSITIONAL AND MATURE MILK

Following the period of colostral milk secretion, a transitional phase toward the production of mature milk is encountered. In this phase of transition, the concentration of immunoglobulins and total proteins diminishes; lactose, fat, and the total caloric value of breast milk increase. Also, in transitional milk the water soluble vitamins begin to increase toward their highest values as in mature milk. After 3–4 months of nursing, intrinsic supplies of iron, vitamin C, and vitamin D become inadequate, and a supplement is necessary. The white color of milk is due to emulsified lipids and to the presence of a casein calcium salt (calcium caseinate). The yellow color of butter fat is due to its content of carotenoids.

c. NUTRITIONAL VALUE OF VARIOUS MILK CONSTITUENTS

Lactation and thus provision of nutrition and protection against infection may be looked upon as the final stage of the reproductive process. In Table 20 the nutritional value of the various milk constituents is outlined.

i. MILK PROTEINS. From about twenty amino acids present in milk proteins, eight are essential for all stages of life because they cannot be synthesized in the body. In some term infants and more so in prematurely born babies, the activity of the liver enzyme cystathionase may be insufficient or absent; therefore, methionine cannot be utilized and the newborn must depend on cystine as the essential sulfur-containing amino acid. Intestinal protein absorption is so efficient that the feces of breast-fed infants contain only amino acids provided by the bacterial flora of the gut or by mucosal secretion. Thus milk proteins are completely digestible in the newborn's intestinal tract. Although casein is the predominant milk protein, its nutritive value for promotion of tissue growth is less than that of the whey proteins (Table 22). In fresh milk, casein, a phosphoprotein of pH 6.6, is present as a complex of calcium caseinate and calcium phosphate. In the process of milk clotting, the stomach enzymes, rennin and pepsin, and the pancreatic chymotrypsin hydrolize casein to a soluble paracasein that, in the presence of calcium, is converted into insoluble casein (calcium paracaseinate, or curd); the residual clear fluid is called "whey" and contains water, electrolytes, albumin, and globulin. Rennin causes clotting of milk at pH 7 by first liberating a glycopeptide from casein; the remaining para-

TABLE 22

Protein Distribution in Human Milk and Cow's Milk [a]

Milk proteins	Human			Cow's mature milk
	Colostrum (1–5 days postpartum)	Transitional milk (6–10 days postpartum)	Mature milk 30 days postpartum and thereafter)	
Total proteins g%	2.7	1.6	1.0	3.3
Casein g%	1.2	0.7	0.4	2.8
Albumin g%	1.1	0.8	0.4	0.4
α-Lactalbumin	—	—	0.1	—
β-Lactoglobulin	—	—	0.3	—
Serum albumin	—	—	0.03	—
Globulin g%	5	0.5	0.2	0.2
Euglobulin	—	—	0.05	—
Pseudoglobulin	—	—	0.15	—
Whey proteins g% (albumins and globulins)	1.5	—	0.6	0.5
Nonprotein nitrogen mg%	47	42	32	21

[a] From Macy and Kelly (1961).

casein then reacts with calcium to form an insoluble paracasein–calcium complex (curd) for further digestion. Rennin is a potent milk-clotting enzyme that is secreted as prorennin and is activated by hydrogen ions into rennin. For the process of milk curdling the following reactions can be observed:

$$\text{Prorennin} + H^+ \rightarrow \text{rennin} \quad (1)$$

$$\text{Casein} + \text{rennin} \rightarrow \text{soluble paracasein} \quad (2)$$

$$\text{Soluble paracasein} + Ca^{2+} \rightarrow \text{insoluble } Ca^{2+} - \text{paracaseinate (curd)} \quad (3)$$

Rennin is present in early infancy only; it appears before pepsin is secreted and is most abundant in young ruminants. In the older infant as well as in children and adults, gastric pepsin and pancreatic chymotrypsin are responsible for the process of milk clotting and milk protein digestion. Human milk proteins provide a digestible curd with a high protein value and a well balanced composition of essential amino acids for the growing infant.

ii. MILK LIPIDS. Fat is the main energy provider in milk. Milk is a natural emulsion and comprises a complex of lipid material including phospholipids, cholesterol, cholesterol esters, cartenoids, a squalen-like compound, and a high-melting glyceride fraction, which forms a membrane around aggregates of triglycerides. Such fat globules vary in size and are suspended in an aqueous system that also contains proteins in colloidal dispersion and milk sugar in true solution. Fatty acids of C_{12} to C_{18} series comprise 80–90% of the total. Among the saturated fatty acids ranging from butyric to arachidic acid, palmitic acid (15%) is predominant; whereas, among the unsaturated fatty acids of milk fat, oleic acid (monounsaturated C_{18} fatty acid) is abundant (32%). Glycerol and short-chain fatty acids up to C_{10} compounds are synthesized in mammary aveolar cells via acetyl coenzyme-A. Palmitic and oleic glycerides reaching the mammary glands through the bloodstream are split here into fatty acids and glycerol. Upon entry into the mammary alveolar cells, respective fatty acids are resynthesized into triglycerides by the glycerol of the alveolar secretory cell. The highly unsaturated fatty acids of milk, linoleic, linolenic (essential fatty acids), and arachidonic acid, are derived from blood plasma; their content in milk depends on the maternal intake. Milk fat is almost completely digestible, and its nutritional value appears superior to that of vegetable oils. In cow's milk the "unnatural" (cis-trans or trans-cis forms) configuration of essential fatty acids has been held responsible for their lower activity. Moreover, these "unnatural" fatty acid isomeres may antagonize the effect of other essential fatty acids contained in the infant's diet. The well-emulsified, small fat globules of human milk contribute to a fine curd that can be digested easily by the newborn. Unsaturated fatty acids of lower molecular weight in milk are well absorbed in the gut, assuring an adequate energy supply for the thriving of the infant. Milk fat is also an important carrier for vitamins A and D needed by the newborn.

iii. MILK LACTOSE. The presence of lactose (milk sugar) as the sole carbohydrate is unique in milk. Lactose provides a beneficial medium for intestinal function by promoting growth of the bacterial bifidus flora and adequate hydrogen ion concentration of the gut, as well as proper intestinal absorption of calcium for bone formation. The one split product of milk sugar, galactose, is an essential component for the formation of galactolipids of various nervous tissues. Hydrolysis of the disaccharide lactose into galactose and glucose occurs probably on the luminal surface of the intestinal mucosal cells, and the resultant monosaccharides are subsequently absorbed into the portal circulation. Initially, in the newborn, the intestinal mucosal cells cannot hydrolyze lactose because they lack the enzyme lactase (β-galactosidase). However, after exposure of the intestinal epithelium to lactose the enzyme β-galactosidase is induced by lactose and hydrolyzes it. Mucosal lactase activity is highest during infancy, and it may disappear in adulthood. Genetic failure or delayed induction of lactase (Paige *et al.,* 1971) will lead to symptoms of lactose intolerance, clinically evidenced by anorexia, vomiting, and diarrhea. Through lactose and another factor of human milk, which is a glycoside of *N*-acetylglucosamine, growth of the intestinal *Lactobacillus bifidus* flora is supported. The feces of breast-fed infants contain predominantly bacilli of the bifidus type, in contrast to the predominance of the coliform bacterial group observed in bottle-fed infants' feces. Bifidus lactobaccilli seem to provide some benefit to the newborn because they can ferment lactose into lactic acid. Thereby, the resorptive capacity of the intestinal mucosa is increased, and the stools become slightly acid (pH 5). Upon milk souring, induced by *Streptococcus lactis* and other microorganisms in milk, lactose is converted into lactic acid, which is responsible for the special taste of "sour milk."

iv. MILK WATER AND ELECTROLYTES. Water is the primary solvent of milk, serving as an inert vehicle for dissolving and emulsifying the various constituents of milk. After ingestion of milk, its water contributes greatly to temperature-regulating mechanisms of the newborn's organism (25% heat loss from the body occurs by evaporation of water from lungs and skin). In human milk the total inorganic constituents are lower (0.25 g%) than in cow's milk (0.7 g%). The varying concentrations of minerals in milk of different species appear to be related to the rate of growth of their young; the young of fast-growing cattle, for instance, require a larger mineral supply from milk than a slower growing infant. Accordingly, the mineral content in human milk is fully sufficient to allow optimal growth. Calcium and phosphorus are already amply provided in human milk, but cow's milk contains 4 times more calcium and phosphorus than human milk. The maternal calcium stores account for 10–15% of the milk calcium, and this is independent of the amount of dietary calcium intake by the mother. Due to iron stores provided during intrauterine life and postnatal supply by breast milk, the average full-term infant has a sufficient supply of iron for about 6 months of neonatal growth without developing anemia. Nevertheless, supplementary iron after the third to the fourth month of nursing is recommended. During the period of early neonatal life, when milk is the only source of food, potassium is as important as the bone minerals. At all stages of postpartum development the infant's body contains more potassium than sodium, and this difference increases with advancing growth.

v. MILK VITAMINS. The vitamins contained in milk are essential for the biological growth mechanisms and the enzymatic systems involved in energy metabolism of proteins, fat, and carbohydrates. Thus vitamins are the initiators and supporters of proper function of the respiratory, digestive, and excretory systems of the newborn.

(a) Fat-soluble vitamins (vitamins A, D, E, K). The supply of vitamin A and its precursors (carotenoids) is adequate in human milk. The main physiologic action of vitamin A is to support proper epithelial growth and function; vitamin A is also a constituent of visual pigments. Because vitamin D content of human milk is inadequate, it needs to be supplemented after the third month of nursing. Premature babies require early vitamin D supplement. Cow's milk is often enriched with vitamin D to provide 400 units per quart. Vitamin D increases the intestinal absorption of calcium and phosphate; it also is related to calcium metabolism and alkaline phosphatase activity in bone formation. The fat-soluble vitamins K and E in milk are not considered significant for the infant's nutrition and thriving.

(b) Water-soluble vitamins (vitamin B-complex, vitamin C). Human milk is an outstanding source for water-soluble vitamins. The proportion of some vitamins (mainly vitamin B_1 or thiamine) in milk increases with the duration of lactation. The content of vitamin B_2 (riboflavin) in milk is most important for the young infant in whom intestinal bacterial synthesis of vitamin B_2 is minimal. In the rumen of cattle, vitamin B_2 is amply synthesized by bacteria. Vitamin B_1 is essential for the utilization of carbohydrates in pyruvate metabolism (cofactor in pyruvic acid decarboxylation) for fat synthesis. B_1-Avitaminosis leads to insufficient carbohydrate oxidation with accumulation of intermediary metabolites such as lactic acid, acetoacetic acid, glucuronic acid, pyruvic acid, and methyl glyoxal. In women with vitamin B_1 deficiency and beriberi disease, it is likely that methyl glyoxal is the neurotoxic agent that is excreted from plasma into milk. It may cause toxic symptoms in the breast-feeding infant and is detectable in the infant's urine. Methyl glyoxal is found neither in plasma nor in milk of healthy women and therefore is not in the urine of their breast-fed infants. Vitamin B_2 is involved in oxidative intracellular systems (constituent of flavoproteins) and thus is essential for protoplasmic growth. The vitamin niacin (nicotinamide) represents an integral part of pyridine nucleotide coenzymes (constituent of $NAD+$ and $NADP+$), thus serving as a part of the intracellular respiratory mechanisms. Another B-complex vitamin, pantothenic acid, is a fundamental part of coenzyme A,

a catalyst of acetylation reactions in various organs. Coenzyme A reacts with acetic acid to form acetyl coenzyme A, which is of pivotal importance in intermediary metabolism. Vitamin B_6 (pyridoxine) forms the prosthetic enzyme group of certain decarboxylases and transminases involved in metabolism of nerve tissues. The B-complex vitamin folacin (folic acid) participates in the conversion of glycine to serine and in the methylation of nicotinamide and homocystine to methionine (coenzyme for "one-carbon" transfer). The daily requirement of folic acid (0.5 mg per diem) and of vitamin B_{12} (2 μg per diem) is not fully met by breast milk. Vitamin C (ascorbic acid) takes part in activities of enzyme and hormone systems as well as in the intracellular chemical reactions. This vitamin is also essential for collagen synthesis. One tablet per day of a mineral–multivitamin preparation (Prenatal, Engran, Mi-cebrin, etc.) will increase the vitamin and mineral content of breast milk sufficiently and will ensure adequate supply for the breast-fed infant. The amount of fluorine in breast milk is insufficient for adequate protection of the breast-fed infant against dental caries. Lactating mothers who drink water containing 0.55 ppm of fluorine have milk levels of only 0.1–0.2 ppm. A daily fluorine supply of 0.25–0.5 mg is recommended for the first 3 years of life; between the age of 3–10 years 1 mg is recommended. In districts where the water supply is not fluoridated the maternal intake of fluorine from water is between 0.6–1.8 mg/day; fluoridation (1 ppm) of water may double these figures and provide sufficient fluorine supply for the nursing infant from breast milk. In areas where the fluorine content of drinking water is below 1 ppm, intake of a multivitamin preparation containing sodium fluoride (i.e., Adeflor Chewable, Mulvidren-F, etc.) is advisable for the nursing mother. However, caution has to be exercised in areas where the fluorine content of drinking water exceeds 1 ppm; here additional fluorine supply by vitamin tablets may cause dental mottling in the breast-fed infant.

vi. MILK ENZYMES. The milk enzymes (lipase, amylase, catalase, peroxidase, alkaline and acid phosphatase, xanthine oxidase, aldolase) are not important for the digestion of milk, although milk lipase splits triglycerides in milk. Some milk enzymes (xanthine oxidase, aldolase, alkaline phosphatase) are associated with the fat globule–membrane proteins, and the others are contained in milk serum. These enzymes enter the alveolar milk from mammary blood capillaries via intercellular fluid or by breakdown of mammary secretory cells. Because some of the enzymes' substrates are not present in milk, and since the enzymes are largely destroyed in the process of milk digestion in the infant's gastrointestinal tract, they seem to have no considerable nutritional significance.

7. Excretion of Drugs into Milk—Potential Hazards to the Breast-Fed Infant

Postpartum, puerperas rather frequently require medication, and the physician is confronted with the question of whether breast-feeding, if desired, is advisable because of hazards to the infant who is ingesting breast milk containing drugs.

a. PRINCIPLES OF DRUG TRANSPORT INTO MILK

Drugs that have entered systemic circulation become protein bound or remain in free solution in the blood plasma. Among the substances extensively bound to plasma proteins are sulfonamides, salycilates, coumarins, and phenylbutazone. Antihistamines and meperidine are to a lesser extent protein attached, and injected oxytocin or ADH remain free in the circulation. In plasma the protein-bound part of a drug represents an inactive reservoir with which the drug molecules in free solution maintain a certain equilibrium. Thus, when the plasma level of the freely dissolved drug falls, drug molecules are freed for replacement from the plasma protein reservoir. Only free ionized and unionized substances of plasma can leave the blood capillaries to appear in the interstitial fluid. Within the extravascular space, the drugs may remain in free solution (ionized or unionized), or they may become bound to proteins contained in the interstitial fluid. Drug attachment to protein receptors of the cell membrane also occurs. During lactation, drugs, appearing in the mammary perialveolar interstitium, may enter the alveolar epithelium and pass into the milk in much the same way that ingested drugs penetrate the intestinal mucosa and reach the bloodstream (Knowles, 1965; Stockley, 1970; Arena, 1970; Traeger, 1970; Wang and Ober, 1971). Pharmacologic agents may cross the cell membrane through water-filled pores, by diffusion and active transport. Therapeutic substances, albeit to a minor extent, may also pass directly into milk via intercellular spaces of the alveolar epithelium (Figs. 31 and 32, Table 23). The main site of drug entrance appears to be the basal cellular part, and here small water-soluble substances (ethanol, urea) may pass through water-filled cellular pores. Most molecules of drugs, especially larger ones, probably penetrate the plasma membrane by diffusion and active transport. Because the cell membrane possesses a phospholipid–protein structure, unionized drugs cross the cellular barrier much better than ionized ones. The liver may chemically alter drug molecules by hydroxylation, sulfonation, and glucuronidation. Thereby their lipid solubility decreases, and the cellular drug uptake is diminished. The

Fig. 32. Mechanisms of drug excretion into milk. Ingested drugs reach mammary blood vessels via the general circulation and may leave the capillary bed (a) through endothelial pores or by reversed pinocytosis. After passing the mucopolysaccharide material of the perivascular space, the basement membrane (b) of the mammary alveolar cells must be penetrated. Then the drug has to pass the trilaminar structure of the plasma membrane (c) to reach the cytoplasm (d) and from there the milk in the mammary alveolar lumen (e) (see also Figs. 29 and 30). Drugs may also enter the milk via intercellular spaces (f) of the alveolar epithelium. The basement membrane (thickness 500–1000 Å) functions as a sieve, and apparently drugs can easily pass through it by passive diffusion. The basement membrane is subjacent to basal surfaces of cells; it serves as support for mammary secretory and myoepithelial cells and resembles in its composition (collagen filaments embedded in a mucopolysaccharide matrix) the plasma membrane's superficial coat ("fuzz"). The plasma membrane (thickness 100 Å) consists of a phospholipid matrix in which proteins and cholesterol are embedded. Some protein molecules appear also to be adsorbed on the surface of plasma membrane, which shows a trilaminar structure: outer layer

(predominantly globular proteins), middle layer (bimolecular leaflet of lipids), and inner layer (proteins). Because the phospholipid molecules' hydrophobic ends oppose each other in the center of the membrane, the permeability is low for water-soluble compounds and high for lipid-soluble agents. The hydrophilic ends of the phospholipids lie next to the outer and inner enveloping protein layers, and here the permeability is high for water-soluble drugs and low for lipid-soluble agents. The external coat of the plasma membrane ("fuzz") makes up the intercellular space (thickness 200 Å); it forms microvilli (Fig. 29) and is composed of polysaccharides associated with other macromolecules (glycolipids, glycoproteins). Because the external coat contains ionized groups (negative sialiac acid residues) on the terminal units of saccharide chains (g), the cell surface is negative, contributing, together with glycolipids and glycoproteins, to a regular cellular spacing of about 200 Å distance. The outer, strongly negative sialiac acid residues interact with ions (ionized drugs) and other macromolecules of the intercellular space. The intact plasma membrane confines the cell as a morphological and functional unit with pores and other transport sites for movement of drugs and metabolic products into the cytoplasm and back from it. The permeability of the plasma membrane is low for ionized and high for unionized and lipid soluble substances. The hydrophobic phospholipid "tails" located in the equator of the plasma membrane and the hydrophobic protein ends (amino acid residues) of the inner part of the membrane appear to inhibit the transport of water-soluble drugs. The various possible mechanisms for membranal attachment and passing of drugs through the plasma membrane into the cell interior are marked with the symbols h to k. h, Lipid soluble unionized drugs (⟨□⟩) may become attached to membrane receptors and subsequently penetrate into the cell interior after dissolving and descending in the lipid phase of the membrane. i, Ionized drugs (□⟩) and other polar substances may become bound to membranal carrier proteins, and by a so-called ionophore transport mechanism the ions are carried in a tridimensional cage across the membrane's lipid barrier. This transport is facilitated by the hydrophobic side chains of the protein carrier. j, Lipophilic (permease) protein molecules of the plasma membrane containing drug binding sites may bring about drug transport through the hydrophobic regions of the cell membrane; some of the embedded membrane proteins may bind drugs and move them into the cell by means of cell-inward directed rotation. Glycophorin molecules (glycoproteins) possessing a hydrophilic portion attached to the protein chain, a hydrophobic middle portion (uncharged amino acids), and hydrophilic carboxyl tails (charged amino acids) can bind drugs and carry them through the plasma membrane. k, Cell-inside transport of some ions and hydrophilic drug molecules also occurs through water-filled membrane channels (pores) that are formed by hydrophilic amino acid molecules. Besides passive diffusion, active drug transport through these protein channels by drug binding to a special protein subunit of the interior surface of the protein channel also appears possible; due to energy-yielding (ATP) enzyme reactions a process of successive inward shifting of subunits enables the drug to reach the cytoplasm. Membranal protein channels also allow water and ethanol to enter the cell by passive diffusion. For that purpose the channel proteins must carry a negative charge and moderate hydration on polar sites because water and ionized drugs are only bound at charged sites causing neutralization of charges and thus a shift toward the hydrophobic state with increased cell-inward flow. Multivalent ions, organic ions, drug ions, or charged macromolecules with low hydration may even more readily attach to and neutralize the electronic state of the channel proteins resulting in accelerated transmembranal transport.

TABLE 23

Passage of Drugs into Breast Milk

1. Mammary alveolar epithelium represents a lipid barrier with water-filled pores and is most permeable for drugs during colostral phase of milk secretion (first week postpartum)
2. Drug excretion into milk depends on the drug's degree of ionization, molecular weight, solubility in fat and water, and relation of pH of plasma (average 7.4) to pH of milk (average 7.0)
3. Drugs preferably enter mammary cells basally in unionized, nonprotein-bound form by diffusion or active transport
4. Only water-soluble drugs of molecular weight below 200 pass through water-filled membranal pores
5. Drugs leave mammary alveolar cells apically by diffusion, active transport, and apocrine secretion
6. Drugs may enter milk via spaces between mammary alveolar cells
7. Most drugs ingested appear in milk; drug concentration in milk usually does not exceed 1% of ingested dosage and is independent of milk volume
8. Drug binding to milk proteins is much less than binding to plasma proteins
9. Drug metabolizing capacity of mammary epithelium is not understood

unionized agents are more lipid soluble and thus pass the membrane by diffusion, i.e., by simple solution within the cell membrane. In this fashion specifically lipid-soluble substances such as barbiturates also enter the cells. In some instances passive drug transport in the form of facilitated diffusion takes place. Here, the active compound associates with a carrier enzyme (protein) and is transported across the cell membrane; the drug is liberated inside. Both simple diffusion and carrier-mediated diffusion are passive transport processes associated with a concentration gradient ("downhill" flow), whereas active transport mechanisms are moving against a concentration gradient ("uphill" transport). Ionized drugs are usually too large to penetrate the small water-filled cellular pores and too lipid insoluble to pass through the plasma membrane by diffusion elsewhere. Here, as in the case of unionized lipid-insoluble drug molecules, active transport mechanisms take place, which necessitate energy provision by intracellular metabolic processes (Fig. 32). It is thought that for active transport a drug-carrier mechanism similar to that for the process of facilitated diffusion is operative, with the difference that active transport occurs against a concentration gradient. Larger drug molecules and proteins may also be transported actively into the mammary alveolar cell by pinocytosis ("cell drinking"); here the substances are ingested in solution. For the process of pinocytosis, drug molecules that are dissolved in the interstitial fluid attach to receptors located at the surface of the cell membrane. At the place of drug attachment the membrane invaginates the drug bringing it inside the cell; by pinching off the invaginated part from the cell surface, the membrane-surrounded drug remains inside the cell. Next, this membrane is removed, probably by lysosomal enzyme activity, and the drug molecules can spread within the cell. After reaching the apical part of the alveolar cell, the drug is extruded into the alveolar lumen during the process of milk secretion. Drugs may pass into milk apically via small pores by diffusion, active transport, and reverse pinocytosis ("cell vomiting," or apocrine secretion). During the process of reverse pinocytosis the apical membrane evaginates after fusion of intracellular membrane-bound secretion granules with the plasma membrane; these granules contain lipids, proteins, lactose, drug molecules, and other cellular constituents. The protruding apical cellular parts together with the plasma membrane are pinched off from the rest of the cell and are thus released into the alveolar lumen (Fig. 30). Through this mechanism of apocrine milk secretion, drugs reach the alveolar milk simultaneously. All these processes, beginning with drug intake and ending when the active compound has entered the milk, are not well understood. The appearance and concentration of drugs in milk depend greatly on the drugs' solubility in fat and water, their degree of ionization ($[H^+][A^-]/[HA] = K_a$), and the extent of active transport mechanisms. Many drugs are weak acids or bases and exist in solution in more or less ionized or unionized forms. When the normal pH of plasma (pH 7.4) or milk (pH 6.8 to 7.3) is altered, the degree of drug ionization changes. In general, drugs of low acidity, such as organic acids, coumarins, phenobarbital, salicylic acid, phenylbutazone, benzyl penicillin, streptomycin, sulfonamides, and diuretics are absorbed better at a lower pH because more drug molecules are in unionized form. The reverse holds true for drugs of low alkalinity, such as lincomycin, erythromycin, alkaloids, antihistamines, isoniazid, amphetamine, chloroquine, imipramine, meperidine, mecamylamine, and theophylline, which are more ionized at a lower pH and thus are absorbed only in small amounts. To what extent absorption and excretion of drugs into milk are influenced by changes of the pH of plasma and milk is not well known. Drug absorption and excretory mechanisms are far better understood regarding the influence of acidic and alkaline pH changes of plasma and urine on the renal excretion of various drugs. Here, clear-cut responses and quantitative assessments are possible; for instance, alkalinization of urine with sodium bicarbonate has been used to accelerate the clearance and excretion of weak acids such as phenobarbital and salicylate. Nevertheless, similar mechanisms seem to apply to milk, too, because the concentration of weak alkaline compounds in milk is often as high as in plasma. Drugs of weak acidity are more ionized in plasma and interstitial fluid than weak alkaline agents.

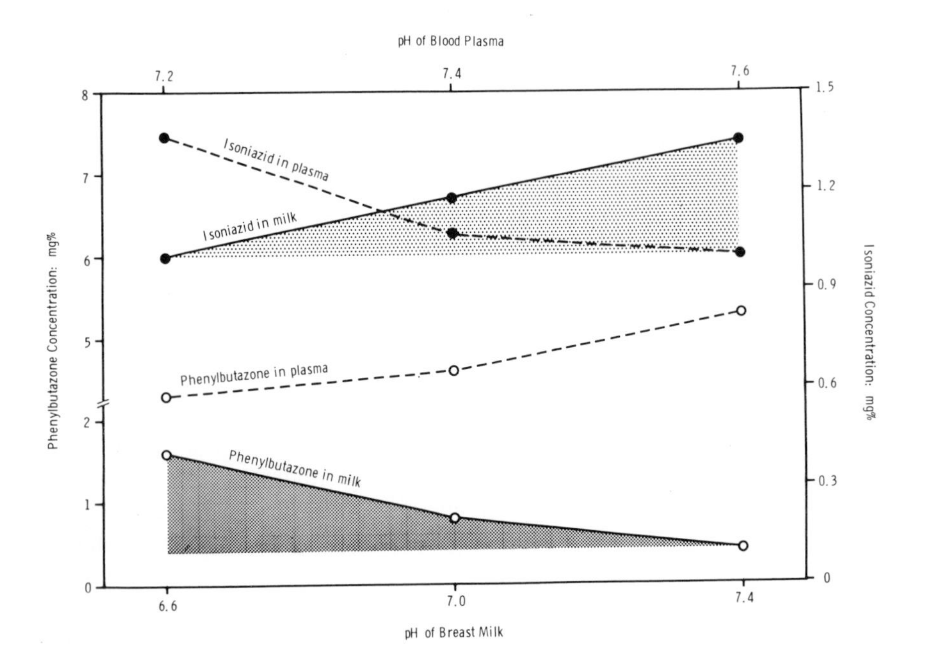

Fig. 33. Passage of drugs into breast milk in relation to pH of pharmacotherapeutic agent, blood plasma, and milk. Weak acidic drugs are ionized to a larger extent in an alkaline milieu (blood plasma and interstitial fluid) than weak alkaline drugs. Because unionized drug molecules penetrate the cell membrane more readily than ionized ones, weak acidic drugs pass into milk to a lesser extent than the mainly unionized, weak alkaline drugs. Accordingly, with decreasing pH of plasma and milk, a weak acidic drug, such as phenylbutazone, becomes less ionized allowing more unionized drug molecules to enter mammary alveolar cells and milk; under this condition plasma drug levels decrease. The reverse applies when pH of plasma and milk increase. Weak alkaline drugs such as isoniazid, are less ionized in an alkaline milieu, and the unionized drug molecules readily enter the mammary milk-secretory cells. With increasing pH values of plasma and milk, more isoniazid molecules are in an unionized state, and isoniazid milk levels increase and plasma levels decrease. Therapeutic doses of phenylbutazone (300–500 mg/day) and of isoniazid (200–400 mg/day) yield plasma levels of 2–5 mg% and 0.6–1.5 mg%, respectively; in the case of isoniazid, milk levels are similar to plasma levels in the normal pH range of plasma (7.4) and milk (7.0). The amount of drugs passing into milk also depends on plasma protein binding of the respective therapeutic agents. Over 90% of phenylbutazone is bound to plasma proteins. This, in addition to the drug's weak acidity, explains why a smaller amount of phenylbutazone passes into milk. The entrance of the weak base, isoniazid, into milk is facilitated by a lower degree of plasma protein binding (50%) and drug ionization allowing more rapid diffusion of isoniazid into the extravascular space and passage of unionized drug molecules through the cell membrane of the mammary epithelium.

Therefore, in the case of weak acid drugs relatively fewer lipid-soluble, unionized drug molecules are available for entering mammary alveolar cells; accordingly, the concentration of weak acidic compounds in milk is lower than in blood plasma (Table 25, Fig. 33). The reverse applies to weak alkaline drugs such as alkaloids, caffeine, theophylline, antihistamines, pyrazolones, meperidine, imipramine, amitriptyline, ephedrine, amphetamines, sulfapyridine, sulfanilamide, trimethoprim, isoniazid, acetanilid, mecamylamine, pempidine, erythromycin, lincomycin, levorphanol. Uptake of weak alkaline drugs by mammary epithelium is increased, equaling or exceeding respective plasma levels because the number of unionized molecules in plasma and interstitial fluid is higher (Table 25). The ratio of ionized to unionized parts of acidic drugs is usually below one at the normal pH of milk; it rises toward one when the pH of milk approaches the pH of plasma. For alkaline drugs the milk ratio of ionized to unionized molecules is one or above one, and it is reduced toward one with a rising pH of the milk.

Nonelectrolytes, such as ethanol and tetracycline, readily enter milk by diffusion through the lipid membrane barrier and may reach the same concentration in milk as in plasma. Because drug concentration in milk is independent of milk volume, a diffusional process of unionized drug fractions into alveolar milk is most likely to occur.

b. CLINICAL ASPECTS OF DRUG INTAKE DURING LACTATION

Although most drugs ingested by the nursing mother are excreted into milk, usually no adverse effects on the breast-fed infant are encountered with the exception of a few agents as documented in a small number of case reports (Table 24). Therapeutic doses of salicylates, potassium iodide, bromide, and caffeine yield milk levels of 1–3 mg%, which is not harmful to the breast-fed neonate. In Table 25 drug levels in maternal blood and milk are presented, and it can be noted that in some instances drug concentrations in maternal plasma and milk are very similar; however, in spite of that, no adverse effects in the breast-fed infant are observed. In general, it can be stated that most drugs (analgesics, antipyretics, narcotics, sedatives, tranquilizers, antihistamines, antibiotics, laxatives, oral antidiabetics) in therapeutic doses pass into breast milk in undetectable, low, or amounts nontoxic to the infant. The reason most drugs are innocuous to the breast-fed infant is that relatively low total amounts are ingested with milk. With daily ingestion of 500 ml breast milk containing 9 mg% sulfanilamide, the baby receives 45 mg of sulfanilamide per day, an

TABLE 24 Drugs Passing into Breast Milk and Potential Effects on the Infant[a]

Agents taken in excessive amounts	Potential drug effects on the breast-fed infant
Alcohol	Vomiting, drowsiness
Barbiturates	Sleepiness, decreased suckling efforts
Diphenylhydantoin	Vomiting, tremors, rash, blood dyscrasia
Dihydroxyanthraquinone	Increased bowel activity, loose stools
Ergot alkaloids	Nausea, vomiting, diarrhea, weakness
Dextroamphetamine	Tremor, insomnia
Bromide	Rash, drowsiness
Thiouracil, iodine	Hypothyroidism, goiter
Atropine	Diminution of milk secretion; inadequate supply of nutrients; rarely: pupillary dilation, tachycardia, constipation, urinary retention
Nicotine (smoking)	Diminution of milk secretion and thus reduced supply of nutrients for the infant
Hexachlorobenzene	Vomiting, diarrhea, fever, papular skin lesions, anemia, dystrophia
Metronidazole	Anorexia, vomiting, blood dyscrasia
Reserpine	Nasal stuffiness with breathing difficulties during suckling

[a] From various literature sources as indicated in the text.

TABLE 25 Concentration of Various Drugs in Blood and Breast Milk[a]

Drugs ingested (therapeutic dosage)	Drug levels: Plasma or serum	Drug levels: Milk	Percentage of ingested drug appearing in milk
Chloral hydrate	0–30 mg%	0–1.5 mg%	0.6%
Chloramphenicol	4.9 mg%	2.5 mg%	1.3%
Erythromycin	0.1–0.2 mg%	0.3–0.5 mg%	0.07–0.12%
Ethyl biscoumacetate	2.7–14.5 mg%	0–0.17 mg%	0.1%
Folic acid	3 μg%	0.07 μg%	0.1%
^{131}Iodine	0.002 μc%	0.13 μc%	4%
Isoniazid	0.6–1.2 mg%	0.6–1.2 mg%	0.5–1%
Lithium	0.2–1.1 mg%	0.07–0.4 mg%	0.15–0.8%
Methotrexate	3 μg%	0.3 μg%	0.01%
Novobiocin	1.2–5.2 mg%	0.3–0.5 mg%	0.1–0.2%
Penicillin	6–120 μg%	1.2–3.6 μg%	0.03%
Phenylbutazone	2–5 mg%	0.2–0.6 mg%	0.4%
Pyrazolones	2.3 mg%	2 mg%	1%
Quinine sulfate	0.7 mg%	0.1 mg%	0.05%
Sulfanilamide	8 mg%	9 mg%	1.5%
Sulfapyridine	3–13 mg%	3–13 mg%	0.5–2%
Tetracycline HCL	80–320 μg%	50–260 μg%	0.01–0.05%
Thiouracil	3–4 mg%	9–12 mg%	4.5–6%

[a] From various literature sources as indicated in the text.

amount that is not harmful. Even with relatively high drug concentrations in milk, the total amount ingested by the infant is low, rarely reaching therapeutic or toxic plasma levels. Moreover, drugs may be excreted into milk as ineffective metabolites (50% of chloramphenicol in breast milk is antimicrobially ineffective) decreasing the risk for adverse effects in the infant. Also backdiffusion of drugs from milk into the bloodstream may occur when the maternal plasma levels decrease (Catz and Giacoia, 1972).

Actually, only a few drugs are known to have caused adverse reactions in infants in isolated instances (Table 24) after ingestion of milk from mothers taking these drugs in excessive amounts. Nevertheless, it is important to be aware of possible dangers to the breast-fed infant when a nursing mother uses drugs. Some drugs passing into milk are very potent, and due to the newborn's insufficient hepatic and renal drug detoxification potentials, seemingly low amounts of these agents may accumulate in the infant's organism and exert pharmacologic, adverse or toxic effects; for instance, even small doses of an ingested sulfonamide may produce hyperbilirubinemia in premature infants by displacing bilirubin from its protein binding sites. Sulfonamides and antibiotics contained in milk may, in some instances, provoke bacterial resistance in the breast-fed baby, thereby enhancing the danger for infants suffering from infectious disease. Because only small amounts of antibacterial agents are secreted into the milk, the intestinal bacterial flora of the infant is usually not impaired. Prolonged medication with barbiturates and diphenylhydantoin is liable to cause hepatic microsomal enzyme induction not only in the mother but also in the baby; this may accelerate the metabolism of other drugs as well as that of barbiturates (Catz and Giacoia, 1972). Such infants may consequently experience therapeutically insufficient effects of a given drug (e.g., an antibiotic), although the therapeutic dosage was correctly compounded according to their body surface area. Maternal treatment with high doses of anticoagulants of coumarin type may result in decreased prothrombin levels (50% of normal) in the infant; in one infant a massive hematoma was observed at the site of operation for repair of an inguinal hernia. Hemolytic drug-induced anemia derived from drugs excreted into the breast milk has been reported in infants with hereditary glucose-6-phosphate dehydrogenase deficient erythrocytes. Drugs excreted into the breast milk may also cause jaundice in the infant through deficient hepatic glucuronyltransferase activity with a handicap for detoxification and excretion of meprobamate, chloramphenicol, morphine, and bilirubin; in infants with deficient erythrocyte-diaphorase activity maternal breast milk containing phenytoin and barbiturates may cause methemoglobulinemia (Catz and

Giacoia, 1972). Sex steroids contained in contraceptive pills may reach the milk in concentrations up to 1% (usually only 0.05%) of the oral dose taken. Whether complications such as thromboembolism and decrease in milk yield occur, seems to depend greatly on the doses of sex steroids taken by the mother. In general, daily doses not exceeding 50 μg of an oral estrogen (ethinyl estradiol or mestranol) and 2.5 mg of a progestin (19-nortestosterone derivatives) are not likely to increase the risk of maternal thromboembolism significantly nor to alter the milk yield (Table 26). It also seems to depend on how soon the oral contraceptive is given postpartum. Diminished lactation is most apt to result when oral contraception is instituted early in the puerperium, especially when he sex steroid doses are rather high. The possibility of inhibitory action is reduced when oral contraceptives are begun in the third to fourth postpartum week; here only high dosages diminish the milk yield. If sex steroids are absorbed in larger amounts from ingested milk into the baby's circulation, they may cause hyperbilirubinemia, since the steroids compete with bilirubin for glucuronic acid binding. Neonatal jaundice has been correlated with the mother's intake of contraceptive pills, the estrogen content of which is thought to inhibit conjugation of bilirubin in the liver (Wong and Wood, 1971). On the basis of the above presented example of drug distribution in maternal plasma and milk with consideration of the final dose appearing in the infant's system, it seems rather unlikely that the estrogen of the pill is the causative agent for the infant's hyperbilirubinemia since the daily maternal dose of estrogen ingested with the pill is rather low. In an earlier study on nursing women taking no oral contraceptives (Arias and Gartner, 1964), the breast milk content of endogenous $3\alpha,20\beta$-pregnanediol was held responsible for causing hyperbilirubinemia in the breast-fed infant, by inhibiting the hepatic glucuronyltransferase. So far, no conclusive evidence has been presented that maternal steroids excreted into milk are the cause of neonatal jaundice, and further well-controlled studies are needed to clarify this subject.

At present it is poorly understood as to what extent endogenous hormones are secreted into breast milk and how their action or interference, if any, may be found in the suckling infant. Because postpartum endogenous estrogens and progesterone drop very rapidly and since during the first 6 weeks of lactation only low sex steroid levels are found in the circulation, it appears unlikely that sex steroids or their metabolic products contained in milk exert any influence on the breast-fed infant. Proliferation of vaginal epithelium has been reported in female breast-fed infants of some mothers taking contraceptive pills. In one case, gynecomastia occurring in a male breast-fed baby was attributed to the estrogen component of the oral contraceptive taken by the mother. Although the estrogen and progestin content of presently applied oral contraceptives has not been proven to interfere with the mother's or the infant's health, the American Medical Association recommended in 1967 not to prescribe hormonal contraceptives to nursing mothers.

Many women take laxatives postpartum for normalization of bowel function. Anorganic cathartics (milk of magnesia), hydrophilic colloids (methyl cellulose), fecal softeners (dioctyl sodium sulfate), and most of the irritant or stimulant laxatives (castor oil, bisacodyl, phenolphtalein) do not enter the milk in measurable amounts. Actually only the anthraquinone or emodin alkaloids (cascara sagrada, senna, rhubarb, aloe) have been suspected of entering the milk in sufficient amounts to cause purgation in the suckling infant. Some investigators, however, could not observe any passage of anthraquinones into the milk, which supports the general belief that laxatives, even anthraquinones, do not affect the infant's bowel motility to a significant extent. Ingested caffeine passes into the milk in amounts up to 1% but no adverse effects on the breast-fed infant were detectible even after the mother had consumed several cups of strong coffee. In the maternal organism, caffeine and theophylline may even be of benefit since they promote pituitary prolactin release for milk secretion. Prolonged maternal medication with thiouracil, ergot alkaloids, metronidazole, and reserpine may cause adverse effects in the infant (Table 24). Nicotine, too, enters the milk, and, indeed, smoking of more than 20 to 30 cigarettes per day may not only decrease the milk yield significantly but also, albeit very rarely, may cause symptoms like nausea, vomiting, abdominal cramping, and diarrhea in the infant; innocuous milk levels of about 0.01–0.05 mg% of nicotine per liter milk are found if the mother smokes 20 cigarettes per day. Nicotine is not readily absorbed in the baby's intestinal tract, and, since it is rather quickly metabolized, signs of nicotine intoxication occur very seldom. Thus mothers who smoke not more than 20 cigarettes per day, especially when they leave large butts, do not encounter the aforementioned problems.

In Turkey, breast-fed infants were poisoned and died after the mother had consumed seed wheat treated with the fungicidal agent hexachlorobenzene. This compound reached the infant's system via breast milk and caused severe poisoning with porphyrobilinogenemia. DDT is contained in amounts of about 0.05–0.1 ppm in breast milk and cow's milk. This concentration is not dangerous to the breast-fed infant. As mentioned before, in a community with a fluorine drinking water content exceeding 1 ppm and eventual fluorine supply by intake of fluorine-containing vitamin tablets, relatively high amounts of fluorine may appear in milk possibly caus-

TABLE 26

The Effect of Contraceptive Steroids on Postpartum Lactation[a]

References	Type, dosage, and duration of administration of contraceptives	Effect on milk yield, milk composition, breast-fed baby, and mother	Comments
Curtis *et al.* (1963)	Enovid: 2.5, 5 mg norethynodrel and 100 μg mestranol per tablet Norlutin: 10 mg norethindrone per tablet (immediately postpartum for at least 3 months)	Successful breast feeding for one month in about 50%, falling to 38% by the end of the third month (same values apply for controls); Norlutin patients had decreased lochial flow of 14 days duration; after 2–3 months on oral contraceptives uterine length and width were 1 cm more than those of controls	Simple controls; in one male infant whose mother received Enovid 5, gynecomastia developed due to the compound's mestranol content and disappeared after discontinuence of breast feeding; "no added maternal hazards"
Rice-Wray (1963)	Anovlar: 4 mg norethindrone and 50 μg ethinyl estradiol per tablet Ortho-Novum: 10 mg norethindrone and 60 μg mestranol per tablet (duration of administration not reported)	Lactation was decreased by 45% with Anovlar and by 58% with Ortho-Novum	No data given regarding time of treatment postpartum and no method of data evaluation reported; no simple or placebo control studies performed; *Note*: a 2½-fold progestin dose, as in the case of Ortho-Novum, results only in a slightly higher percentage of decreased milk yield when compared to Anovlar
Rice-Wray *et al.* (1963)	Ortho-Novum (Norinyl): 2, 5, 10 mg norethindrone and 100, 60, 60 μg mestranol per tablet, respectively; from 4 weeks postpartum on for over a year	"Lactation declined or stopped altogether in approximately 1 out of 3 cases on norethindrone 10 mg" (not reported whether lower progestin doses influenced milk yield too). During treatment there was an initial transitory secretory phase which was followed by glandular regression; breakthrough bleed-	No placebo controls; no data presented as to how effect of norethindrone on milk yield was evaluated

		ings of 10–25% were observed during the first 3–4 cycles and thereafter the rate was only 2–3%; fertility significantly increased after discontinuation of 10 mg norethindrone medication	
Ferin *et al.* (1964)	Ethyl-estrenol: 5 mg daily for 15 days and thereafter 2.5 mg daily; from 6th to 10th week postpartum	After 4 weeks' treatment: no impairment of milk yield or infant's weight gain as compared to placebo controls; these progestogenic 19-nortestosterone derivatives transform the proliferative endometrium, as most frequently seen in untreated controls, into an atrophic or resting stage	Placebo controls
Pincus (1965)	Enovid: 2.5, 5, 10, 20 mg norethynodrel and 100 μg mestranol per tablet; during lactation	No effect on maternal breast size; no breast discomfort; with progestin doses of 2.5 mg or less no change in milk volume; daily progestin doses of 5 mg or more decreased milk volume by 40–80%	In this textbook report no details of study methods used for evaluation of the drug effects on lactation were given
Semm (1966)	Lyndiol: 2.5 mg lynestrenol and 75 μg mestranol per tablet; group a: given from day 1 to day 10 after delivery; group b: given from day 10 to day 31 after delivery	Group a: no change in milk yield Group b: slight increase (21%) in milk yield as compared to the placebo group	Placebo controls; period of treatment was too short
Kaern (1967)	Norinyl 1 + 50: 1 mg norethisterone and 50 μg mestranol per tablet; from 2nd to 8th day after delivery	In norinyl group complementary feeding was needed in 12% of cases against 3.5% in placebo group; febrile breast engorgement slightly lower in norinyl group; no significant changes in weight gain as compared to placebo group	Placebo controls: "measurable diminution in quantity of milk" in norinyl group (no amounts reported); period of treatment was too short.

TABLE 26 *(continued)*

<table>
<tr><th>References</th><th>Type, dosage, and duration of administration of contraceptives</th><th>Effect on milk yield, milk composition, breast-fed baby, and mother</th><th>Comments</th></tr>
<tr><td>Frank *et al.* (1969)</td><td>Ovulen: 1 mg ethynodiol diacetate and 100 μg mestranol per tablet; treatment from 3 days postpartum on for up to 18 months</td><td>"Nursing function was occasionally impaired;" lochia rubra subsided more rapidly; 4 weeks postpartum atrophic appearance of endometrium; no thromboembolic phenomena occurred; no pregnancy occurred during treatment; after ovulen withdrawal, fertility was not impaired</td><td>No placebo controls</td></tr>
<tr><td>Kamal *et al.* (1969b)</td><td>Lyndiol 1.0: 1 mg lynestrenol and 100 μg mestranol
Lyndiol 2.5: 2.5 mg lynestrenol and 75 μg mestranol
Lynestrenol 0.5 mg (given continuously)
Deladroxate: 150 mg dihydroxyprogesterone and 10 mg estradiol enanthate (i.m. cyclically)
(from 8th to 32nd week postpartum)</td><td><table>
<tr><th>Agents applied</th><th>Infants' age in weeks when additional feeding needed</th><th>Diminished milk yield in % (subjective maternal impression)</th><th>Measured milk volume</th></tr>
<tr><td>Placebo</td><td>15</td><td>14</td><td>—</td></tr>
<tr><td>Lyndiol 1.0</td><td>12</td><td>30</td><td rowspan="3">No difference as compared to placebo controls</td></tr>
<tr><td>Lyndiol 2.5</td><td>14</td><td>45</td></tr>
<tr><td>Lynestrenol 0.5</td><td>11</td><td>33</td></tr>
<tr><td>Deladroxate</td><td>18</td><td>43</td><td>Slight decrease in milk yield</td></tr>
</table></td><td>Study performed on a group of postpartum nursing women of a low socioeconomic condition; a 20–40% decrease in milk protein, milk fat, and inorganic milk constituents (Ca^{2+}, Mg^{2+}, Na^{+}, K^{+}, P) has been observed, after using contraceptives, however, the growth rate curves of the babies in all groups were within the normal range</td></tr>
<tr><td>Kates *et al.* (1969)</td><td>C-Quens: 2 mg chlormadinone acetate and 80 μg mestranol (sequential); from 5th day</td><td>C-Quens does not interfere with lactation; "27/55 lactating mothers had no difficulty with lactation;" reduced amount and dura-</td><td>No placebo controls; "None of the mothers taking the pill discontinued breast feeding because</td></tr>
</table>

	postpartum throughout lactation	tion (14 days) of lochia; in 9% breakthrough bleeding; no abnormalities regarding Papanicolaou cytology, no impairment of pituitary-ovarian function on discontinuance of the pill; no pregnancy occurred; no change in infants' genitals; normal bone growth	of inadequate milk supply;" in some patients "increase in a requirement for supplemental feedings"
Kora (1969)	Ovulen: 1 mg ethynodiol diacetate and 100 μg mestranol per tablet; from 4th to 24th week postpartum	After 5 weeks of ovulen administration the mean milk volume obtained at a test feeding was decreased by 30% and the mean weekly infants' weight gain was diminished by 26% as compared to nontreated controls	No placebo controls
Gambrell (1970)	Norethindrone 2 mg/mestranol 100 μg Norgestrel 0.5 mg/ethinyl estradiol 50 μg Norethindrone acetate 2.5 mg/ethinyl estradiol 50 μg Ethynodiol diacetate 1 mg/mestranol 100 μg Norethindrone 1 mg/mestranol 50 μg Norethynodrel 2.5 mg/mestranol 100 μg Dimethisterone 25 mg/ethinyl estradiol 100 μg (sequential) (per tablet; from 5th day postpartum on for 6 weeks)	25% of pill users stopped breast feeding after 6 weeks of lactation versus 13% of controls (difference is not significant); in pill users slightly less lochial flow and lower incidence of uterine subinvolution (1.2% in pill users and 3.3% in controls); postpartum weight loss was 14½ lbs. in pill users and 17 lbs. in nonpill users; no pregnancy occurred in pill users	Simple controls; "patient acceptance was excellent and no major problems were encountered;" "side effects were minimal"
Miller and Hughes (1970)	ORF: 1 mg norethindrone and 80 μg mestranol per tablet Group a: ORF given from 14th	Group a: Infants' weight gain was 23% less than that of placebo controls at 5th week postpartum; "breast feeding success" de-	Placebo controls; these authors contradict themselves in stating that Semm (1966) administered

TABLE 26 (*continued*)

References	Type, dosage, and duration of administration of contraceptives	Effect on milk yield, milk composition, breast-fed baby, and mother	Comments
Miller and Hughes (1970)—*continued*	day to 12th week postpartum Group b: ORF given from 6th to 12th week postpartum	clined and at 12 weeks postpartum only 21% were still nursing (controls: 73%); at 6 weeks postpartum lochia ceased in 50% (placebo controls: 25%); at 3 weeks postpartum the vaginal smear showed, besides a few basal cells, mainly intermediate and superficial cells (in placebo group in 70% an atrophic smear was observed); at 5 weeks postpartum decreased endometrial glandular activity Group b: Lactation began to decline at 8th week postpartum; at 12 weeks postpartum only 52% were still nursing (controls: 73%); ORF medication did not effect the infants' breast size	a higher daily mestranol dose than they did
Borglin and Sandholm (1971)	Lyndiol 2.5: 2.5 mg lynestrenol and 75 μg mestranol per tablet Ethinyl estradiol: 50 μg per tablet Mestranol: 80 μg per tablet (for a period of 7 days during lactation)	Lyndiol: after administering it for 7 days, approximately 70% reduction in milk volume as compared to placebo group Ethinyl estradiol or mestranol alone: no effect on milk volume	Placebo controls; accurate measurement of milk volume by using an electric breast pump; but too short a period of investigation
Karim *et al.* (1971)	Medroxyprogesterone acetate (150 mg i.m. every 3 months) Norethisterone enanthate (200 mg i.m. every 84 days) (for up to 18 months)	Milk yield was not altered; growth (no changes in bone epiphyses as judged by x-ray examination) and weight gain of infant not impaired; no pregnancy occurred during treatment	No placebo controls

Koetsawang *et al.* (1972)	Ovulen: 1 mg ethynodiol diacetate and 100 μg mestranol per tablet C-Quens: 2 mg chlormadinone acetate and 80 μg mestranol (sequential) (treatment from 6th to 16th week postpartum)	After 4 weeks of medication: Ovulen: Milk volume decreased by 32% compared to nontreated control group; no additional feeding necessary C-Quens: Milk volume decreased by 20% compared to nontreated control group; additional feeding necessary in 20% of cases After 10 weeks of medication: Ovulen: Milk volume decreased by 55% and infants' weight was 27% less compared to nontreated control group; additional feeding necessary in 5% of cases C-Quens: Milk volume decreased by 32% and infants' weight was 10% less compared to nontreated control group; additional feeding required in 70% of cases No changes in milk composition; no signs of impaired health or breast engorgement of the infants	No placebo controls; authors report a 55% decrease in milk volume after 10 weeks intake of ovulen but despite that, additional feeding was necessary in only 5% of cases
Ramadan *et al.* (1972)	Ovosiston: 3 mg chlormadinone acetate and 100 μg mestranol per tablet; administration started around 3–4 months postpartum for 4 months	"Sufficient milk yield—infants were satisfied;" no effect on milk constituents (proteins, lactose, chloride, ash); 30% increase in milk fat	No placebo controls

[a] From Vorherr (1973).

ing dental mottling in the breast-fed baby. Protein allergens may appear in breast milk, due to excessive maternal intake of eggs, fish, and cow's milk, but no related pathology has been observed in the newborn. Obviously, such allergens as well as maternal rhesus and diphteria antibodies in milk are not absorbed in the infant's gastrointestinal tract. A hypersensitivity reaction (fever, rash) has been described in a breast-fed infant whose mother received 2.4 million units of penicillin (Catz and Giacoia, 1972). Staphylococci derived from cracked nipples and inflamed mammary tissue admixed to milk are usually not infectious to the breast-fed baby.

In conclusion, it may be stated that our knowledge of drug excretion into milk and the potentially acute and chronic harmful effects on the breast-fed baby is still limited. Although drug excretion into breast milk rarely exceeds 1% of the total dose ingested, maternal intake of drugs over a prolonged period of time may be hazardous to the breast-fed baby. Bottle-feeding should be recommended for postpartum women taking potentially dangerous drugs such as diuretics, steroids, reserpine, ergot alkaloids, coumarins, diphenylhydantoin, barbiturates, antithyroid agents, laxatives (anthraquinone and emodin type), metronidazole, and anticancer agents. If drug treatment is necessary when nursing is strongly desired, the lactating woman should receive drugs only in well-controlled doses over a limited period of time; careful observation of mother and infant is mandatory.

CHAPTER IV

Lactation and Maternal Changes

Lactation sets in within 2 to 4 days following delivery due to postpartal withdrawal of placental lactogen and luteal and placental sex steroids (Figs. 8 and 25). Because during the last 4 weeks of gestation, in the absence of lactation, prolactin blood concentrations are even higher than those observed postpartum (Fig. 26), it appears that during pregnancy mainly sex hormones and probably placental lactogen antagonize the secretory prolactin endorgan effect on the mammary epithelium. With postpartum prolactin stimulation, the breasts become extremely full and tender, and the superficial mammary veins become greatly dilated and blood filled. At the same time the puerpera experiences a sensation of heaviness and "milk coming in," and many distended milk-filled alveoli are visible in the histiologic mammary section of such a breast (Fig. 27).

For the first 5 days of lactation, so-called colostrum, or colostral milk, is secreted; this sticky, yellowish fluid may also be secreted premenstrually, during pregnancy, and during the time of weaning. Colostrum contains corpuscles of Donné, which are composed of leukocytes, histiocytes, lymphocytes, and desquamated glandular cells; colostral proteins are coagulated by heat in contrast to the proteins of mature milk. In Table 21 the major components of colostrum milk, mature human milk, and cow's milk are listed. At first, colostrum appears as a thin serous fluid then changes into one of a sticky, yellowish consistency during the following 2–5 postnatal days. During the subsequent 2 weeks, some sort of transitory milk is produced that soon acquires the composition of mature breast milk. Colostrum milk may contain up to 13 g% lactoglobulin, which is identical with the euglobulin of the blood and thus carries immune bodies. It is not known to what extent these immune bodies can pass the gastrointestinal barrier of the newborn to cause effective passive immunization of the neonatus. Since most of the maternal immune substances against infectious bacterial and viral diseases, with the exception of pertussis, cross the placental barrier, the newborn does not seem to depend on immune bodies supplied by breast milk. The lactalbumin content of colostrum is higher than that of mature breast milk; lactalbumin is considered to be the most valuable protein for promotion of growth in the neonatus. The caloric value of colostrum is slightly less than that of mature breast milk (Table 21).

During the initial phase of postnatal lactation, a decline of serum calcium and magnesium is observed in the mother due to the withdrawal of luteal and placental estrogens; whereby calcium binding to serum albumin is decreased. Increased excretion of calcium into the milk also occurs during lactation. At the same time, more calcium is eliminated in the urine through enhanced postpartum diuresis. In dairy cattle, true hypocalcemia is sometimes observed at the onset of lactation, resulting in muscular paresis, a condition also called "milk fever." Even though hypocalcemia is encountered postpartum in humans, no such muscular paresis has been observed (Bach and Messervy, 1969; Ramberg *et al.*, 1970; Rice *et al.*, 1969).

The plasma free fatty acids increase during the second half of gestation, which is probably due to the physiologic "diabetogenic" state of pregnancy; plasma free fatty acids return rapidly to normal after delivery (Gürson and Etili, 1968).

As a rule, an oral intake of 2–3 g of food proteins is required for the production of 1 g of milk protein. Accordingly, an adequate protein supply of 85 g/day on the average is recommended (Table 27). Besides sufficient amounts of carbodydrates and fat, a proper supply of iron and vitamins is necessary. A daily diet plan of 3000 calories, including a regular intake of cow's milk, cheese, eggs, orange juice, and whole wheat bread in adequate amounts, will provide optimal conditions for lactation. However, even with a well-balanced maternal diet, the amount of iron in human milk is not sufficient. Breast feeding without supplement will lead to iron deficiency in the newborn as soon as the prenatally stored iron reserves of the liver are exhausted. The breast milk content of vitamins B and D is also inadequate for the newborn. Therefore, the breast-fed infant requires an additional supply of iron and vitamins B and D (Zilliacus, 1967; Taylor, 1966; this subject will be discussed in detail in Chapter V, Section B.).

TABLE 27

Daily Caloric Allowances and Dietary Recommendations for Pregnant and Lactating Women [a]

Groups of women	Calories (kcal)	Protein (gm)	Fat (gm)	Carbohydrate (gm)	Calcium (gm)	Phosphorus (gm)	Iron (mg)	Magnesium (mg)	Iodine (μg)	Vit A (IU)	Vit B_1 (mg)	Vit B_2 (mg)	Vit B_{12} (μg)	Vit C (mg)	Vit D (IU)
Pregnant women during first trimester of pregnancy	2100	55	60	330	0.8	0.8	15	300	100	5000	0.8	1.3	5	55	200
Pregnant women during second and third trimester of pregnancy	2300	65	65	360	1.2	1.2	20	450	125	6000	1.0	1.6	8	60	400
Lactating women	3000	85	80	480	1.3	1.3	20	450	150	8000	1.4	2.0	6	60	400

[a] From Vorherr (1972b).

A. MAINTENANCE OF LACTATION AND HORMONAL MECHANISMS INVOLVED

Regular removal of milk, adequate diet (Table 27), appropriate psychologic attitude toward nursing, and an intact hypothalamic–pituitary axis regulating prolactin and oxytocin secretion are key factors for successful onset and maintenance of lactation. The process of lactation is comprised of (1) prolactin-induced milk synthesis and release of milk into the mammary alveoli and smaller milk ducts, and (2) oxytocin-induced ejection of milk from alveoli and smaller ducts into lactiferous sinuses. Failure to remove milk will result in reduction of milk synthesis because the milk-dilated alveoli obstruct the mammary capillary blood flow. Moreover, in the absence of the suckling stimulus no adequate pituitary prolactin or oxytocin secretion takes place, resulting in cessation of lactation. Basal prolactin levels (Fig. 26) with additional spurts of maternal prolactin and oxytocin release induced by the infant's suckling are necessary for maintenance of lactation. Within 4 weeks postpartum the basal prolactin blood levels in lactating women decline gradually. During the infant's suckling, however, spurts of prolactin release bring about a temporary tenfold rise in the basal prolactin blood levels (Fig. 26). In nonlactating puerperas prolactin blood levels diminish to about 10 ng/ml or below by the fourth to seventh week postpartum. Hypothalamic–pituitary insufficiencies due to pathologic lesions will result in nursing failure when either prolactin or oxytocin synthesis and/or release are impaired. It can be shown in rabbit experiments that endogenous prolactin depletion by active immunization with human placental lactogen prevents lactation (El Tomi *et al.,* 1971). In our studies active immunization of rabbits against oxytocin resulted in formation of oxytocin antibodies. Despite high oxytocin antibody titers, the rabbits could conceive and carry their pregnancy to term, but they failed to lactate postpartum, which indicates a selective endogenous depletion of oxytocin (H. Vorherr, unpublished observations, 1971).

In anxious and fearful postnatal women with pelvic pains, increased activity of the sympathetico–adrenal system may cause nursing difficulties through catecholamine antagonism of oxytocin-induced milk ejection. Also injection of epinephrine can inhibit oxytocin-induced milk ejection; thus catecholamine antagonism toward oxytocin effects the mammary myoepithelium. Through catecholamine-induced stimulation of myoepithelial β-adrenergic receptors, the mammary myoepithelium is rendered inexcitable to oxytocin. In addition, α-adrenergic receptor stimulation of mammary blood vessels by norepinephrine and epinephrine leading to vasoconstriction and decreased mammary blood flow plays a role because not enough oxytocin can reach its target cells for initiation of milk ejection (Vorherr, 1971). This problem of catecholamine-induced inhibition of milk ejection can be easily corrected by intramuscular injection of 5–10 units of oxytocin. It is not well known as to what extent sympathetic stimuli or other functional mechanisms can influence the secretion or release of prolactin or its action on the mammary secretory epithelium in humans; under the situation of stress increased prolactin levels have been reported (Fig. 47). Recently, *in vitro* incubation studies with rat pituitaries showed that catecholamines may inhibit prolactin secretion into the incubation medium (MacLeod *et al.*, 1969). In similar studies it was found that theophylline stimulates prolactin secretion by a direct effect on the pituitary prolactin cell, an action that cannot be influenced by the hypothalamic prolactin-inhibiting factor (Parsons and Nicoll, 1971). Treatment of hypogalactia due to pituitary prolactin deficiency or impaired releasing mechanisms has not been reported yet. It appears, however, that administration of either highly purified prolactin, synthetic thyrotropin-releasing hormone, or theophylline may be clinically helpful for treatment of hypogalactia due to insufficient prolactin secretion.

B. NEUROHUMORAL REFLEX PATHWAYS FOR PITUITARY RELEASE OF PROLACTIN AND OXYTOCIN

Sensory nerve endings, located mainly in the breast areola and nipple, are stimulated by suckling or breast manipulation, and an afferent neural reflex pathway for release of prolactin and oxytocin becomes operative (Fig. 34). In studies on goats, ewes, rabbits, rats, and guinea pigs, the various components of this afferent neural pathway have been traced. Through the stimulus of suckling or other breast manipulations, impulses that travel in sensory nerve fibers and enter the spinal column through its dorsal routes may be elicited. From there, the neural impulses ascend via dorsal, lateral, and ventral spinothalamic tracts to the mesencephalon. In the mesencephalon and lateral hypothalamus the neurons responsible for the stimulation of prolactin release, probably through depression of a hypothalamic prolactin-inhibiting factor (PIF), seem to separate from those inducing posterior pituitary release of oxytocin and its synthesis in hypothalamic cell bodies (nucleus paraventricularis mainly). From the mesencephalon, stimuli are transmitted via diencephalic routes to the hypothalamic areas regulating release of prolactin and oxytocin. Similar reflex pathways probably also exist in man (Tindal *et al.*, 1969; Richard, 1970; Tindal, 1972).

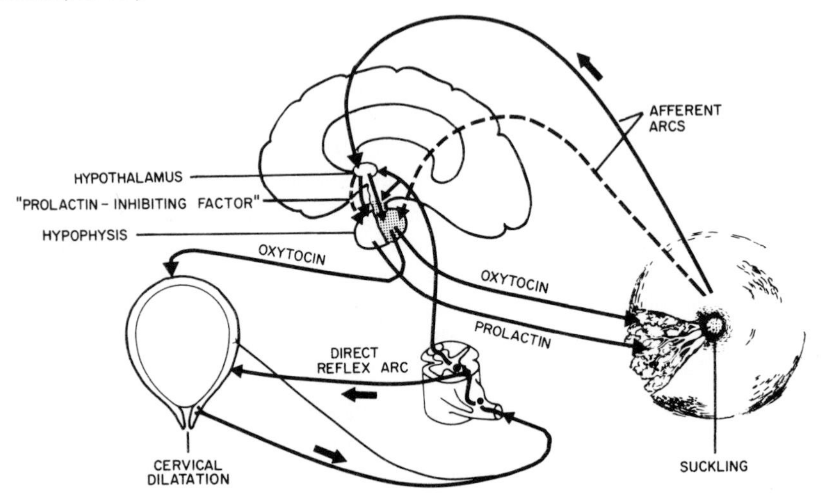

Fig. 34. Afferent neural reflex pathways for humoral secretion of oxytocin and prolactin (from Vorherr, 1968). Breast stimulation by suckling elicits afferent neural impulses reaching the spinal column. Within the spinal cord the impulses are transmitted in spinothalamic tracts to the mesencephalon, as traced in animal experiments. In the mesencephalon the neural fibers bifurcate, and stimuli travel separately via different diencephalic routes to the hypothalamic areas involved with secretion and release of prolactin and oxytocin. Adenohypophysial prolactin release is most likely due to hypothalamic suppression of the tonic secretion of a prolactin-inhibiting factor (dashed line); the possibility exists that secretion of a hypothalamic prolactin-releasing factor(s) is also triggered by the stimulus of suckling. Prolactin released into the systemic circulation induces mammary milk synthesis and milk secretion into the alveoli. Afferent impulses reaching the paraventricular and supraoptic nuclei of the hypothalamus via the dorsal diencephalic route induce oxytocin synthesis and its secretion along neurosecretory axons into the posterior pituitary lobe for storage. Simultaneously, release of oxytocin from posterior pituitary neurovesicles is brought about by hypothalamic impulses traveling down along secretory neurones of the hypothalamo-hypophysial tract to the posterior lobe (see Fig. 37). It seems unlikely that afferent stimuli can bring about release of oxytocin due to suckling by direct stimulation of the posterior pituitary neurovesicles (dashed line). Liberated oxytocin rapidly enters the circulation, causing ejection of milk from mammary alveoli and smaller milk ducts into larger lactiferous ducts and lactiferous sinuses ("milk letdown") and evoking myometrial contractions in support of postpartum uterine involution. Cervix dilation or vaginal stretching, as occurring during labor and delivery, may elicit primarily uterine contractions via a direct lumbar spinal reflex; also an ascending afferent neural reflex pathway (Ferguson reflex) has been suggested to be responsible for neurohypophysial oxytocin release in rabbits.

1. Prolactin Synthesis and Secretion

a. PROLACTIN SYNTHESIS AND MOLECULAR STRUCTURE IN RELATION TO HUMAN PLACENTAL LACTOGEN AND GROWTH HORMONE

Synthesis of prolactin occurs on ribosomes of the rough surfaced endoplasmic reticulum in erythrosinophilic lactotrophs (prolactin cells, Fig. 23) of the anterior pituitary gland. Prolactin synthesis and release is induced by female sex steroids (estrogens mainly) and by afferent neurogenic and psychogenic stimuli reaching the anterior pituitary via median eminence and other hypothalamic centers (Meldolesi *et al.,* 1972; Turkington 1972d). Electron microscopy reveals that prolactin cells contain many membrane-limited secretory granules; an enlarged Golgi apparatus; well-developed, rough-surfaced endoplasmic reticulum; and many mitochondria. These are all attributes indicating active intracellular hormone synthesis. It seems, however, well established that only neurohumoral stimuli, causing impairment of hypothalamic prolactin-inhibiting factor secretion, are capable of inducing transmembranal prolactin release from adenohypophysial lactotrophs into the capillaries of the hypophysial portal system. During pregnancy, estrogens and progesterone effectively stimulate prolactin synthesis. Prolactin release into the circulation is rather high toward the end of gestation and may exceed the basal prolactin levels at the onset and during the period of lactation (Fig. 26, Table 28). The existence of prolactin as a different entity from human growth hormone is now well established (see below). In Table 29 the biochemical and biological characteristics of human prolactin and growth hormone are presented and compared with those of placental lactogen and sheep prolactin. The prolactin molecule (Fig. 35) resembles that of growth hormone and human placental lactogen (HPL), which is also called "human chorionic somatomammotropin" (HCS). The molecular weight of human prolactin (HPr) has been estimated at 23,500 (isoelectric point: 5.1) and is similar to that of sheep prolactin (24,000). The amino acid sequence of HPr has been found to be in close homology with sheep prolactin (Li *et al.,* 1969; Hwang *et al.,* 1971b). Although human prolactin carries the amino acid leucine at the NH_2-terminus and the NH_2-terminal amino acid of sheep prolactin is threonine, the homology between the two prolactins is rather striking with 16 out of 23 terminal amino acid residues matching. A homology of only 2 out of 23 terminal amino acid residues is observed between HPr and HPL. Recently, human pituitary prolactin has been greatly purified (ammonium acetate-NaOH extraction, ethanol precipitation, Sephadex gel filtration, and DEAE-cellulose chromatography), but no completely pure

TABLE 28

Prolactin Plasma Levels in Nonpregnant, Pregnant, and Lactating Women, and in Galactorrhea Patients

			Hormone levels			
References	Prolactin assay	Standard	Nonpregnant women	Pregnant women	Lactating women	Galactorrhea patients
Simkin and Goodart (1960) Simkin and Arce (1963)	Pigeon crop	Luteotrophin (Squibb) and NIH sheep prolactin	0.02–0.045 IU/ml blood (first 5 days of cycle and second half of cycle)	—	Approximately 1 IU/ml blood	—
Canfield and Bates, (1965)	Pigeon crop	No standard reported	<5 mU/ml plasma	—	—	5–27 mU/ml plasma
Herlyn *et al.* (1969)	Pigeon crop	No standard reported	—	—	8–15 IU/100 ml plasma	—
Kurcz *et al.* (1969)	Pigeon crop	Human purified prolactin (Apostolakis) and ovine prolactin 15 IU/mg (Paulitar) and 22 IU/mg (Ferring)	No activity	—	No activity	No activity
Berle and Apostolakis, (1971	Pigeon crop	Sheep prolactin NIH-P-S3	—	2–7 IU/ml plasma (second half of pregnancy)	4–8 IU/ml plasma (1–4 days postpartum) <1 IU/ml plasma (after 6th day postpartum)	—
Berle *et al.* (1971)	Pigeon crop	Sheep prolactin NIH-P-S3	End proliferation phase: 0.31 IU/100 ml plasma End secretion phase: 0.41 IU/100 ml plasma		2–7 IU/100 ml plasma (1–4 days postpartum)	—
von Berswordt-Wallrabe *et al.* (1971)	Pigeon crop	Bovine prolactin 20 IU/mg	<2 IU/100 ml plasma	—	14 IU/100 ml plasma (3rd day postpartum)	—
Forsyth *et al.* (1971)	Rabbit mammary tissue *in vitro*	Sheep prolactin NIH-P-S6, 25 IU/mg	No activity	—	100–200 ng/ml plasma (during suckling)	100–200 ng/ml plasma 50–2000 ng/ml plasma
Kleinberg and Frantz (1971)	Mouse mammary gland *in vitro*	Sheep prolactin NIH-P-S8, 28 IU/mg	<0.4 mU/ml plasma	—	0.6–4.5 mU/ml plasma (1–30 days postpartum)	0.4–3.5 mU/ml plasma
Bryant *et al.* (1971)	Radioimmunoassay	Human prolactin 2.5 IU/mg as compared to standard NIH-OP-S6	30–37 ng/ml plasma	—	87 ng/ml plasma (during nursing) (Presumably pre-suckling value = 40 ng/ml plasma)	—

Hwang *et al.* (1971a, b) Marcovitz and Friesen (1971)	Radioimmunoassay	Postpartum prolactin serum sample containing 161 ng/ml sheep prolactin equivalents (Dr. Frantz)	0–28 ng/ml serum Menstrual cycle: 8–14 ng/ml (no ovulatory prolactin peak)	1st Trimester: 30 ng/ml 2nd Trimester: 55 ng/ml 3rd Trimester: 200 ng/ml	1st Day postpartum: 140 ng/ml 7th Day postpartum: 35 ng/ml Before suckling: 20–120 ng/ml After suckling: 200–400 ng/ml	Pituitary tumor, drugs = 40–400 ng/ml serum
Tyson *et al.* (1972a, b)	Radioimmunoassay	Not reported (probably human prolactin 20 IU/mg)	0–28 ng/ml plasma (mean: 6 ng/ml plasma)	18 Weeks: 16 ng/ml plasma 35 Weeks: 100 ng/ml plasma At term: 207 ng/m' plasma	32 ng/ml plasma (postpartum)	—
Friesen *et al.* (1972)	Radioimmunoassay	Postpartum prolactin serum sample containing 161 ng/ml sheep prolactin equivalents (Dr. Frantz)	2–10 ng/ml serum	—	—	Forbes-Albright syndrome: 1800–1900 ng/ml serum
Turkington (1972a)	Mouse mammary gland *in vitro*	Sheep prolactin NIH-P-S9, 30 IU/mg	<2 ng/m' serum	—	—	Forbes-Albright syndrome: 65–980 ng/ml serum
Jacobs *et al.* (1972)	Radioimmunoassay	Human prolactin from male patient with large pituitary tumor and galactorrhea: 60 mU/ml (2000 ng/ml) plasma as compared to sheep prolactin	9 ng/ml plasma	1st Trimester: 56 ng/ml 2nd Trimester: 117 ng/ml 3rd Trimester: 251 ng/ml	—	Forbes-Albright syndrome: 107–1642 ng/ml plasma; drug-induced (chlorpromazine): 236 ng/ml plasma
Tyson *et al.* (1972a, b)	Radioimmunoassay	Human prolactin 20 IU/mg as compared to sheep prolactin	12 ng/ml serum (first half of cycle)	10 Weeks: 15 ng/ml serum 20 Weeks: 50 ng/ml serum 30 Weeks: 100 ng/ml serum 40 Weeks: 200 ng/ml serum	24 Hours postartum: 80 ng/ml First week postpartum: 124 ng/ml (before suckling) 163 ng/ml (30 min. after suckling) Second week postpartum: 30 ng/ml (before suckling) 280 ng/ml (30 min. after suckling)	—

TABLE 29

Biochemical and Biological Characteristics of Human Prolactin and Growth Hormone

1. Separation of HPr from HGH and purification of HPr
   - Acrylamide gel electrophoresis
   - HGH antibodies coupled to a sepharose column to remove residual HGH contained in HPr-extract solution
   - Final HPr preparation is contaminated only by 0.5–1% of HGH allowing a rather specific antibody production against HPr and iodination of HPr; antibody cross-reaction with HGH is minimal. HPr cross-reacts with monkey and sheep prolactin
   - Commercial HGH preparations contain an impurity of 0.2% of HPr
2. HPr as a distinct molecule different from HGH
   - Different mobilities of HGH and HPr on acrylamide gel electrophoresis at the same pH; isoelectric point of HPr is slightly higher than that of HGH
   - Hereditary ateliotic dwarfs lack HGH but are able to lactate successfully (presence of HPr)
   - Increase in serum HPr (bioassay and radioimmunoassay) during pregnancy, puerperal and nonpuerperal lactation, conditions under which serum HGH levels are normal or slightly decreased
   - Different molecular weight and amino acid sequence
   - NH_2-terminal amino acid residue is leucine for prolactin and phenylalanine for growth hormone
3. Molecular weights of HPr, HGH, and HPL
   - Human prolactin (HPr) 23,500 (similar to monkey and sheep prolactin); $t/2 = 15$ to 40 minutes
   - Human growth hormone (HGH) 21,500; $t/2 =$ 20 to 30 minutes
   - Human placental lactogen (HPL) 22,000; $t/2 =$ 10 to 20 minutes
4. Homology between amino acids of sheep prolactin, HGH, HPL, and HPr molecules
   - Sheep prolactin and HGH about 40%
   - HGH and HPL 80 to 90%
   - HGH and HPr about 20%
5. Ratio of HGH to HPr in adenohypophysis is 100:1 to 500:1 (ratio HGH to HPr is reversed in patients with prolactin-secreting pituitary tumor). Blood levels of HGH and HPr are similar under normal conditions

natural or synthetic prolactin is available at present. Human pituitaries obtained at autopsy and extracted contain 40–50 μg of HPr, whereas the HGH content is 10–30 mg/gland. Recently, HPr concentrations of about 800 μg per pituitary gland of nonpregnant women and of men have been reported (Sinha *et al.,* 1973c). Currently available human, monkey, and sheep prolactin preparations are 80% pure. The main contamination of about 20% is due to other proteins, and only 0.5–1% of the impurity is caused by admixed growth hormone (Hwang *et al.,* 1972). Such preparations contain 20–30 IU of prolactin activity per milligram powder (Friesen, 1972). The sequence of about 198 amino acids of sheep prolactin has been identified (Li *et al.,* 1969; Fig. 35); prolactin contains 10 more amino acids and one more disulfide bridge than human growth hormone (HGH). Sheep prolactin possesses six half-cystine and seven methionine residues that help to determine the shape of the molecule in three-dimensional space. The half-life of HPr has been estimated at between 15–43 minutes (Malarkey *et al.,* 1971; Friesen, 1972; Ehara *et al.,* 1973), which is higher than that of HPL ($t/2$ = 10–20 minutes). HPr appears to be a single homogenous protein and more highly charged than HGH. Serum prolactin obtained from lactating women is physically, biochemically, and biologically indistinguishable from serum prolactin found in male patients with pituitary adenoma and galactorrhea (Chrambach *et al.,* 1971). Synthesis of HGH occurs in separate acidophilic adenohypophysial cells (orangeophilics, Fig. 23), and the molecular weight of HGH at 21,500 is different from that of prolactin. In situations of acromegaly in which high blood levels of HGH are found (20–100 ng/ml), serum prolactin levels are not above normal; the reverse applies for the galactorrhea patient (Tables 28 and 30). HGH and prolactin resemble HPL, which is produced in the syncytial trophoblast in increasing amounts as pregnancy advances (Table 31). The HPL concentration is higher in retroplacental and intervillous blood than in peripheral maternal circulation; only minor amounts of HPL are found in cord blood (less than 1% of the maternal serum concentration). The molecular weight of HPL has been estimated at 22,000. The physiology of HPL is not understood. Besides supporting mammary development during pregnancy a potentiation of the HGH action by HPL has been proposed. A luteotrophic effect of HPL on the ovary resulting in increased synthesis of progesterone and thus potentiating the effect of HCG has been suggested. In addition, stimulation of placental sex steroid production by HPL in synergism with HCG is possible. Also a maternal metabolic effect of HPL regarding storage of nitrogen and mobilization of body fat has been considered. Furthermore, HPL may stimulate aldosterone secretion and thus contribute to the increased production of

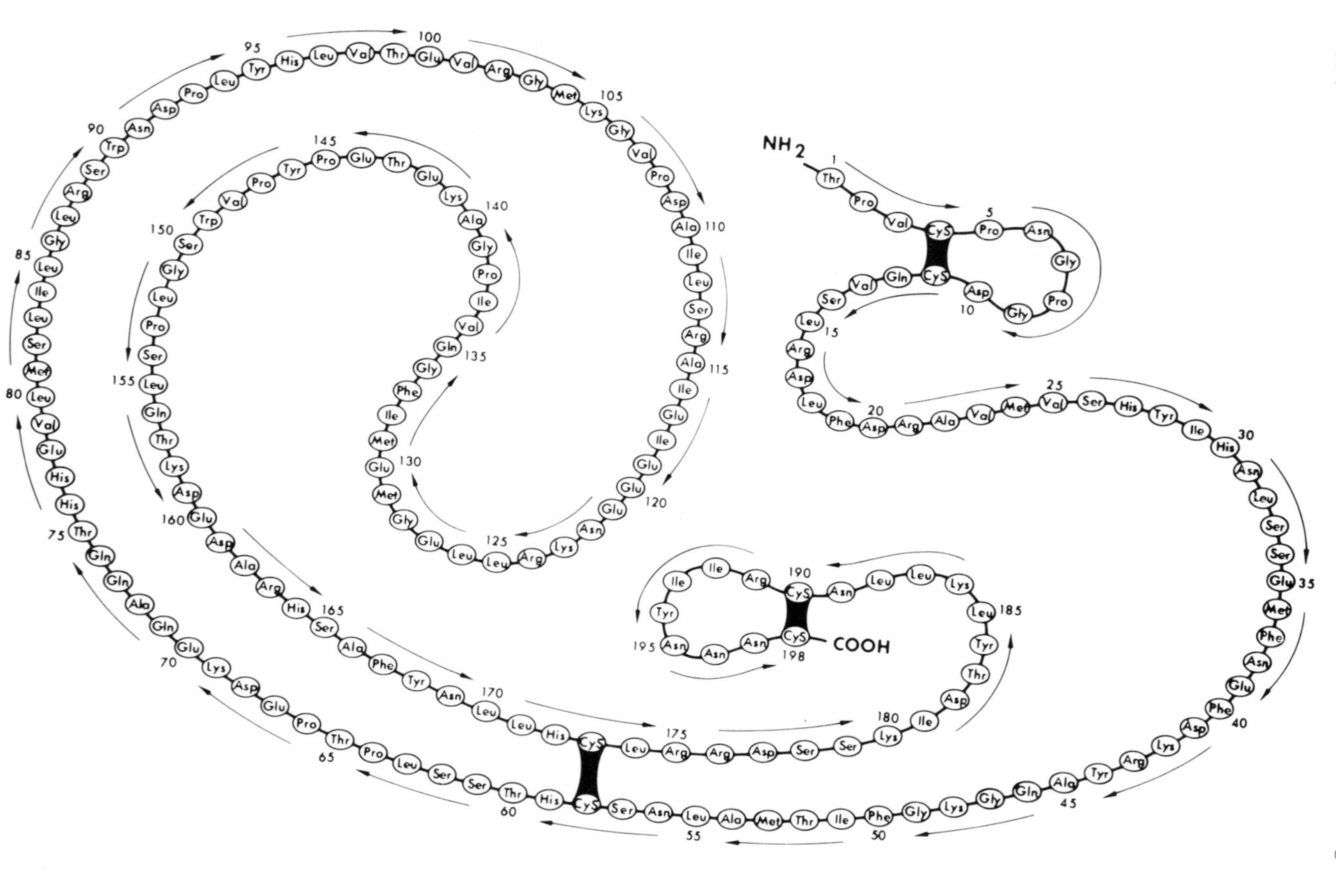

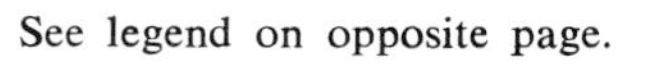
See legend on opposite page.

aldosterone during pregnancy. It seems that both HPL (during the last trimester of pregnancy) and prolactin (during early lactation) may suppress the adenohypophysial output of HGH to some extent. It is also assumed that HPL inhibits prolactin release or its effect on the mammary epithelium during gestation (Spellacy and Buhi, 1969). HPL has growth hormonelike and prolactinlike effects, but immunologically it resembles HGH much more than it does prolactin. In radioimmunoassy systems cross reactions are observed between HPL, HGH, and HGH-contaminated prolactin preparations. The HGH molecule also possesses a slight intrinsic biologic HPr activity and vice versa. HPL is first measurable in circulation around the fifth week of gestation (about 0.5 μg/ml serum), and approximately 10–15 μg/ml serum (200–500 μg/24 hour urine) are found near term (Table 31). Human placental lactogen enters amniotic fluid by passage through amnion and chorion; near term amniotic fluid contains about 1 μg/ml of HPL. In fetal cord blood the HPL concentration is approximately 15 ng/ml, in newborns' urine it is around 0.5 ng/ml. During the last weeks of pregnancy 1–10 g of HPL are produced per day (1000 times the production rate of HGH). The HPL production rates and blood levels seem to correlate with placental weight and function. As mentioned, the physiology of HPL is unknown; however, recently its participation in maternal anabolic processes (HGH is anabolic too) has again been suggested (Spona and Janisch, 1971). HGH and HPL possess two disulfide bonds that are located at the same position in the polypeptide chain of both hormones and that are not essential for these hormones' small lactogenic effect. The disulfide bonds that connect half-cystine residues at positions 53 and 164 and at positions 181 and 188, however, are required for the HGH's and HPL's growth hormone activity. Disruption of disulfide bonds of HPL and HGH markedly alter their immunoreactivity, which is almost completely lost, whereas the hormones' lactogenic potency is not decreased

---

Fig. 35. Structure of ovine (sheep) prolactin (by courtesy of C. H. Li). The prolactin molecule resembles that of placental lactogen and human growth hormone (HGH). The molecular weight of sheep prolactin has been estimated at 24,000, and the amino acid sequence of 198 has recently been clarified (Li *et al.*, 1969). Sheep prolactin contains 10 more amino acids and one more disulfide bond than HGH. Ovine prolactin possesses 6 half-cystine and 7 methionine residues. The amino acid sequence of human prolactin has been found to be in close homology with sheep prolactin carrying threonine at the NH_2-terminus. Both sheep prolactin and human prolactin appear to be a single homogenous protein, more highly charged than HGH. Human prolactin has been greatly purified recently allowing a rather specific antibody production against it, thus providing a reliable homologous radioimmunoassay procedure for the human lactogenic hormone.

TABLE 30

Growth Hormone Plasma Levels in Nonpregnant, Pregnant, and Lactating Women and in Galactorrhea Patients [a]

References	Standard	Hormone levels			
		Nonpregnant women	Pregnant women	Lactating women	Galactorrhea patients
Greenwood *et al.* (1964) Hunter and Greenwood (1964)	Purified human growth hormone, I.C.R.F., batch 3 (Raben)	0.55 ng/ml plasma	20 ng/ml plasma (8th week, thereafter no valid measurements possible)	4.5 ng/ml plasma (2nd–8th day postpartum) no valid measurements before 2nd day postpartum	7 ng/ml plasma
Beck *et al.* (1965)	Human growth hormone, Armour (Fisher) 50% of potency of Wilhelmi's standard HS612A	—	19 ng/ml plasma (at term)	—	—
Kaplan and Grumbach (1965)	HGH 475A (Wilhelmi)	—	4 ng/ml serum	5 ng/ml plasma (immediately postpartum)	—
Board (1968)	NIH-GH-HS722A	0.9–10.6 ng/ml plasma	—	1–8 ng/ml plasma (during 1st–5th day postpartum)	—

Spellacy *et al.* (1968, 1970)	NIH-HS612B and NIH-624Bb 3N (Wilhelmi)	2–15 ng/ml plasma	—	8.4 ng/ml plasma	3–8 ng/ml plasma
Benjamin *et al.* (1969)	NIH-GH-HS840FA (Wilhelmi)	3.9 ng/ml plasma	—	4.8 ng/ml plasma	1.5–2.5 ng/ml plasma
Spellacy and Buhi (1969)	Lot HS612B and lot 624Bb 3N (Wilhelmi)	—	7 ng/ml plasma (36–40 weeks)	6 ng/ml plasma (2nd–6th day postpartum) 9 ng/ml plasma (6 weeks post-partum)	—
Shearman and Turtle (1970)	Human growth hormone NIH-GH-HS968C	—	—	—	1–5 ng/ml plasma
Bryant *et al.* (1971)	I.C.R.F. (Raben)	<1–3.2 ng/ml plasma	—	—	—
Forsyth *et al.* (1971)	NIH-GH-HS-612A	1–12 ng/ml plasma	—	<2 ng/ml plasma	<2–9 ng/ml plasma
Kleinberg and Frantz (1971)	HS1103C (Wilhelmi)	1 ng/ml plasma	—	1.5 ng/ml plasma (during days 1–30 postpartum)	0.3–2.6 ng/ml plasma
Turkington (1972a)	Not reported (probably standard Wilhelmi)	—	—	—	<10 ng/ml serum (Forbes-Albright syndrome)
Friesen *et al.* (1972)	NIH lot no. HS1216C	—	—	—	<5 ng/ml serum

[a] Radioimmunoassay used for measurement of human growth hormone.

TABLE 31

Placental Lactogen Plasma Levels in Nonpregnant, Pregnant, and Lactating Women [a]

References	Standard	Hormone levels			
		Nonpregnant women	Pregnant women	Lactating women	Galactorrhea patients
Beck *et al.* (1965)	Purified placental lactogen; final product was 150–200 times more potent than the crude KCL extract	—	25 μg/ml plasma (36–40 weeks) 39 μg/ml plasma (during labor)	—	—
Kaplan and Grumbach (1965)	Purified placental lactogen CGP 10–5	—	—	5.6 μg/ml serum (immediately postpartum)	—
Frantz *et al.* (1965)	Purified placental lactogen (Friesen)	—	0.1 μg/ml plasma (6th week of pregnancy); 2–9 μg/ml plasma (last trimester)	—	—
Josimovich *et al.* (1969)	Purified placental protein (Lederle)	—	10–12 μg/ml serum (36–40 weeks)	—	—
Bryant *et al.* (1971)	Purified placental lactogen (Grumbach and Kaplan standard; Lederle standard 4508 C75)	<4 ng/ml plasma	—	—	—
Spellacy and Buhi (1969)	PPP 438 7C 125 (Lederle)	—	10 μg/ml serum (36–40 weeks)	No measurable activity after second day postpartum and thereafter	—
Spona and Janisch (1971)	Purified placental lactogen (Friesen)	—	10 μg/ml serum (36–40 weeks)	—	—

[a] Radioimmunoassay used for measurement of human placental lactogen.

(Handwerger *et al.,* 1972). The three disulfide bonds of HPr are essential for the hormone's lactogenic action. The lactogenic potentials of HGH and HPL are identical and are only a fraction (5%) of the HPr's lactogenic potency; the somatotrophic activity of HPL is only 0.1–1% of that of HGH. HGH and HPL molecules display an 85% homology of amino acids. HGH shows remarkable resistance to acid, alkali, and urea denaturation, whereas HPL can be much more easily denatured by these agents (Handwerger *et al.,* 1972; Aloj *et al.,* 1972).

b. PROLACTIN: PHYSIOLOGY, SECRETION, AND METABOLISM

Under conditions such as stress, pregnancy, lactation, and in the neonatus, prolactin secretion is enhanced. Nevertheless, the physiologic effect of prolactin appears to be of vital importance only during pregnancy and lactation (Table 32). Prolactin combines with hormone-specific receptors in the plasma membrane of mammary alveolar cells and thereby stimulates intracellular milk synthesis via induction of messenger and transfer RNA. Thereby polymerase activity and cellular transcription are enhanced, resulting in selective phosphorylation of histones and formation of milk proteins. Prolactin also stimulates the activity of galactosyltransferase, lactose synthetase, and α-lactalbumin (Turkington, 1972f). It appears that prolactin promotes cyclic AMP-stimulated protein kinase activity, which phosphorylates the specific milk proteins in the mammary secretory cells (Turkington, 1972d). An extramammary effect of prolactin on the kidney has been observed in rats, cats, and humans. Here, the lactogenic hormone

TABLE 32

Physiology of Prolactin

1. During pregnancy: Stimulation and preparation of mammary alveolar epithelium for secretory activity
   During lactation: Synthesis and secretion of milk
2. Potential participation of prolactin in regulation of water and salt transport across amniotic membranes (increased prolactin levels of amniotic fluid) and in water and salt regulation in the early neonatal period (increased serum prolactin levels of the newborn)
3. Possibly slight increase in prolactin release during the secretory phase of menstrual cycle; this increase probably corresponds to the increased serum levels of luteal sex steroids. With decline of luteal function around the time of menstruation, prolactin induces minor secretory mammary changes
4. Potential anabolic effect of prolactin which is increasingly released in situations of stress to counteract the catabolic action of cortisol
5. In males prolactin together with testosterone appears to participate in the stimulation of prostatic growth and function

decreases the urinary excretion of sodium, potassium, and water, which is followed by a rise in serum sodium and potassium. A direct proximal tubular action and/or distal tubular aldosterone-like prolactin effect (retention of water and salt) has been linked with the development of edema as found in the premenstrual tension syndrome and in patients during the last trimester of pregnancy (Turkington, 1972d). In contrast to HGH, prolactin has no metabolic effect (Noel *et al.,* 1972). In animals, prolactin can stimulate growth and function of testis, prostate, and seminal vesicles (Meites, 1973; Table 32).

In both animals and humans, the stimulus for anterior pituitary prolactin release seems to require primarily a depression of hypothalamic prolactin-inhibitor factor secretion and/or secretion of a hypothalamic prolactin-releasing factor(s) (Fig. 25). One of these hypothalamic prolactin-releasing factors (PRF) in man appears to be thyrotropin-releasing hormone (TRH) since after i.v. injection of 200–500 μg of TRH a five- to tenfold rise of serum prolactin occurs; a similar rise is observed during the infant's suckling (see Table 33). Thyrotropin-releasing hormone most likely acts directly on pituitary lactotrophs to bring about increased HPr secretion. After a given dose of TRH, prolactin serum levels rise twice as high in females as in males. Increased prolactin blood levels return to normal within 2 hours following the TRH injection. Intravenous administration of TRH always leads to elevation of serum prolactin (peak levels of HPr after 10–20 minutes of TRH injection) whereby milk volume and concentration of milk fat become increased; in TRH-treated lactating puerperas, increased thirst and fluid intake were also observed (Tyson *et al.,* 1973). It is not clear whether TRH itself is the prolactin-releasing factor or whether it is chemically related to a PRF complex.

TABLE 33

Stimulation of Adenohypophysial Prolactin Secretion in Normal Adults

Drug and dose regimen (references)	Serum prolactin: Pretreatment level	Serum prolactin: Posttreatment level
Chlorpromazine 50 mg i.m. (Turkington, 1972b)	<2 ng/ml	20–300 ng/ml (peak values 2 hours after injection)
Thyrotropin-releasing hormone 200 μg i.v. (L'Hermite *et al.*, 1972)	75 mU/ml	450 mU/ml (peak values 5–10 minutes after injection)
Thyrotropin-releasing hormone 500 μg i.v. (Sachson *et al.*, 1972)	12 ng/ml	110 ng/ml (peak values 15 minutes after injection)

Since surgical transection of the human pituitary stalk induces chronic prolactin secretion, it is assumed that under normal conditions adenohypophysial prolactin release is regulated by the hypothalamus, which exerts a continuous inhibitory influence upon transmembranal pituitary prolactin secretion. In humans, PIF has not yet been identified; in rats it seems that the PIF action is manifested only when sufficient extracellular calcium is available to support spontaneous pituitary prolactin secretion (Parsons and Nicoll, 1971). Increased activity of hypothalamic dopamine neurons or enhanced hypothalamic dopamine levels following L-dopa treatment counteract transmembranal prolactin secretion by increasing PIF activity (Table 34). Dopamine and other local monoamines (catecholamines, serotonin) are considered as mediators controlling the discharge of some of the hypothalamic prolactin-releasing and inhibiting factors. With decreased PIF function and/or release of PRF, transmembranal prolactin release from the pituitary lactotrophic cells probably occurs through exocytosis (Meldolesi *et al.*, 1972; Shiino *et al.*, 1972) (see also Fig. 37 illustrating

TABLE 34

Influence of Drugs on Prolactin Secretion

Drug	Effect on serum or plasma prolactin concentration	Mechanism of drug action
L-Dopa	Decrease	L-Dopa increases hypothalamic dopamine-catecholamine levels leading to enhanced activity of prolactin-inhibiting factor (PIF)
Ergocornine Ergocryptine	Decrease	Direct inhibition of adenohypophysial prolactin secretion; possibly also increase of hypothalamic PIF activity (continued PIF function)
Thyrotropin-releasing hormone (TRH: pyroglutamyl-histidyl-prolinamide)	Increase	Direct stimulation of adenohypophysial lactotrophs for increased prolactin secretion
Phenothiazines (chlorpromazine)	Increase	Decrease of hypothalamic dopamine-catecholamine levels leading to diminution of PIF activity
Amphetamine	Increase	Decrease of hypothalamic dopamine-catecholamine levels leading to diminution of PIF activity
α-Methyldopa	Increase	Decrease of hypothalamic dopamine-catecholamine levels leading to diminution of PIF activity
Theophylline	Increase	Directly stimulating adenohypophysial prolactin release from lactotrophs

the mechanism of oxytocin release by exocytosis), and this is usually prevented by the membrane-stabilizing effect of continuously secreted hypothalamic PIF. When PIF secretion is impaired through neurogenic or psychogenic stimulation, prolactin is not only promptly released from the adenohypophysial lactotrophs but intracellular prolactin synthesis is also induced. In rat experiments *in vitro* a time lag of 45 minutes was observed until intracellular synthesized prolactin contained in secretion granules was transported from the cell interior to the cell membrane and then released outside of the cell (Meldolesi *et al.,* 1972). Also *in vitro* experiments showed that catecholamines directly inhibit pituitary prolactin release by such a PIF-like membrane-stabilizing mechanism (MacLeod, 1969). *In vivo,* however, it appears that local catecholamines inhibit prolactin release via stimulation of continued hypothalamic PIF function rather than by exerting a membrane-stabilizing mechanism on lactotrophs.

The extent of prolactin metabolism in liver and eventual prolactin excretion via the bile are not understood. Because HPr serum levels are enhanced in patients with renal failure (HGH is not increased), it appears that besides decreased PIF activity the kidney also plays a role by influencing prolactin serum levels, its metabolism and/or its urinary excretion. Physiologically, a negative feedback mechanism could also play a part in the balance of prolactin blood levels.

c. FACTORS AFFECTING PROLACTIN SECRETION

In rats a diurnal pattern for prolactin secretion seems to exist with peak values of 18 ng/ml serum at 11 PM; under conditions of stress peak levels of 33 ng/ml serum at 5 PM were observed (Dunn *et al.,* 1972). A diurnal rhythm of prolactin secretion appears to occur in cows (Tucker *et al.,* 1973), and in humans, a similar diurnal prolactin pattern has been found indicating a rise in plasma prolactin levels of 10 ng/ml (waking period) to 20–40 ng/ml during sleep (Sassin *et al.,* 1972); nocturnal rise of HPr is due to sleep and not related to the time of day. Whereas a circadian periodicity in nonpregnant women has been observed (peak values at 1–5 AM), in pregnancy no diurnal pattern of serum HPr appears to exist (Nokin *et al.,* 1972). In humans slight stress exerted by venipuncture did not influence the serum prolactin levels when compared with values (6–9 ng/ml) obtained in blood drawn from an indwelling needle (Jacobs *et al.,* 1972). In situations of anesthesia and operative stress, blood prolactin levels of 20 to >400 ng/ml serum in humans (Noel *et al.,* 1971, 1972) and up to 500 ng/ml in rhesus monkeys have been observed. In patients with Hand-Schüler-Christian disease (reduced

PIF function, HGH deficiency) and in those with renal failure (defect in PIF function, defective prolactin metabolism and/or excretion by the kidneys), increased plasma prolactin levels are observed. Moreover, strenuous exercise, psychic stress (patients awaiting elective surgery), and hypoglycemia result in two- to fifteenfold increased prolactin release (Sassin *et al.*, 1972; Noel *et al.*, 1972). In females after exercise, hypoglycemia, coitus with orgasm, and operative stress, basal HPr levels of 9 ng/ml plasma were increased to 22, 129, 97, and 170 ng/ml, respectively; also plasma HGH rose during hypoglycemia and surgery from baseline values of 0.5–2 ng/ml to 19 and 9 ng/ml, respectively (Noel *et al.*, 1972). Besides increasing HPr secretion, thyrotropin-releasing hormone also brings about enhanced release of TSH and FSH-LH, whereas the HGH blood levels remain unchanged and those of ACTH decline.

d. DRUGS AFFECTING PROLACTIN SECRETION

L-Dopa (0.5–1 g orally) penetrates the blood–brain barrier and increases, via its transformation into dopamine and norepinephrine, the catecholamine levels in the brain; locally enhanced catecholamine concentrations will bring about increased hypothalamic PIF activity resulting in decreased prolactin secretion. Also secretion of hypothalamic corticotropin and FSH-LH and TSH releasing hormone are diminished by enhanced local catecholamines (Edmonds *et al.*, 1972; Boden *et al.*, 1972; Rapoport *et al.*, 1973). Because local catecholamines promote hypothalamic secretion of growth hormone-releasing hormone, pituitary HGH secretion is increased after L-dopa from 2 to 10 ng/ml plasma within 1–2 hours (Boden *et al.*, 1972). With 0.5–1 g L-dopa orally HGH plasma levels rose from 2 to 15 ng/ml whereas plasma FSH-LH and TSH concentrations remained unchanged (Saito *et al.*, 1972). Furthermore, in a patient with a pituitary eosinophilic adenoma secreting both HPr and HGH, L-dopa was found to suppress not only HPr but also HGH secretion (Guyda *et al.*, 1973). Drugs that lead to a decrease in hypothalamic catecholamines, such as chlorpromazine and reserpine, inhibit hypothalamic growth hormone-releasing hormone activity and PIF function, resulting in decreased HGH and increased prolactin secretion (Sherman and Kolodny, 1971; Edmonds *et al.*, 1972). Amphetamine intake may raise serum levels of HPr to 80–90 ng/ml. This effect is probably due to amphetamine-induced blockade of dopamine re-uptake of synaptic neuronal endings in the hypothalamus whereby norepinephrine secretion is reduced (Ehara *et al.*, 1973).

According to Bryant *et al.* (1971) baseline serum prolactin levels of

30–40 ng/ml in normal nonpregnant females increased to 80–90 ng/ml after stimulation of prolactin release with fluphenazine. In 1972, Turkington reported that prolactin was not measurable (<2ng/ml serum) in healthy subjects (1972b); after phenothiazine stimulation, however, prolactin concentrations of 30–300 ng/ml serum were found. In patients with panhypopituitarism, the prolactin levels were only slightly elevated not exceeding 10 ng/ml after 50 mg of phenothiazine i.m., and in 2 patients who were unable to nurse, no increase in serum prolactin was measurable after phenothiazine administration. Moreover, no breast engorgement or lactation occurred in a postpartum patient with Sheehan's syndrome and serum prolactin levels less than 10 ng/ml (Tyson *et al.,* 1973). Thus, an isolated deficiency of prolactin secretion seems to exist. In these cases either a hypothalamic inability to achieve temporary relief of the continuous function of PIF or selective deficiency of functional adenohypophysial prolactin cells may be responsible for the lack of prolactin secretion (Turkington, 1972b); isolated deficiency of pituitary basophilic cells to produce FSH-LH and of eosinophilic cells to secrete HGH have been reported in the past (Ewer, 1968). Accordingly, administration of chlorpromazine or thyrotropin-releasing hormone can be utilized as tests for hypothalamoadenohypophysial function regarding prolactin secretion (Table 33). A positive response to chlorpromazine indicates an intact hypothalamoadenohypophysial function with respect to prolactin secretion, whereas a positive response to thyrotropin-releasing hormone discloses the presence of functional adenohypophysial lactotrophs. A negative chlorpromazine prolactin test suggests either an abnormal PIF function or nonfunctional lactotrophs; the latter abnormality also exists when the prolactin stimulation test with thyrotropin-releasing hormone is negative. L-Dopa can also be utilized for testing pituitary control of prolactin secretion. Because L-dopa inhibits prolactin secretion by increasing the hypothalamic PIF activity, a negative test result indicates that adenohypophysial lactotrophs are functioning autonomously, i.e., they are no longer under hypothalamic control. Since pretreatment with L-dopa can significantly inhibit TRH-stimulated pituitary prolactin secretion, a direct effect of L-dopa on adenohypophysial lactotrophs (decreased prolactin release) appears possible. Inhibition of lactation by a prolactin-antagonistic L-dopa effect on mammary alveolar cells has not been ruled out either.

Theophylline, a stimulant of cyclic AMP formation, may promote prolactin release through a direct action on the lactotrophic pituitary cell and does not seem to affect the control of the prolactin-inhibiting factor (Parsons and Nicoll, 1971, Table 34). Ergocornine may inhibit prolactin release in rats (it also depresses FSH-LH secretion) through a direct effect on lactotrophs of the adenohypophysis; an indirect effect of ergot alkaloids on the hypothalamus through increased PIF activity also seems possible. Ergocornine also counteracts the stimulatory effect of estrogen on prolactin release (Lu *et al.*, 1971); in ewes estradiol stimulated prolactin secretion from basal levels of 10–15 ng/ml plasma to 100–500 ng/ml (Fell *et al.*, 1972). Estrogens are also potent stimulants for prolactin secretion in man. In postpartum women and in patients with galactorrhea, bromergocryptine has been found to be an efficient agent for the depression of serum prolactin and thus for the inhibition of normal and abnormal lactation (del Pozo *et al.*, 1972); bromergocryptine appears to decrease HPr secretion below normal serum levels by exerting a direct inhibitory effect on adenohypophysial lactotrophs. In Table 34 the influence and mode of action of various drugs on adenohypophysial prolactin secretion is outlined. Recently, also L-dopa was found to inhibit most effectively prolactin secretion in some patients with the Chiari-Frommel syndrome. Administration of daily oral doses of 2 g of L-dopa lowered these patients' serum prolactin levels from upper values of 1000 ng/ml to less than 2 ng/ml and thus proved to be an effective cure of galactorrhea (Turkington, 1972c). In some galactorrhea patients with pituitary tumors (Forbes-Albright syndrome), L-dopa failed to reduce effectively increased serum prolactin (for more details see Chapter VIII).

e. METHODS FOR MEASUREMENT OF PROLACTIN; PROLACTIN LEVELS IN BLOOD AND URINE

Neither pituitary staining, immunofluorescence, nor prolactin assays of adenohypophysial tissue allow estimates of prolactin secretion rates; the latter can only be determined by measuring prolactin blood levels and disappearance of prolactin ($t/2$ = half-life) from the bloodstream. Most earlier data regarding plasma levels of prolactin in humans are derived from bioassays using the stimulating effect of prolactin on pigeon crop tissue. The prolactin levels measured in plasma vary considerably among investigators, and the results are by and large inconclusive (Kurcz *et al.*, 1969). Only more specific bioassay procedures or radioimmunoassays can be expected to yield reliable data regarding human prolactin plasma levels under physiologic and pathologic conditions. Recently an *in vitro* bioassay for prolactin has been described using breast tissue from midpregnant mice. The mammary tissues were incubated together with human plasma in organ culture, and the degree of secretory tissue transformation served as an indicator for the plasma's prolactin concentration. A mammary secretory growth hormone effect was excluded in this assay procedure by preincubation of the plasma samples with growth hormone antiserum, which specifically eliminated the lactogenic activity due to growth hormone. With this rather specific assay technique, plasma prolactin activity was found to be less than 0.42 mU/ml (less than 10–20 ng/ml) in normal men and women; nursing mothers had elevated plasma prolactin activity (Kleinberg and Frantz, 1971; Table 28). Using a purified human prolactin preparation, containing less than 0.5% of growth hormone, and an amino acid sequence with a great homology to sheep prolactin (Friesen and Guyda, 1971), serum prolactin levels determined by radioimmunoassay ranged between 0 and 25 ng/ml in normal adults of both sexes. Recently a homologous radioimmunoassay for human prolactin was reported using a human prolactin preparation containing 30 units per milligram powder (Sassin *et al.*, 1972; Sinha *et al.*, 1973a, b). With this assay midmorning plasma prolactin levels averaged 14 ng/ml in male and female subjects, and no cross reaction to human growth hormone or placental lactogen occurred. In Sassin's radioimmunoassay a prolactin amount as low as 1 ng/ml plasma can be measured. In children with HGH deficiency, plasma HPr levels were 10 ng/ml, double the level of normal children (Kaplan *et al.*, 1972). Prolactin levels in urine appear to correspond to those measured in the blood plasma or serum (Manaro *et al.*, 1971; Table 35). More recently (Sinha *et al.*, 1973 a, b), urinary HPr levels were estimated to be 3–6 ng/ml in males and 2–5 ng/ml in females; the respective 24 hour urinary HPr excretion was 5 μg and 3 μg. During the menstrual cycle HPr serum levels ranged between 10 to 25 ng/ml and no cyclic rhythm comparable to the midcycle FSH-LH peak could be detected. Under the condition of

TABLE 35

Prolactin Levels in Blood and Urine [a]

Subjects	Prolactin concentration	
	Plasma or serum (ng/ml)	Urine (ng/ml)
Children	5	10
Men	6	10
Women		
Menstruating	10	12
Pregnant (near term)	200	30
Lactating	50	30
Menopausal	10	8

[a] The values presented in this table are derived from various literature data as referred to in the text.

amphetamine induced high HPr serum levels (80–90 ng/ml) ovarian function (ovulation) was not impaired (Ehara *et al.,* 1973).

f. PROLACTIN BLOOD LEVELS DURING PREGNANCY AND LACTATION

With bioassay techniques, during the first 20 weeks of gestation, no prolactin was measurable in blood plasma (Berle and Apostolakis, 1970); thereafter, an increase of biologic prolactin activity until term was observed. Most recently Jacobs and co-workers (1972) measured prolactin levels of nonpregnant and pregnant women by radioimmunoassay (Table 28). In normal nonpregnant females prolactin in amounts of 9 ng/ml serum were found (normal males: 6 ng/ml serum). During gestation the serum prolactin concentration increased to 50, 125 and 200 ng/ml in the first, second, and third trimesters, respectively (Friesen *et al.,* 1971; Fig. 26; similar values were found by Jacobs *et al.,* 1972; Table 28); prolactin concentration of amniotic fluid appears to exceed that of maternal serum 100- to 200-fold; HPr levels of amniotic fluid range between 1 to 10 μg/ml. In a similar study (Tyson *et al.,* 1972a, b) the serum prolactin concentration was measured to be 12 ng/ml in nonpregnant women; the corresponding values for pregnant women were 15, 50, 100, and 200 ng/ml serum at 10, 20, 30, and 40 weeks of pregnancy, respectively (Table 28). The fetal cord blood levels of HPr are similar to those of maternal serum, returning to normal values by the sixth neonatal week. Whereas in some studies a postpartum drop in plasma prolactin levels has been reported (Berle and Apostolakis, 1970; Tyson *et al.,* 1972a, b), in others, prolactin was found at its highest levels during postpartum lactation (Herlyn *et al.,* 1969; Forsyth, 1969; Forsyth and Myres, 1971; Kleinberg and Frantz, 1971; Bryant *et al.,* 1971) (Table 28). The plasma levels of growth hormone usually do not increase in lactating women (Board, 1968; Spellacy *et al.,* 1970; Bryant *et al.,* 1971). During the period of lactation, pituitary synthesis and release of prolactin seem to follow a more discrete pattern, and only an instantaneous additional pituitary discharge seems to occur shortly after the beginning of suckling (Bryant *et al.,* 1971). These authors measured a prolactin plasma level of 87 ng/ml 5 minutes after suckling. In lactating mothers during the first week postpartum, the basal prolactin levels were estimated to be 120 ng/ml, rising to 200 ng/ml within 15 minutes of suckling. Between weeks 2–4 postpartum, the basal prolactin levels dropped to 25 ng/ml and increased tenfold (250 ng/ml) during the baby's suckling (Friesen, 1972; Fig. 26). Gautvik and co-workers (1973) found an eightfold increase in HPr levels (from 30 to 250 ng/ml) during suckling but no change in TSH plasma concentrations (0.5–2.5 μU/ml) occurred; intravenous injection of 10 μg of TRH could raise HPr plasma levels from 10 ng/ml to 40–50 ng/ml within 10 minutes. In nonlactating puerperas, HPr levels return to normal around the seventh postpartum week. In cows, postpartum prolactin plasma levels of 20 ng/ml were measured by radioimmunoassay, rising rapidly within 5–10 minutes to 100–300 ng/ml during the stimulus of milking (Johke, 1969). Exogenous oxytocin injected in amounts of 80 units had no influence on cow's prolactin levels during milking (Schams, 1972).

The regulation of prolactin release during lactation seems to be closely related to the strength and frequency of neural stimulation. However, the possibility exists that plasma prolactin in rats may exert a negative feedback on hypothalamic prolactin-inhibiting factor activity or directly act on the adenohypophysial cells reducing its own secretion (Spies and Clegg, 1971; Sud, 1971). Prolactin infusion in cows was found to reduce synthesis and systemic release of endogenous prolactin during milking, probably due to an "auto" or "short-loop" feedback mechanism (Tucker *et al.*, 1973).

g. PROLACTIN: A MOLECULE DISTINCT FROM GROWTH HORMONE AND PLACENTAL LACTOGEN

Because in former studies rather impure prolactin preparations for antibody production were used, cross reaction occurred with HPL and HGH in radioimmunoassays, casting doubt in the past on the existence of a prolactin molecule as an entity distinct from HGH and HPL. Although at present synthetic prolactin is not available, highly purified human prolactin preparations can be obtained for antibody production and for measurement of prolactin in homologous radioimmunoassays. Such antisera against human prolactin produced in rabbits show no substantial cross reaction with HPL or with HGH. There is also clinical, biochemical, and histological evidence to justify the assumption that a distinct prolactin molecule is synthesized in the human pituitary: (1) Hypophysial ateliotic dwarfs, lacking endogenous growth hormone, lactate normally and show a rise in plasma prolactin during lactation. (2) Human growth hormone levels have not been found elevated during lactation, whereas the plasma growth hormone activity is considerably increased with insulin-induced hypoglycemia. (3) Some nonpregnant women with or without pituitary tumors develop galactorrhea; plasma prolactin levels in these patients were increased whereas growth hormone levels were normal, and acromegaly symptoms were lacking. In acromegalic patients, however, the increased plasma lactogenic activity was inhibited after incubation with growth hormone antiserum, indicating growth hormone as the cause for enhanced lactogenic activity. In one of the galactorrhea patients a prolactin-rich pituitary tumor (chromophobe adenoma) was identified by selective erythrosine staining of lactotrophic cells, electron microscopy, and measuring the tumor tissue for prolactin activity, which was 40-fold increased in comparison with normal pituitary tissue. (4) A remarkable increase in prolactin cells during pregnancy and lactation can be demonstrated through selective staining of anterior pituitary cells with carmoisine L and erythrosine; the number of growth hormone cells decreases during pregnancy relative to the increase of pituitary lactotrophs. (5) Selective immunofluorescent studies performed on tissues of monkey pituitary glands showed that antibodies against ovine prolactin and human growth hormone were localized in erythrosinophilic prolactin cells and orangeophilic growth hormone cells, respectively. This indicates that secretion of prolactin and growth hormone from different types of acidophilic cells is possible. (6) Electron microscopic studies of the human pituitary gland proved that prolactin cells could be distinguished from growth hormone cells by size and shape of intraplasmatic granules (Sherwood, 1971; Board, 1968; Peake *et al.*, 1969) (Tables 11 and 14). (7) Further evidence for the existence of a prolactin molecule as distinct from growth hormone has been obtained by using greatly purified human prolactin extracts for immunization procedures, allowing the production of specific prolactin antibodies. Measurement of prolactin, HPL, and HGH in serum of nonpregnant, pregnant, and lactating women, as well as of galactorrhea patients with rather specific radioimmunoassay, showed (Tables 28, 30, and 31) that (a) no HPL is measurable in nonpregnant women and men; (b) in pregnant, lactating, and galactorrhea patients, prolactin blood levels are increased, and HGH concentrations are normal; and (c) HPL disappears from circulation within a few hours postpartum.

Hopefully, in the near future, the complete chemical analysis of the molecular structure of human prolactin and its synthesis will provide ultimate proof of its existence.

2. Oxytocin Synthesis and Secretion

a. LOCATION AND MODE OF SYNTHESIS

Although oxytocin is found in relatively high amounts in the posterior lobe (10–16 IU per lobe) of the neurohypophysis, this octapeptide is actually synthesized in the cell bodies of neurons located mainly in the nucleus paraventricularis and in smaller amounts in the nucleus supraopticus of the hypothalamus (Bargmann and Scharrer, 1951; Adamsons *et al.*, 1956; Bargmann, 1968). The neurohypophysis consists of an infundibular recess of the third ventricle, a median eminence of the tuber cinereum, an infundibular stem, and a posterior pituitary lobe. The biosynthesis of oxytocin is similar to that of arginine vasopressin; both are octapeptides and differ by two amino acids only. Their molecular weight is 1007 and 1084, respectively. Accordingly, octapeptide synthesis takes place in the cell body of the hypothalamic neuron, whereby messenger RNA and transfer RNA seem to trigger the synthesis of peptides on ribosomes. During the first stage of synthesis, the hormone precursor is probably biologically inactive, and, in a further step, the active octapeptide appears in a place distant from that of synthesis. Hormone precursor and active peptide are thought to be incorporated into neurosecretory granules, which are possibly formed in the Golgi apparatus (Sachs, 1967) (Fig. 36). The neurosecretory material is then transported along the axons into the posterior lobe for storage.

b. HYPOTHALAMO-NEUROHYPOPHYSIAL FUNCTION

In addition to producing neurosecretory material and transporting it along the axons into the posterior lobe for storage, the hypothalamic neurons transmit nervous impulses along the axons to the posterior pituitary, leading to immediate hormonal release. Thus the cells of the hypothalamo-neurohypophysial system fulfil the three criteria for classification as neurosecretory neurons (Scharrer, 1959): (1) reception of stimuli from other neurons, (2) production of secretory material that can be stained by special histological techniques, and (3) humoral release of active substances affecting distant organs.

The neurons of the hypothalamo-hypophysial tract consist mainly of unmyelinated nerve fibers, and intermingling with these fibers are pituicytes with plasmic processes that are several hundred microns long. The pituicytes serve as supportive glial elements providing the mechanical framework for the parenchymal neural tissues. The axons of the hypothalamo-neurohypophysial tract begin in the hypothalamic nuclei and end in the posterior pituitary lobe (neural lobe); the axonal downward flow of the neurosecretory osmiophilic material is probably 2–3 mm/day. Within the neural lobe, both axons and pituicyte processes form an arborization pattern and then lie close to the pericapillary connective tissue. Under the electron microscope, the axonal endings show small vesicles (25 mμ, synaptic-like vesicles) that appear empty and larger vesicles (100–300 mμ in size) that can be stained by chromalum-hematoxilin and with osmium te-

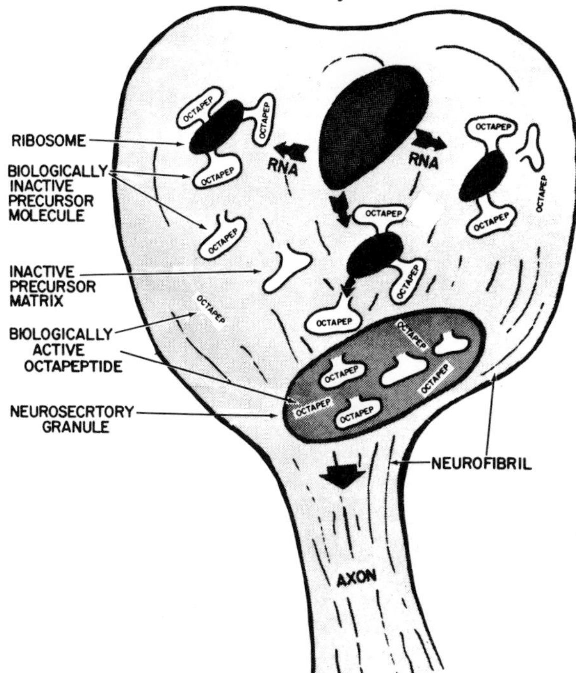

Fig. 36. Diagram of octapeptide synthesis and formation of neurosecretory granules in the cell body of hypothalamic paraventricular and supraoptic neurons (from Kleeman and Vorherr, 1969). Afferent stimuli elicited by suckling, osmotic changes, hypovolemia, or pains and converging on the hypothalamic nuclei appear to exert a triple function: (1) initiation of hormone synthesis in the nucleus paraventricularis (mainly synthesis of oxytocin) and in the nucleus supraopticus [mainly synthesis of vasopressin (ADH)]; (2) hormone transport along the axons of the hypophysial stalk into the posterior pituitary lobe for storage; (3) posterior pituitary release of oxytocin (suckling) and vasopressin (dehydration, hypovolemia, pains). Through the stimulus of suckling, for instance, synthesis of oxytocin is initiated predominantly at the hypothalamic cell bodies of the nucleus paraventricularis. The first stage of octapeptide synthesis on ribosomes of the endoplasmic reticulum in the hypothalamic cell body is triggered by nuclear messenger and transfer RNA. It is believed (Sachs, 1967) that the biologically inactive hormone is formed on a precursor-protein matrix. In the second stage of hormone production the biologically active peptide separates from its protein precursor matrix and appears in a place removed from that of synthesis. In the cisternae of the Golgi apparatus, probably, both hormone precursor-protein matrix and free protein matrixes are united with the biologically active octapeptide, oxytocin, and are incorporated into the neurosecretory granules (100–300 mμ in diameter). Neurosecretory granules are transported along the axons of the hypophysial stalk into the posterior pituitary lobe for storage, and the neurosecretory material can be identified in the neurovesicles of Herring bodies. Simultaneously with these processes, upon initiation, hypothalamic stimuli travel via the axons of the hypothalamo-neurohypophysial tract down to the posterior pituitary lobe causing depolarization of neurovesicles and thus rapid release of oxytocin into the circulation (see also Fig. 37).

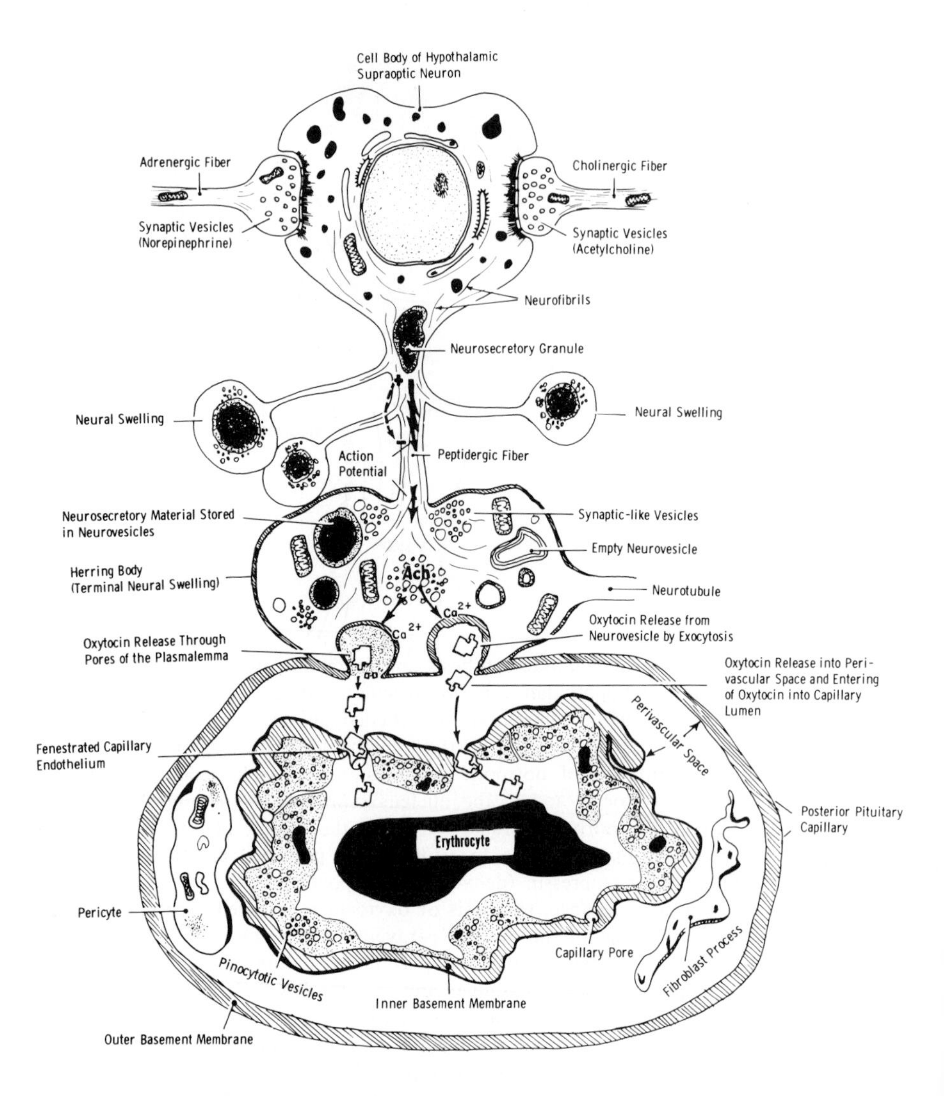

Fig. 37. Diagram of neurohypophysial hormone release into the bloodstream upon stimulation. Cholinergic and adrenergic fibers impinge, via synaptic knobs, the cell bodies of hypothalamic neurons. Afferent stimuli, as elicited for instance by suckling, cause excitation of cholinergic fibers and, via synaptic acetylcholine action, synthesis of oxytocin in the hypothalamic cell bodies of paraventricular (predominantly) and supraoptic hypothalamic nuclei. Simultaneously, acetylcholine-induced membrane action potentials travel down the axons of the hypothalamo-neurohypophysial tract bringing about immediate release of oxytocin from its neurovesicular storage into the circulation. It is thought that the action potentials arriving at the dilated axon terminals, the Herring bodies (unmyelinated, true nerve terminals of about

troxide (Fig. 37). The larger vesicles seem to correspond to osmiophilic neurosecretory granules; the osmiophilic substance is membrane bound or centrally accumulated in the neurovesicle and consists of octapeptides bound to their carrier protein, neurophysin. For each neurohypophysial octapeptide, oxytocin and arginine vasopressin, the structure of the respective neurophysin (Van Dyke protein, about 10,000 MW) is slightly different in the rat (Burford *et al.*, 1971). A number of small and large neurovesicles is contained within a Herring body, which is about 5 μm in size. Herring bodies are considered to be true dilated nerve terminals

---

5 μm in diameter), induce release of acetylcholine from the clear synaptic-like vesicles (20–75 mμ in diameter). Acetylcholine in turn causes membrane depolarization of neurovesicles (150–300 mμ in diameter) and vesicular entrance of Na^{+} and Ca^{2+}. Thereby, calcium-induced dissociation of oxytocin from the stored octapeptide-neurophysin complex leads to the release of oxytocin from the neurovesicles, which correspond to neurosecretory granules. At this stage of acetylcholine-provoked neurovesicular membrane depolarization, the dissociated oxytocin appears in the neuroplasm and is released in molecular dispersion through temporary increased permeability of the plasmalemma (porous secretion), which attaches very close to the outer condensed capillary basement layer. Also, by means of exocytosis (frank rupture of plasmalemma) or by reversed pinocytosis, oxytocin may be released either directly into posterior pituitary capillaries or into the communication network of neurotubules (20 mμ in diameter) of the Herring bodies; neurotubules may act as a sponge for temporary hormone storage. Before reaching the capillary lumen, oxytocin released into the axoplasm (neuroplasm) has to pass the plasmalemma, the outer condensed cement layer of the basement membrane, the mucopolysaccharide material of the perivascular space, the inner condensed layer of the basement membrane, and the capillary endothelium. The endothelial barrier is thought to be traversed by pinocytosis (endothelial pinocytotic vesicles of 30–80 mμ in diameter are abundant in posterior pituitary capillary endothelium). Also, the hormone may enter the capillary through fenestration (pores) of the endothelium (30–80 mμ in diameter). Upon hormone release neurovesicles lose their osmiophilicity. As observed in lactating rats (Monroe and Scott, 1966) a reduction in the number of neurovesicles and other organelles of the Herring bodies occurs through suckling-induced hormone release. The remaining empty neurovesicles (clear neurovesicles) become pleomorphic and possibly break down into smaller vesicles or flatten out and merge with neurofibrils. Simultaneously, synaptic-like vesicles increase and accumulate along terminal axon membranes. Hormone repletion and restoration of the ultrastructural picture to presuckling conditions takes about 1 hour. Gomori-positive material containing octapeptides may be delivered to the terminal neuronal swellings from other neurohypophysial storage depots; also from hypothalamic cell bodies which show signs of increased hormone synthesis (dense sheets of rough endoplasmic reticulum with dilated cisternae; polysomal clusters) neurosecretory granules are transported down the axons for replacement. Some small osmiophilic neurovesicles (50–100 mμ in diameter) are thought to contain catecholamines, which may exert an inhibitory effect on octapeptide release following adrenergic stimulation.

(swellings) and contain neurosecretory granules and mitochondria; they are surrounded by a single membrane and are connected with each other by neural tubules of 20 mμ in diameter. Upon a secretory stimulus, i.e., suckling, hypothalamic synthesis of oxytocin is probably initiated after a time lag of an hour or more, whereas stimuli from the hypothalamic cell body travel immediately down the hypothalamo-neurohypophysial tract and bring about instantaneous release of oxytocin from the neural lobe. Experiments on lactating rats showed that after suckling extensive octapeptide synthesis occurred in the cell bodies of the hypothalamic paraventricular neurons; this is evidenced by the abundance of rough endoplasmic reticulum and polysomal clusters. Secretion and release of oxytocin and ADH are brought about by cholinergic nerve fibers, while excitation of adrenergic fibers, also reaching the hypothalamic cell bodies via synapses, seems to inhibit octapeptide release. Stimuli transmitted during suckling via cholinergic synapses to hypothalamic cells of the supraoptic nucleus thus elicit action potentials that are conducted down the axons of the hypothalamo-hypophysial tract and release oxytocin. At the axonal terminals, synaptic-like small neurovesicles seem to trigger membrane depolarization through their transmitter substance, acetylcholine (Fig. 37). Thereby, calcium, sodium, and chlorine enter the neurovesicular membrane, and potassium moves out. It is thought that calcium is essential for facilitating release of oxytocin (vasopressin) from the neurophysin-octapeptide complex (Ginsburg and Ireland, 1966). Accordingly, the larger neurovesicles lose their osmiophilicity after octapeptide release (Monroe and Scott, 1966). Regarding the releasing mechanism, it is not yet clear whether oxytocin or vasopressin are separated completely from their neurophysin carrier and released alone, or whether part of the octapeptide-neurophysin complex may enter the circulation, too. Within the neurovesicle's membrane one molecule of neurophysin is thought to bind six octapeptide molecules.

c. DIFFERENTIAL RELEASE OF OXYTOCIN

Many stimuli are known to cause release of neurohypophysial hormones. Animal experiments usually indicate that stimuli such as pain or anesthetics release oxytocin and vasopressin simultaneously (Berde, 1959), but recently evidence of their independent release has been obtained. In humans oxytocin is predominantly liberated from the neurohypophysis during suckling (Cobo *et al.*, 1967); after an osmotic stimulus, carotid occlusion, or hemorrhage, however, vasopressin is released in greater quantity, if not exclusively (Ginsburg and Heller, 1953). Accordingly, impulses due to various stimuli that appear to have converged on the hypothalamic nuclei probably travel from there in separate fibers of the hypothalamo-neurohypophysial tract to cause differential release of the hormone from the neurovesicles located within the neuronal terminals (Herring bodies). Since the hormones may be released independently (for instance, oxytocin by the stimulus of suckling), some of the terminals must contain neurovesicles loaded largely with oxytocin. Labor pains predominantly elicit vasopressin release while oxytocin secretion seems minimal or nonexisting (Vorherr, 1972c).

The morphology of the neurohypophysis permits a ready release of oxytocin or vasopressin into the circulation because the neurovesicles lie very close to the basement membrane of the perivascular space. Thus, the released hormones can easily reach it and enter the capillary bed through its thin fenestrated endothelial layer (Fig. 37). Nevertheless, before entering the hypophysial capillaries, the released octapeptide must first penetrate the neurovesicular membrane; then the hormone passes through the envelope of the Herring body to reach the pericapillary connective tissue space from which it subsequently crosses the basement membrane and the endothelium in a molecular dispersed form. All the same, hormone release once initiated allows appearance of the active peptide in the bloodstream within seconds. Since processes of neurosecretory cells may reach the third ventricle, a direct release of oxytocin and vasopressin into the cerebrospinal fluid (CSF) is also possible. After administration of stimuli such as anesthesia or hemorrhage, oxytocin and vasopressin are measurable in the CSF of rabbits and dogs (Vorherr *et al.,* 1968; Bradbury *et al.,* 1968). The bloodstream, however, is the main pathway for neurohypophysial octapeptide release.

The afferent reflex pathway for oxytocin release and milk ejection can be traced in animals. Fibers reach the dorsal roots of the spinal cord from sensory nerve receptors located in the nipple and areola mammae, and impulses travel in spinothalamic fibers to the brain stem. They then relay via collaterals in the mid brain reticular formations, and thus impulses reach the hypothalamic neurosecretory nuclei via the tegmento-hypothalamic pathways (Urban *et al.,* 1971). Upon afferent stimulation, discharge of action potentials down the hypothalamo-neurohypophysial tract to the posterior lobe brings about release of oxytocin (Eayrs and Baddeley, 1956; Cross, 1961). The efferent component of the milk-ejection reflex is a humoral one, the liberated oxytocin enters the circulation and leads to contraction of myoepithelial cells in the mammary gland (milk ejection).

d. PHYSIOLOGY OF OXYTOCIN

Oxytocin released under the stimulus of suckling has at least one definite physiologic action: it causes contraction of the myoepithelial cells surrounding the mammary alveoli, resulting in milk ejection. Thereby, milk is ejected from alveolar and smaller milk ducts into the larger lactiferous ducts and sinuses (milk ejection effect of oxytocin) from which it can be removed by suckling. A galactopoetic effect has also been observed in animal experiments after application of pharmacologic doses of oxytocin. This may be explained by an intensified oxytocin-induced milk ejection, allowing better and more extensive removal of milk during suckling or milking. It is not clear whether oxytocin can stimulate pituitary prolactin release and thereby enhance the process of galactopoesis. In the absence of oxytocin, e.g., after hypophysectomy, or when the oxytocin action on the mammary myoepithelium is blocked by injected or endogenously released catecholamines, as in the state of hyperfunction of the sympathetico-adrenal system, women fail to lactate (Vorherr, 1971). In sheep and goats, however, oxytocin seems not to be essential for lactation because in these species myometrial cells contract in response to the mechanical stimulus of suckling or milking. Moreover, severence of afferent neurogenic pathways blocked lactation in rats, rabbits, and cats but not in sheep or goats (Kolodny *et al.,* 1972). The phenomenon of milk ejection is specific to oxytocin, the only naturally occurring substance known to cause it under physiologic conditions. In lactating women an intravenous dose as little as 1–2 mU of oxytocin elicits milk ejection (Fish *et al.,* 1964). In lactating animals (rabbits, guinea pigs, rats) local injections into an artery supplying the mammary gland show a high sensitivity to oxytocin, so that this procedure has now become the preferred bioassay for oxytocin (Tindal and Yokoyama, 1962; Bisset *et al.,* 1967; Vorherr, 1971). In lactating rats a dose as low as 1–2 μU of oxytocin can be detected when milk ejection is recorded through measurement of intraductal mammary pressure, and the hormone is injected into the femoral artery allowing the bolus to proceed only to the lower mammary glands supplying artery (Fig. 38) (Vorherr, 1971). Most recently a radioimmunoassay also has been applied for measuring oxytocin. Although such a procedure (see Figs. 39 and 40) detects oxytocin specifically, the antibody binding with metabolic inactive fragments of the oxytocin molecule may lead to erroneous results because the fragments do not represent the biologically active oxytocin concentration in the circulation. Because up to 50% of biologically inactive hormone metabolites may be measured by a radioimmunoassay for oxytocin or any other hormone, a false conclusion regarding physiology, metabolism, and excretion of a hormone may be drawn. Therefore, only the simultaneous application of an oxytocin radioimmunoassay in combination with a bioassay procedure for such studies of oxytocin levels and oxytocin metabolism will yield optimal results. In analogy to the prolactin

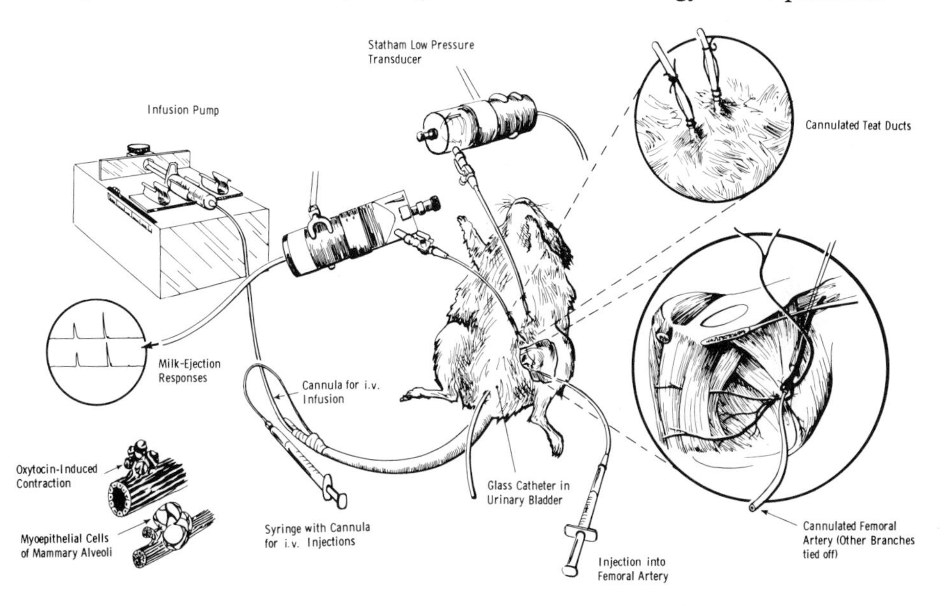

Fig. 38. Oxytocin milk-ejection assay. Experimental set up of a lactating rat under pentothal-ethanol anesthesia. Two abdominal teat ducts are cannulated for measurement of milk-ejection pressure with Statham low pressure transducers. Milk-ejection responses are recorded on a Beckman Dynograph direct writing system. For continuous intravenous infusion with dextrose, anesthetic, etc., a tail vein is cannulated with #24 gauge needle which is connected via a #50 PE tubing with a syringe in the infusion pump. A second tail vein cannula connected with a tuberculin syringe serves for intravenous injections of oxytocin and test solutions with a high oxytocin content. A high sensitivity with milk-ejection responses to 1–2 μU of oxytocin is achieved by cannulation of an ipsilateral femoral artery 3–5 mm distal from the origin of its arterial branch supplying the abdominal mammary glands. All emerging branches of the femoral artery supplying other tissues are ligated, and a loose silk thread #00 is placed 2–4 mm proximal to the femoral origin of the mammary arterial branch. Before a test solution is injected intrafemorally, the silk thread is lifted slightly and twisted; thus, the intraarterial bolus proceeds exclusively into the mammary branch. This injection technique is essential for the achievement of the high sensitivity of this bioassay. Test solutions with an expected high oxytocin content such as neurohypophysial extracts will be injected first intravenously; with this way of screening, amounts of oxytocin that exceed 100 μU will be detected. Then appropriate dilutions for intrafemoral testing are prepared. A thin glass catheter attached to #160 PE tubing is inserted via the urethra into the urinary bladder to prevent distension and to allow control of the animal's fluid balance.

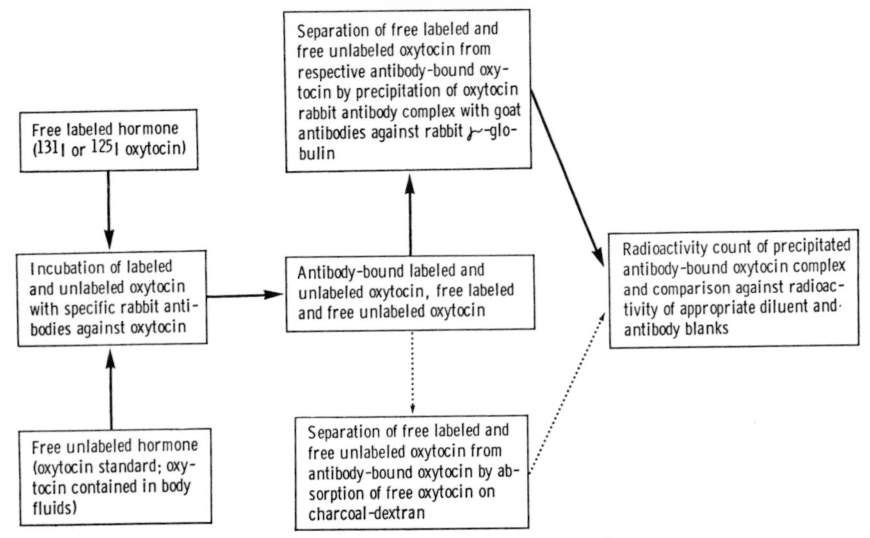

Fig. 39. Scheme of the oxytocin radioimmunoassay. Essential for the radioimmunoassay of oxytocin are rather specific oxytocin antibodies (serum from actively immunized rabbits) and radioactive oxytocin (iodine or tritium labeled). When labeled oxytocin and unlabeled oxytocin (oxytocin standard or oxytocin contained in extracts of body fluids) are incubated with oxytocin antibodies, formation of an antibody-bound oxytocin complex takes place. The amount of labeled oxytocin and of oxytocin antibodies used for the radioimmunoassay is titrated in such a way that only 40–50% of the labeled oxytocin can be bound by the available antibodies (endpoint of binding reaction should lie around 50%). According to the law of mass action, labeled and unlabeled oxytocin compete with the specific oxytocin antibody for binding sites, but some of the labeled and unlabeled oxytocin will not be able to combine with the oxytocin antibody. A dose response curve is obtained when the radioactivity of the antibody-bound labeled oxytocin complex is measured after incubation with blanks (blank diluent and serum from nonimmunized rabbits) and is compared with the radioactivity count obtained after similar incubations with various concentrations of unlabeled oxytocin standard solutions. Because with increasing doses of oxytocin standard solutions increasing amounts of labeled oxytocin are displaced from the antibody, a decreasing radioactivity count of antibody-bound labeled oxytocin is observed. When instead of the oxytocin standard solution an unknown test solution (extract of body fluids or neurohypophysial tissues) is incubated with a solution containing labeled oxytocin and oxytocin antibodies, the oxytocin content of the unknown test solution will determine to what extent the labeled oxytocin will be displaced from its antibody binding site. Thus the counting of antibody-bound labeled oxytocin provides a method for measuring the oxytocin content of an unknown test solution. Because the unbound (free) labeled and unlabeled oxytocin in the incubation solution will interfere with the measurement of the labeled antibody-bound oxytocin, the fraction of free oxytocin needs to be separated from that of antibody-bound oxytocin. The antibody-bound oxytocin complex may be removed from the incubation solution by precipitating it with goat antibodies against rabbit γ-globulin (double antibody radioimmunoassay technique), and thereafter the radioactivity is counted in the precipitate. Use of ammonium sulfate for precipitation of the antibody-bound oxytocin complex or removal of free labeled and unlabeled oxytocin by absorbing them on charcoal coated with dextran are other possible methods for isolation of the antibody-bound oxytocin complex (dotted lines). *Note:* Similar principles as outlined above apply also to the radioimmunoassay of ADH.

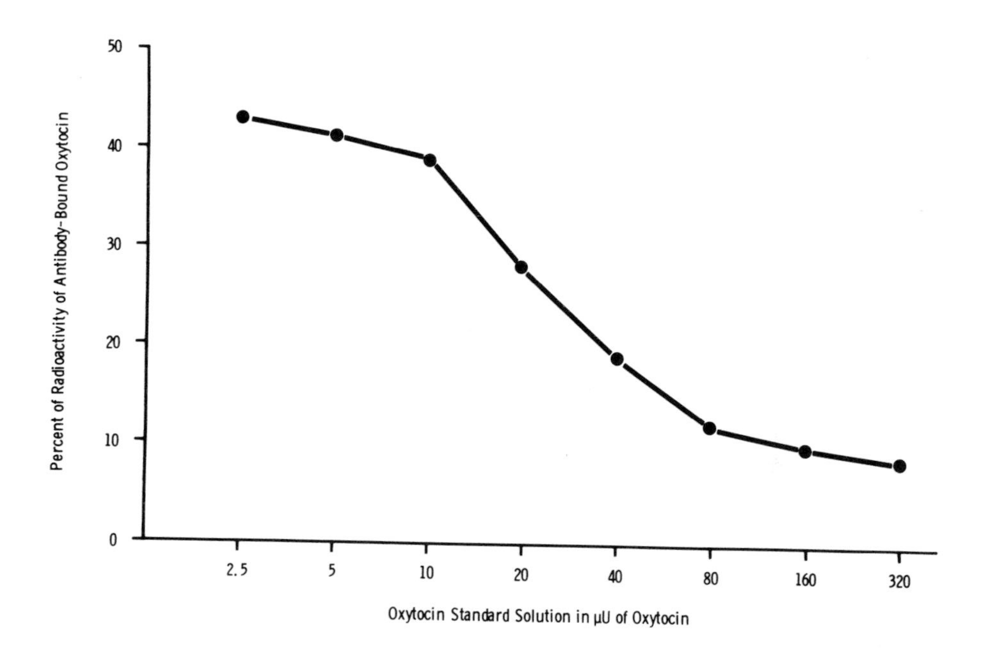

Fig. 40. Oxytocin dose response curve obtained by radioimmunoassay. When nonradioactive oxytocin of the standard solution is incubated with radioactive oxytocin and oxytocin antibodies, molecules of the oxytocin standard solution compete with and displace radioactive oxytocin from its bond with the antibody. Because with increasing concentrations of oxytocin standard solution increasing amounts of radioactive oxytocin are displaced from the antibody, a decreasing radioactivity count of antibody-bound radioactive oxytocin is observed. This displacement curve of antibody-bound radioactive oxytocin in linear between doses of 10–100 μU of oxytocin standard. The lowest amounts of oxytocin detectable by radioimmunoassay are 2.5–5.0 μU per dose. When instead of the oxytocin standard solution an unknown test solution (extract of body fluids or neurohypophysial tissues) is incubated with a solution containing radioactive oxytocin and oxytocin antibodies, the oxytocin content of the unknown test solution will determine to what extent the radioactive oxytocin will be displaced from its antibody binding site. Thus, the counting of antibody-bound radioactive oxytocin provides a method for measuring the oxytocin content of an unknown test solution. Since the chemically similar peptide, vasopressin, can displace radioactive antibody-bound oxytocin only in insignificant amounts, this radioimmunoassay is specific for the measurement of the octapeptide oxytocin.

measurement by bioassay, it also has to be demonstrated for the determination of oxytocin that the activity measured is indeed due to the respective hormone and not due to other extraneous substances in the extract solution causing similar biologic effects. In incubation studies with oxytocin and vasopressin and their respective antisera, it was shown that antisera diluted 1:10 in physiologic saline do not cross react; i.e., diluted oxytocin antiserum cannot destroy the biologic activities of vasopressin and vice versa. Such antisera are therefore highly suited for the identification of neurohypophysial hormonelike activities (milk ejection, antidiuresis) of test solutions, since they allow a determination of whether the measured

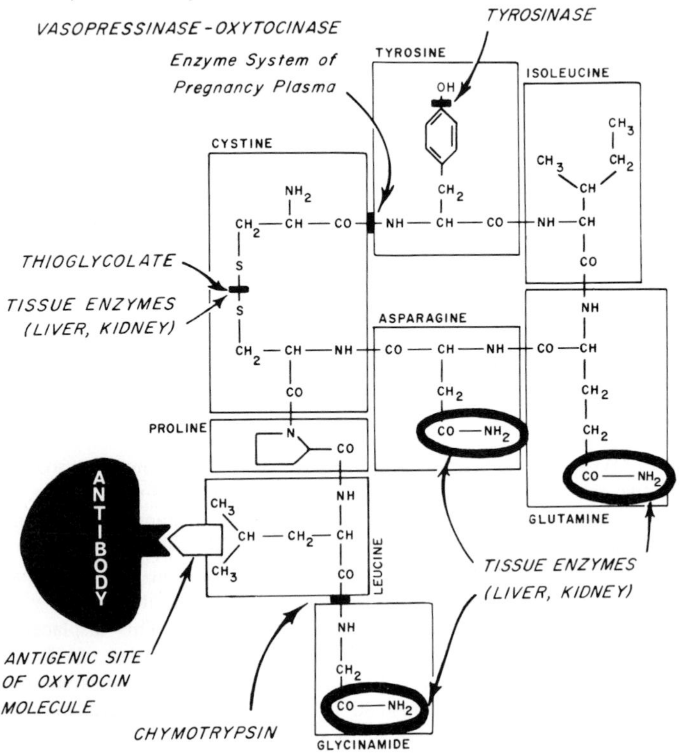

Fig. 41. Biologic identification of oxytocin. Substances such as histamine, 5-hydroxytryptamine, acetylcholine, bradykinin, and ADH may mimic the biologic effects (milk ejection, uterine contraction) of oxytocin. Therefore, oxytocin-like activities measured by bioassay in body fluids have to be identified. Active immunization of rabbits against oxytocin results in antibody formation. Because oxytocin antibody solutions, diluted 1:10, cannot destroy the structurally similar vasopressin (see Fig. 42) or the chemically even closer vasotocin (arginine^{8}—oxytocin, or isoleucine^{3}—vasopressin) it seems that a key portion of the antigenic site of the oxytocin molecule is located at the amino acid leucine. The biologic activity of oxytocin but not that of vasopressin is destroyed by chymotrypsin, which acts at the glycinamide-leucine bond. The biologic activities of both, oxytocin and vasopressin, are destroyed by the oxytocinase-vasopressinase system of pregnancy plasma, tyrosinase, thioglycolate and enzymes of liver and kidneys (see also Fig. 42).

activity is due to either oxytocin, vasopressin, or other unknown substances (Figs. 41 and 42).

e. OXYTOCIN IN THE CIRCULATION

Not much is known about the physiologic amount of oxytocin released into the circulation in the nonpregnant and pregnant state and postpartum despite many literature reports (Table 36). Indirect studies in humans

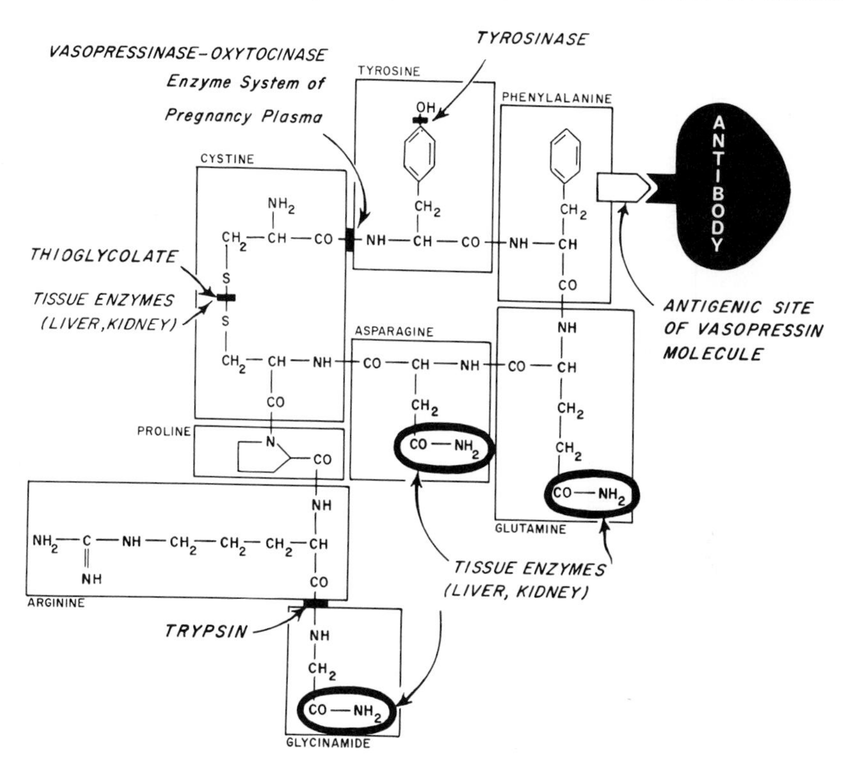

Fig. 42. Biologic identification of vasopressin (ADH). Substances such as angiotensin, catecholamines, 5-hydroxytryptamine, histamine, and oxytocin, may mimic the antidiuretic effect of vasopressin. Therefore, ADH-like activities measured by bioassay in body fluids have to be identified. Active immunization of rabbits against ADH results in antibody formation. Because vasopressin antibody solutions, diluted 1:10, cannot destroy the biologic activity of the structurally similar oxytocin (see Fig. 41) or the chemically even closer vasotocin (isoleucine[3]—vasopressin, or arginine[8]—oxytocin), it seems that a key portion of the antigenic site of the vasopressin molecule is located at the amino acid phenylalanine. The biologic activity of ADH, but not that of oxytocin, is destroyed by trypsin, which acts at the glycinamide-arginine bond. The biologic activities of both vasopressin and oxytocin are destroyed by the vasopressinase–oxytocinase system of pregnancy plasma, tyrosinase, thioglycolate, and enzymes of liver and kidneys (see also Fig. 41).

TABLE 36

Oxytocin Blood Levels in Man

References	Assay method	Identification of activity measured	Oxytocin levels reported
Hawker and Robertson (1957)	*In vitro* rat uterus	Thioglycolate	Peripheral venous blood during pregnancy: 950–43,000 μU/ml plasma
Fitzpatrick (1961a, b)	Rabbit *in vivo* milk-ejection assay	Thioglycolate	During labor in peripheral blood: 100–200 μU/ml blood
Hawker *et al.* (1961)	Rat uterus *in vitro*	Thioglycolate	Peripheral venous blood: Nonpregnant women: 120 μU/ml blood; Pregnant women: 1100 μU/ml blood; Parturients: 1390 μU/ml blood; Lactating women: 640 μU/ml blood
Rorie and Newton (1964)	Guinea pig uterus *in vitro*	Thioglycolate	Peripheral venous blood: Males: 640 μU/ml plasma; Females: 600 μU/ml plasma
Coch *et al.* (1965)	Rabbit *in vivo* milk-ejection assay	Oxytocinase	First stage of labor: Jugular vein blood: 20–200 μU/ml plasma; Peripheral venous blood: 50–100 μU/ml plasma
			Second stage of labor: Jugular vein blood: 200–900 μU/ml plasma; Peripheral venous blood: 100 μU/ml plasma
			3–5 Days postpartum: Jugular vein blood: 100–275 μU/ml plasma; Peripheral venous blood: 100–200 μU/ml plasma
Ahmed *et al.* (1966)	Rat uterus *in vitro*	None	Peripheral venous blood: Men: 300–2100 μU/ml plasma; Nonpregnant women: 450–2400 μU/ml plasma; Patients with diabetes insipidus: 1000–6000 μU/ml plasma

Caldeyro-Barcia and Méndez-Bauer (1966)	Rabbit *in vivo* milk-ejection assay	Paper chromatography and electrophoresis	Jugular vein blood during nursing: 10–20 μU/ml plasma
Bashore (1967)	Radioimmunoassay	Specific radioimmunoassay	Peripheral venous blood during labor: 7500 μU/ml plasma
Cobo *et al*. (1967)	Measurement of milk-ejection activity in one breast during suckling of baby on other breast and matching milk-ejection activity with exogenous oxytocin doses	Indirect	During nursing: 5–25 μU/ml peripheral blood (estimation from oxytocin matching doses by H. Vorherr)
Coch *et al*. (1968)	Rabbit *in vivo* milk-ejection assay	Paper chromatography and electrophoresis	Internal jugular vein blood: 12–25 μU/ml plasma during suckling
Chard *et al*. (1970)	Radioimmunoassay	Specific radioimmunoassay	Nonpregnant women, peripheral venous blood: <1 μU/ml plasma Venous blood during labor: <1 μU/ml plasma Fetal venous cord blood: 24 μU/ml plasma Fetal arterial cord blood: 45 μU/ml plasma
Vorherr (1968, 1972c)	Rat *in vivo* milk-ejection assay, injections of test solution via femoral artery into arterial branch supplying the lower abdominal mammary glands	Incubation of blood extracts with specific oxytocin and vasopressin antisera from rabbits	Peripheral venous blood: Nonpregnant women: <1–2 μU/ml plasma Pregnant women and parturients: <1–2 μU/ml blood Nursing women: 5–15 μU/ml blood

suggest that during suckling oxytocin is discharged into the bloodstream (Cobo *et al.,* 1967). While the baby was suckling one breast of the mother, changes in milk-ejection pressure were recorded by means of a fluid filled plastic cannula previously inserted into the nipple of the other breast. By this method intramammary pressure increases of 20–25 mm Hg could be observed repeatedly during the period of suckling. This mammary intraductal pressure pattern, as recorded during suckling, was matched by intravenous administration of various oxytocin doses. Accordingly, during a 15 minute nursing period, an endogenous release of 70–380 mU of oxytocin could be calculated (Cobo *et al.,* 1967). Based on this information, temporary plasma levels of oxytocin of about 5–25 μU/ml can be estimated. In our recent studies, maternal blood levels of 5–15 μU/ml of oxytocin were measured during the infant's suckling (H. Vorherr, unpublished observations, 1971). From this it follows that plasma levels of oxytocin must be lower when no stimulation for its release occurs. Although human plasma taken from nonpregnant and pregnant women, and gravidas in labor was used in a 20-fold concentration, no oxytocin was detectable after trichloroacetic acid extraction and chromatography on CG 50 cation exchange resin. From the recovery of oxytocin after the extraction procedure and from the sensitivity of the bioassay, it can be concluded that the levels of oxytocin in peripheral blood are below 1–2 μU/ml under the indicated conditions (Vorherr, 1968, 1972a, c). Similarly, no oxytocin (< 1 μU/ml) was measurable in urine samples obtained from normal men, nonpregnant and pregnant women, and patients during labor (Vorherr, 1968, 1972c). Recently these bioassay results were confirmed by radioimmunoassay determinations of oxytocin (Chard *et al.,* 1970).

Oxytocin circulates in its monomeric form, and it does not seem to be bound to plasma proteins, for when added to plasma it is completely ultrafiltrable. Oxytocin can penetrate into extravascular spaces. The plasma half-life of oxytocin is very short, and within 3–5 minutes 50% of the oxytocin disappears from circulation (Fabian *et al.,* 1969). When a dose of 500 μU of oxytocin was injected i.v. into ethanol anesthetized rats under water diuresis, 10% of the hormone appeared in the urine within 15 minutes (Kleeman and Vorherr, 1969).

f. METABOLISM OF OXYTOCIN

i. INTRACELLULAR DESTRUCTION. The kidneys and liver are mainly responsible for the metabolism of oxytocin. After injection of tritiated oxytocin, the kidneys retain far more radioactivity than the liver (Aroskar *et al.*, 1964). Most of the oxytocin seems to be removed from the blood at the peritubular side of the tubular cells of the nephron. Nevertheless, other target organs such as uterus and mammary gland may trap and thus remove considerable amounts of oxytocin from the circulation. The mechanisms of intracellular hormone destruction probably involve reduction of —S—S— bonds and proteolytic enzymatic attack on the amide groups that are necessary for its biologic activity (Fig. 41). Because natural oxytocin seems to circulate in its monomeric form and is not bound to plasma proteins, its enzymatic destruction in the plasma is unlikely.

ii. OXYTOCINASE–VASOPRESSINASE OF PREGNANCY PLASMA. Normal plasma does not destroy oxytocin *in vitro* (Kleeman and Vorherr, 1969). In contrast, pregnancy plasma of humans and anthropomorphous apes contains an enzyme system, oxytocinase–vasopressinase, formed in the syncytiotrophoblast of the placenta, which is able to inactivate relatively large amounts of oxytocin and vasopressin *in vitro* (Fekete, 1930; Tuppy, 1960; Semm and Waidl, 1962). However, *in vivo* this enzyme does not seem to inhibit the physiologic activity of the hormones to a major extent because exogenously administered oxytocin or vasopressin show similar antidiuretic effects in pregnant and nonpregnant women (Vorherr and Friedberg, 1964). In animals the oxytocinase–vasopressinase of human pregnancy plasma does not significantly influence the effect and metabolism of neurohypophysial hormones (Kleeman and Vorherr, 1969). This might be explained by supposing that injected oxytocin is quickly removed from the circulation and escapes enzymatic destruction and/or that the foreign human enzyme–protein system cannot be operative in the bloodstream of the experimental animal. Indeed, the oxytocinase–vasopressinase of pregnancy plasma cannot be active in the homologous system under *in vivo* conditions, because, if it were, pregnant women would develop symptoms of diabetes insipidus, which has never been observed. The mechanisms by which the enzyme is activated in serum or plasma after venous withdrawal of pregnancy blood are not understood; the physiologic importance of the appearance of the enzyme during pregnancy is not clear either. Suggestions that the enzyme prevents premature oxytocin-induced labor or that a decrease of its activity at term allows oxytocin to evoke contractions have not been substantiated. It appears that the enzyme, locally present in high concentration in the syncytial layer of the placental villi, may have a protective function by preventing maternal neurohypophysial hormones from entering the fetal circulation.

iii. OXYTOCIN AND VASOPRESSIN ANTIBODIES. Vasopressin antibodies have been detected in humans with diabetes insipidus following vasopressin treatment (Roth *et al.,* 1966), but no antibodies have yet been observed after injections of oxytocin. Recently, antibodies against oxytocin (Gilliland and Prout, 1965) and vasopressin (Permutt *et al.,* 1966) have been produced in rabbits. Rabbit antisera against oxytocin or vasopressin are very effective in blocking the biologic activity of the hormones *in vitro* as well as *in vivo,* and they are generally more potent inactivators than the oxytocinase–vasopressinase of pregnancy plasma (Kleeman and Vorherr, 1969). Also, experimental animals (rabbits, rats) actively immunized against oxytocin and vasopressin may develop diabetes insipidus and display impaired lactation, respectively (H. Vorherr, unpublished observations, 1971).

g. BIOLOGIC IDENTIFICATION OF NEUROHYPOPHYSIAL HORMONES

Serum antibodies against oxytocin or vasopressin produced in rabbits, when incubated in a 1:10 diluted solution, do not cross react. Such specific antisera are therefore very well suited for the differentiation of neurohypophysial octapeptides from other substances also displaying antidiuretic, milk-ejection, or oxytocic activities; such substances may be contained in extracts of body fluids or tissues to be measured for their oxytocin and vasopressin content. The antisera method of octapeptide identification is superior to all those previously reported such as inactivation by thioglycolate or pregnancy plasma (Vorherr and Munsick, 1970). However, often the antibody titers actively produced against oxytocin or vasopressin do not seem to be high enough or too much bound to plasma proteins to inhibit greatly the endogenous hormone production of the immunized animals, because diabetes insipidus, for instance, cannot be readily provoked in rabbits immunized against ADH. In addition, when respective rabbit antisera in low amounts (0.1 ml) are injected into bioassay rats, they usually fail to block the activity of subsequently administered oxytocin or vasopressin; here immune mechanisms, elicited by contact of species different proteins, may prevent activity of foreign rabbit donor antibodies in the circulation of rats. Nevertheless, efficient active or passive immunization of experimental animals against oxytocin and vasopressin may provide a research tool to determine further the effects of selective oxytocin and ADH depletion on reproductive performance and fluid metabolism, respectively.

CHAPTER V

Feeding the Newborn

A. TECHNIQUES AND PRINCIPLES OF BREAST FEEDING

In order to stimulate hypophysial prolactin and oxytocin secretion (Fig. 34) necessary for the process of lactation, the newborn is put to the mother's breasts a few hours postpartum, and 3 minutes of suckling each breast at 3- to 4-hour intervals is allowed. Thereafter, each individual suckling period is increased by 1 minute every day. Toward the end of the first postpartal week, regular 4 hour breast-feeding intervals, eventually interrupted by an 8 hour nursing-free period at night, are generally recommended. Initially, about 75% of all newborns cry during the night demanding nursing, which, preferably, should be granted. Only 50% of infants will need night feeding after 4 weeks, and 75% of all neonates will sleep through after 8 weeks postpartum.

After 7 to 10 days, the nursing process is usually well established, and thereafter not more than 10 minutes of nursing at each breast should be allowed because within 4 to 7 minutes of normal suckling the baby removes practically all the available milk. Suckling periods exceeding 10 minutes at each breast will only cause undue mechanical areola and nipple irritation and thus promote the development of nipple fissures and mastitis. During the first week of lactation the newborn often cannot empty the breasts adequately, and additional manual expression or pumping off of milk is necessary in order to maintain and increase the process of milk secretion. Pumping off also relieves breast tenderness and pains caused by the engorged milk-filled mammary glands. After the first week the growing baby suckles better, consumes more milk, and empties the breasts satisfactorily. When the baby is put to the breast, nipple erection through sensory stimulation and contraction of nipple smooth musculature ("nipple sphincter") occurs; the nipple erection is supported by venous stasis and hyperemia of the nipple base. As soon as the baby has seized the nipple and starts suckling, the mother usually experiences a sensation of "drawing" in the breasts due to posterior pituitary release of oxytocin, inducing mammary myoepithelial contraction and thus milk ejection; occasionally, some milk may drip from the nipple of the other breast. Maternal emotional stimulation on seeing the baby or on hearing the infant cry may elicit nipple erection (sympathetic stimulation) and "drawing" in the breasts (milk letdown) even before the infant is laid to the breast for suckling. It therefore can be assumed that respective emotional stimuli may bring about secretion of pituitary prolactin and possibly oxytocin. Also, due to smooth muscle contraction of mammary areola and nipple, the lactiferous sinuses may be partly emptied, i.e., milk drips from the nipple. During suckling the nipple, the infant's lips, gums, and tongue create a vacuum with an underpressure of about -50 cm H_2O (Nagai and Yamada, 1971) in the oral cavity, and by additional peristalsis-like chewing movements milk from lactiferous sinuses can pass the nipple sphincter and squirt into the baby's mouth (Vorherr, 1972d).

At each nursing period, both breasts should be suckled. In this way, the infant is more easily fed, receives more milk, and less breast engorgement and pain occur. Before and after nursing, the areola and nipple of the breast should be cleaned with a nonirritant sterile solution, and between nursing periods the nipple area should be covered with a sterile gauze pad.

Under normal conditions the following daily amounts of milk are available for the infant during the first postpartum week: second day, 120 ml; third day, 180 ml; fourth day, 300 ml; end of first to early second week, 500 ml.

Effective and sustained milk production is achieved 10 to 14 days postpartum, and, after that time, the milk yield is 120–180 ml per feeding. Daily amounts of 800 ml of breast milk or more may be produced after 2–3 weeks of lactation. Women nursing, without supplementary feeding, produce 10–20 ml milk/kg/day (500–1000 ml per day), whereas in lactating rats for instance 50–120 ml milk/kg/day (15–36 ml per day) are secreted. For nursing women the total energy output for milk is 30–50 kcal/day/kg. Some women have given as much as 4–6 liters of milk per day, necessitating a higher caloric intake than lactating women with average milk yield.

Early diminution of milk secretion may be caused by incomplete emptying of the breasts during nursing due to inadequate efforts of the baby, especially in cases where complimentary feeding is given; psychologic problems of the mother with an aversion toward nursing can also cause a decreased milk yield (Table 37). Chlorpromazine, 25 mg 2 to 3 times per day, will not only relax anxious and fearful women who are afraid that they do not have enough milk for their baby, but the drug also increases pituitary prolactin secretion severalfold and thus greatly stimulates milk production. Because ergot alkaloids reduce pituitary prolactin secretion (del Pozo *et al.*, 1972) lactating puerperas should not receive methyl ergonovine or another ergot derivative for support of uterine involution but rather 5–10 units of oxytocin i.m. In contrast to former beliefs that high fluid intake is important for the nursing mother's milk yield, recent investigations (Buckman *et al.*, 1973) have disclosed that a water load of 1–1.5 liters greatly suppresses pituitary prolactin secretion and thus inhibits mammary milk production; accordingly, excessive fluid intake during lactation should be avoided. On the other hand, theophylline can increase pituitary prolactin secretion and thus tea or coffee may be valuable for nursing mothers. Also TRH has been found to increase milk volume and milk fat concentration in nursing puerperas by means of stimulating pituitary prolactin secretion (Tyson *et al.*, 1973), and perhaps synthetic TRH or prolactin may become a treatment modality for lactating mothers with insufficient milk production due to insufficient pituitary prolactin secretion. After administration of synthetic FSH-LH releasing hormone, prolactin plasma levels remain unaltered. Lactating women taking contraceptive pills with larger sex hormone amounts or those becoming pregnant in the early period of lactation may experience diminishing milk secretion. It seems, however, that small amounts of estrogens and progestins as contained in present contraceptive pill preparations (Ovral, Loestrin) do not interfere

TABLE 37

Early Diminution of Breast Milk Secretion Postpartum

A. Failure of adequate milk removal
   1. Inadequate efforts of the suckling infant
   2. Increased sympathetic-adrenal activity in fearful and anxious puerperas resulting in catecholamine antagonism of oxytocin-induced milk ejection
   3. Aversion toward nursing
B. Failure of adequate milk formation
   1. Inadequate food and fluid intake
   2. Early return of menstruation and ovulation postpartum
   3. Early pregnancy postpartum
   4. Intake of contraceptive pills containing relatively high amounts of sex steroids

with the milk yield during nursing. Because contraception is important for the puerpera, it should be pointed out here (see also under Chapter VI) that (1) postpartum contraception is best provided immediately after delivery for puerperas who might not come back for a clinical checkup 4 to 5 weeks after delivery; (2) insertion of an intrauterine device postpartum is accompanied by a rather high expulsion rate; (3) intramuscular injection of a 3 month long acting progestin (150 mg medroxyprogesterone) will provide effective contraception; (4) oral contraceptive pills with low estrogen and progestin content or oral progestin alone ("minipill") do not seem to impair milk yield, puerperal uterine involution, or the health of the breast-fed infant and mother (Vorherr, 1973) (see also Table 44).

After a nursing period of 4 to 6 months the proteins supplied by the breast milk no longer satisfy the needs of the growing child adequately. For this reason, and because the insufficient milk yield necessitates complimentary feeding anyway, the period of lactation does not have to be extended beyond 6 months postpartum (Zilliacus, 1967; Benson, 1968; Taylor, 1966; Kamal *et al.*, 1969a; Vorherr, 1972d).

B. NURSING VERSUS BOTTLE FEEDING

Apart from receiving insufficient instruction and encouragement in regard to breast feeding, today many puerperas do not wish to put up with the "inconveniences" of nursing. All the same, not only from a financial standpoint, important to women from low socioeconomic classes, but for many other reasons, breast feeding still has its merits (Table 38; Vorherr, 1972d).

Although available milk formulas are adequate in their composition and are quite satisfactory, cow's milk proteins contained in formulas may occasionally cause milk allergy (eczema, ulcerative colitis, see below) in the newborn. Circulating cow's milk antibodies have been found in the majority of bottle-fed infants and unexpected sudden deaths ("crib deaths") in such infants have been connected with an anaphylactic reaction to cow's milk proteins. Also, albeit in rare incidences, milk formulas may be contaminated by bacilli such as *Mycobacterium tuberculosis, Brucella, Corynebacterium diphtheria, Shigella,* and *Salmonella* bacteria; pasteurization of milk does not kill all the bacteria contained in it. Moreover, milk formulas easily become contaminated with bacteria when terminal sterilization is not practiced; in homes of low socioeconomic groups up to 50% of feeds showed contamination with enteric organisms (Harrison and Peat, 1972). When a milk formula is prepared and stored for more than 24

TABLE 38

Breast Feeding: Advantages and Disadvantages

Breast feeding	Mother	Child
Advantages	1. Feeling of security; psychologically satisfying experience 2. Better myometrial involution; less problems with uterine subinvolution and endometritis postpartum 3. Incidence of breast cancer is lower in mothers who nurse their babies than in nonnursing women 4. Cosmetically favorable for women with small breasts	1. Breast milk is an ideal food containing immune substances; it is easily digestible; readily available at the right temperature, and free from bacterial contamination; no errors in formula are possible 2. Protection against infection provided; avoidance of milk allergy and anaphylactic reactions due to foreign cow's milk proteins ("crib deaths") 3. Superior biologic value of human milk proteins 4. Lower rate of gastrointestinal infections 5. Superior intestinal absorption of human milk fat with better uptake of fat soluble vitamins 6. More physiological supply of minerals 7. Decreased infection and mortality rate in premature infants
Disadvantages	1. Limiting social and professional activities 2. Contraindicated in Mastitis, TB, and other infectious diseases, as well as conditions requiring intake of large drug dosages 3. Increased incidence of puerperal mastitis in nursing mothers 4. Cosmetically unfavorable in women with large breasts	1. None, if mother is healthy, takes no drugs, and has enough milk supply and skill to properly nurse the baby

hours, contamination by *E. coli* may be as high as 38% of samples. In contrast, breast milk is free of bacteria and no denaturation of milk proteins and enzymes, as caused by the heating process of cows' milk, takes place. The close biochemical relation of human milk proteins to serum proteins facilitates digestion and increases its biologic value. The smaller amounts of casein present in breast milk allow better milk curdling, shorter stay of the curd in the stomach, and faster gastrointestinal digestion. Breast milk contains a higher amount of whey proteins than cow's milk, providing more essential amino acids for the newborn's growth and development (Table 22). The proteins of milk consist of casein (caseinogen), albumins, globulins, and proteose–peptone. Caseinogen, usually termed "casein," is a phosphoprotein present in milk as calcium caseinate. When acid (lactic acid, citric acid) is added to milk reaching its isoelectric point at pH 4.6, the caseinogen becomes insoluble, and the milk curdles. Upon longer standing, milk becomes "sour" and curdles because bacteria *(Streptococcus lactis)* in milk grow and multiply and produce lactic acid which induces curdling. While the globulins are insoluble in a half-saturated ammonium sulfate or in a saturated magnesium sulfate solution, the milk albumins are soluble in the latter solutions. With ingestion of milk, the stomach enzymes rennin (not present in the child and adult), pepsin, and chymotrypsin, bring about coagulation (clotting) of milk whereby caseinogen is converted into casein, which forms an insoluble calcium caseinate–calcium phosphate complex (curd). This curd then represents a coagulum of casein protein fibers that enmeshes the fat and most of the whey proteins. After standing, the clot contracts, expressing a clear fluid known as whey, which contains albumins and globulins. In cow's milk 82% of protein nitrogen is derived from casein, but in man only 39%, indicating that the whey proteins are the predominant amino acid providers in man. Casein is one of the relatively few naturally occurring proteins that contains phosphorus, acting as a reservoir of metabolizable phosphorus. When milk is boiled, the milk proteins are denaturated, but no precipitation takes place until the pH is brought down to 4.6. At pH 4.6 and after 20 minutes boiling, the whey proteins co-precipitate with casein, and only the heat-stable proteose–peptone remains in solution. Heat denaturation of milk proteins leads to a progressive weakening and ultimate rupture of cross linkages of polypeptide chains with increased sensitivity to precipitation. Finally after prolonged heating denaturation and precipitation occur. The immune globulins of milk are the most sensitive to heat denaturation, followed by β-lactoglobulin, serum albumin, and α-lactalbumin, which is the least heat sensitive protein. From the milk albumin fraction (α-lactalbumin, β-lactoglobulin, serum albumin), a β-lactoglobulin has been crystallized; it is almost identical with lactalbumin and constitutes the major portion of whey (Table 22). The amino acid composition of β-lactoglobulin is quite different from that of any plasma protein, whereas the milk immune globulins of colostrum are very similar in composition to their plasma counterparts. Casein and β-lactoglobulin are not found in other tissues and bear no relationship to any of the plasma proteins, albeit they are the most nutritive milk proteins (Table

22). The immunoglobulins of milk are closely related to the α-globulins of serum; they are found in very high amounts in colostrum and are responsible for transmission of antibodies in ungulates. A marked decrease of immunoglobulins occurs with the change of colostrum into transitional milk. Although human colostrum contains relatively high amounts of immune globulins, their physiology is not well understood. They seem to be of none or only of secondary importance to intrauterine maternal passive immunization of the fetus. Milk albumins provide biologically the most valuable protein nutrition, whereas the globulin fraction of milk is the carrier of immune bodies that are especially high in colostrum and essential for the survival of the newborn of some animal species. In infants some immune globulins of maternal colostrum may be absorbed undigested by pinocytosis in the intestine. If during the early postpartum period foreign proteins (cow's milk) are absorbed into the infant's circulation by intestinal pinocytosis, formation of antibodies is provoked, and an antigen-antibody reaction occurs upon subsequent entry of the same protein. However, intestinal mucosal cells of infants and young animals seem to lose the ability to take up unsplit proteins very rapidly. Some infants may become allergic to protein foods (cow's milk, eggs), but the allergy usually disappears as they grow older. Nevertheless, some adults may develop an allergy toward certain foods, and in these individuals intestinal absorption of whole proteins or larger split products probably occurs. The biologic value of protein is measured and expressed as the percent ratio of food nitrogen retained in the organism to the total nitrogen intake. In general, a high biologic value of protein indicates that essential amino acids are available and adequately provided in human milk. The nonprotein milk constitutents (urea, uric acid) are derived from plasma and probably have filtered into the alveolar milk via intercellular spaces. Breast milk proteins decrease the gastric acidity to a much lesser degree than cow's milk proteins, which have a higher casein content. Furthermore, the quantitative and qualitative supply of milk fat and vitamins through breast milk is superior to that of cow's milk. As indicated in Table 39, breast milk supplies energy mainly from fat and carbohydrates, and less calories are contributed by proteins. Cow's milk preparations or commercial formulas derived from it contain higher amounts of proteins than breast milk. The buffer capacity of the casein contained in cow's milk is 4 times higher than that in human milk. Thus, more gastric acid is needed for curdling of ingested cow's milk, but the gastric contents remain less acidic than with the ingestion of breast milk. With decreased gastric acidity the bactericidal effect of gastric juice is inadequate, and this may be one of the reasons why bottle-fed children have a higher incidence of gastrointestinal infec-

TABLE 39

Caloric Distribution in Breast Milk and Cow's Milk [a]

Type of feeding	Percent of calories derived		
	Protein	Fat	Carbo-hydrates
Breast milk	7%	50%	43%
Marketed cow's milk	20%	52%	28%
Cow's milk—diluted 2:1 with water and addition of 5% sugar	15%	42%	43%
Commercial milk formulas—diluted 1:1 with water and addition of 5% sugar	10%	48%	42%

[a] From Vorherr (1972b).

tions than those who are breast fed. Cow's milk per se is more acidic (pH 6.8) than breast milk (pH 7.0), and the *E. coli* stool counts of breast-fed babies are much lower than those fed with cow's milk (Harrison and Peat, 1972). Transport and digestion of cow's milk curd from the stomach into the small intestine creates another problem, because here the higher acidity of the curd requires neutralization through increased intestinal secretion of alkaline juice. Thus, the intestinal digestive work load is enhanced, and the propulsion of the intestinal contents may become accelerated resulting in diminished absorption of nutrients.

The curd of cow's milk is coarse-flakey, whereas the curd of breast milk is fine-flakey. Accordingly, the cow's milk curds are more difficult to digest, and this may lead to a decreased utilization of milk proteins, increased loss of nitrogen, and the development of gastrointestinal dyspepsia. Boiling and homogenization of cow's milk with addition of citric or lactic acid will provide a more fine-flakey and better digestible curd.

In breast milk, the emulsion of milk fat is more intensive, and the milk lipases are more active than in cow's milk. Thus, the digestion and utilization of breast-milk fat is superior to that of cow's milk. Heating or boiling of cow's milk also destroys to some extent the lipases contained in it and decreases the degree of fat emulsion, leading to a less adequate absorption of fat in the intestine. Since milk fat is the carrier of vitamins A and D, their absorption and that of calcium, folic acid, and cyanocobalamine is diminished at the same time. The lipolytic milk enzyme resembles in many aspects pancreatic lipase; it may play a part in the hydrolysis of the milk glyceride tributyrin (Jubelin and Boyer, 1972). Furthermore, cow's milk formulas contain a higher proportion of saturated fatty acids, which are not as well absorbed in the intestinal tract of the neonatus as unsaturated fatty acids.

The lactose content of breast milk is higher than that of cow's milk (some commercial formulas contain only sucrose). Lactose facilitates the growth of the intestinal *Lactobacillus bifidus* flora, which produce lactic acid. Intestinal hydrolysis of lactose, besides glucose and lactic acid, leads to formation of galactose, which plays an important role in the synthesis of the cerebrosides of myelin and the glycoproteins of collagen. Oligosaccharides of human milk possess bifidus factor activity, which is practically absent from the milk of ruminants. Deproteinized human milk contains oligosaccharides such as galactose, glucose, fucose, and *N*-acetylglucosamine. Also cow's milk contains a series of oligosaccharides but no fucose. It may, therefore, be assumed that fucose is a contributing factor for the establishment of the intestinal bifidus flora early in the neonatal period. In addition, meconium provides an essential factor for the growth and development of intestinal *Lactobacillus bifidus* in the originally sterile intestinal lumen of the newborn infant (Ling *et al.*, 1961).

The concentration of minerals is 3 to 4 times higher in cow's milk than in breast milk. Consequently, the cow's milk has to be diluted with water, because during the first months of neonatal life the infant's kidneys are not ready to handle larger amounts of salt. In experiments on newborn rats with high salt intake, hypertension was observed. Cow's milk contains 4 times more calcium and 5 to 8 times more phosphorus than breast milk (Table 21). The relatively higher concentration of phosphorus in cow's milk and thus its enhanced intestinal absorption rate and its increased serum level in the newborn may lead to a decrease of serum calcium, thereby the danger of neonatal tetany exists, which occurs almost exclusively in formula-fed infants. Tetany due to low serum calcium may be observed in babies fed with cow's milk during the first 2 weeks after birth. Because the newborn's kidneys are unable to excrete the large amounts of intestinally absorbed phosphorus, the serum phosphorus level increases, inducing a concomitant fall in serum calcium. This explains why despite an abundant supply of calcium from cow's milk, neonatal tetany may occur.

Cow's milk contains about 40% less iron, vitamin A, and vitamin D than breast milk. Therefore, anemia or vitamin A and vitamin D deficiencies will develop faster in an infant receiving cow's milk than in a breast-fed baby. Consequently, iron, vitamin A, and vitamin D are added to milk formulas to supply the recommended amount of 0.6 mg of iron, 500–1000 units of vitamin A, and 200–300 units of vitamin D per day. The vitamin

C content of breast milk is adequate, whereas that of cow's milk (Table 21) is insufficient and is further decreased by about 50% through the milk heating process. Accordingly, from the second postnatal week on, the infant receiving cow's milk needs an additional supply of vitamin C. Because in newborns the intestinal tract is sterile and thus no bacterial vitamin K synthesis takes place, a transient hypoprothrombinemia with occasional hemorrhagic manifestations is observed between the second to sixth day postpartum. Beginning with the second week of neonatal life, the intestinal bacterial flora corrects this "physiologic" vitamin K deficit without any exogenous supply. Prevention of this hemorrhagic disease, encountered early postpartum, is accomplished by providing the pregnant or nursing mother with a supplement of vitamin K or, most effectively, by giving it directly to the newborn. Since prothrombin activity is detectable in the newborn's plasma, it seems that placental transport of vitamin K occurs. When vitamin K is given to the mother immediately postpartum, no appreciable improvement (shortening) of the breast-fed infant's prothrombin time occurs during the first 3 days after delivery. This may be explained by the low vitamin K excretion in colostrum and the only minimal consumption of colostrum by the newborn during the first 3 postpartum days.

Puerperas with a drug intake should be evaluated carefully with regard to breast feeding because drugs may be secreted into the breast milk in amounts that may cause adverse effects in the organism of the suckling baby. Potentially dangerous drugs in the newborn are diuretics, steroids, reserpine, atropine, coumarines, diphenylhydantoin, antithyroid drugs, anthraquinones, metronidazol, and anticancer drugs (see Chapter III, Section D, 7).

Although there has been much speculation about the transfer of maternal immune bodies via colostrum and milk to the suckling newborn, it has not yet been determined whether, and if so to what extent, immunoglobulins contained in human milk are able to pass the intestinal mucosa of the neonatus causing passive immunization. In horses, pigs, cattle, dogs, cats, and rodents antibody transmission in milk and thus provision of passive immunization postpartum is possible. From the various immunoglobulins (IgA, IgG, IgM, IgD) present in human colostrum, IgA is the most important regarding concentration of antibodies and immunological properties. Milk IgA is most likely synthesized in mammary alveolar cells from two molecules of serum IgA linked by disulfide bonds. The IgA content of human colostrum is high during the first days postpartum; thereafter, it declines rapidly and disappears around 2 weeks after delivery. The following immunoglobulin levels were measured for 4 days after delivery and are expressed in milligrams per 100 ml colostrum (Michael *et al.*, 1971):

First day:	600 IgA,	80 IgG, and	125 IgM
Second day:	260 IgA,	45 IgG, and	65 IgM
Third day:	200 IgA,	30 IgG, and	58 IgM
Fourth day:	80 IgA,	16 IgG, and	30 IgM

Serum and milk immunoglobulins contain antibodies against viruses, rickettsiae, bacteria (*E. coli, Salmonella,* diphtheria, etc.), and protozoa, and they are also potential carriers of rhesus antibodies. As mentioned before, the transmission of antibodies via milk immunoglobulins must be very minimal in humans; otherwise hemolytic disease would have been observed in rhesus incompatible situations when immunized mothers nurse their babies.

What is the significance of human milk globulins? *In utero,* the human fetus is amply supplied by transplacental passage of immune bodies (IgG fraction mainly, insignificant amounts of IgA and IgM). Because immunoglobulins of the IgG type readily pass the human placenta, infants born to mothers having an immunity against scarlet fever, diphtheria, and measles are also immune against these diseases (passive immunity). In the newborn, immunity against pertussis and also antibodies against *E. coli* are lacking. Although immunoglobulins in human milk are not absorbed by the gastrointestinal mucosa in significant amounts, these antibodies most likely provide local intestinal protection against viruses (poliomyelitis, etc.) or bacteria (*E. coli,* etc.), which may infect the intestinal mucosa or which may try to enter the organism from the gut. IgA is the predominant immunoglobulin in colostrum, where it is found in higher levels than in blood. The lower intestinal content of coliform bacteria in breast-fed babies is thus attributable to the IgA activity of colostrum. Aerobic and anaerobic lactobacilli are found in greater amounts in breast-fed than in bottle-fed infants. Postpartum the colostral immunoglobulins decrease rapidly as evidenced by the declining agglutinating and bactericidal effect of colostrum against *E. coli.* The antimicrobial effect of colostrum seems to be mainly due to IgA immunoglobulin. Also, early postpartum the saline extracts of stools from breast-fed infants contained a substantial bactericidal activity toward *E. coli;* no such germicidal effect was demonstrable in stool extracts from bottle-fed babies (Michael *et al.,* 1971). Indeed it appears that colostral immunoglobulins protect against coliform septicemia, because in bottle-fed infants higher rates of neonatal infections (gastroenteritis, urinary tract infections, meningitis) have been observed than in breast-fed babies (Winberg and Wessner, 1971).

Besides the immunoglobulins of milk, other factors providing resistance to infection seem to be operative. Breast-fed infants are less susceptible to gastrointestinal disorders (nonbacterial and bacterial diarrhea) and other diseases (respiratory infections, otitis media) than formula-fed babies. This is also confirmed by the observation that early weaning and higher infant morbidity and mortality are correlated. Also during weaning and in the subsequent weeks an increased incidence of gastrointestinal disorders is encountered ("weaning diarrhea"). The exact factors and mechanisms leading to the enhanced resistance of breast-fed newborns against infections are not clear. Human milk may act beneficially in this respect by providing specific antibodies (*E. coli* antibodies) against infective agents or by inhibiting pathogenic intestinal microorganisms through nonspecific antimicrobial factors. A "bifidus factor" activity has been described in human milk, exceeding that in cow's milk by 40 to 100 times. This bifidus growth factor activity is due to a group of N-containing carbohydrates and other saccharides of human milk (lactose, galactose, fucose, *N*-acetylglucosamine, and *N*-acetylneuraminic, or sialiac, acid), leading to growth and development of the intestinal *Lactobacillus bifidus* flora. Accordingly, by the end of the first postpartum week the bacterial bifidus flora represents 95% of all culturable microorganisms (streptococci, bacteroides, clostridia, micrococci, enterococci, *E. coli*) in the fecal smear of the breast-fed infant. With cessation of nursing the nonbifidus flora increases progressively in number. Bifidobacteria metabolize milk saccharides thus producing acetic acid, lactic acid, and trace amounts of formic and succinic acid, which are all responsible for the low pH (pH 5.3) of the feces of breast-fed babies (infants fed with cow's milk formulas have neutral or alkaline stools). The bifidus factor substances of human milk are also considered as "building blocks" for blood group substances. It appears that this bifidus factor in human milk together with the intestinal bifidus flora provide resistance mechanisms against intestinal infections with *Staphylococcus aureus, Shigella,* and protozoa, causing general antagonism toward certain pathogens. Sialiac acid of human milk, for instance, may inhibit influenza virus activities. The low stool counts of *E. coli* and the high lactobacillus counts in breast-fed babies are contrasted by the reversed condition in bottle-fed infants. These low *E. coli* counts in breast-fed babies (human milk has a bacteriostatic effect on *E. coli*) have been connected with a better weight gain of such infants as compared to those who are formula-fed, by providing protection against neonatal gastoenteritis (Harrison and Peat, 1972). Lactoferrin as contained in a concentration of about 4 mg/ml milk was found to exert a bacteriostatic effect on *E. coli.*

Human milk contains another nonspecific antimicrobial factor, lysozyme (muramidase), in concentrations of about 0.2 mg/ml. This thermostabile enzyme is bacteriolytic toward enterobacteriaceae and gram-positive bacteria; lysozyme is found in large amounts in feces of breast-fed newborns but not in those of babies fed with cow's milk formula. Complement and interferon are also present in human milk, but their role in the newborn's resistance against infections is not understood. Colostrum and milk contain cells with immunological properties ("immune cells" carrying IgA); these are thought to be lymphocytes and macrophages with the capacity to phagocytose.

Acrodermatitis enteropathica, recognized as a vesicular dermatitis around the body orifices and the distal parts of the extremities, multiple paronychia on hands and feet, and digestive dysfunctions (diarrheal attacks) are rarely observed in breast-fed infants. This disease is probably due to an inborn metabolic error, caused by abnormally functioning intestinal flora. Here the pathogenic bacterial products may be accumulating in the gut, and when absorbed the infant's organism may be unable to detoxify them. In addition, a defect in interconversion of unsaturated fatty acids in the baby's organism has been considered as contributory to the disease. Acrodermatitis enteropathica can be cured by human milk diet; the therapeutic success confirms again the superior value of human milk for the infant's nutrition and well being (György, 1971; Mata and Wyatt, 1971).

In a clinical trial, it was found that infection and mortality rates in premature infants, fed with breast milk, were lower than in those receiving cow's milk formulas. Therefore, even proponents of milk formula feedings agree that premature infants should receive breast milk and should nurse at the breast as soon as they are able. Although no clear statistics are available for term infants demonstrating that breast milk is a significant factor for increased infant survival and better growth, it seems reasonable to assume that human milk is superior to any feeding formula. Recently, obesitas of infancy due to formula feeding has been related to obesity observed in childhood and adulthood with sequalae such as degenerative vascular and metabolic diseases. According to a recent report, coronary artery disease was found only in teenagers who had been bottle fed after birth (Zilliacus, 1967; Benson, 1968; Ordway, 1970; Silverman, 1961; Meyer, 1968, Baum, 1971; Babson, 1971).

About 5–6% of all women in the United States contract breast cancer during their life (Miller and Fraumeni, 1972; Anderson, 1972). Breast cancer rates increased as nursing declined, and, according to one statistic, mothers who breast feed their babies are 2 to 3 times less in danger of contracting breast cancer than those who do not nurse their infants (Kessler, 1968). Whether this holds true for future respective surveys remains to be seen. For several months during lactation the ovaries may be quiescent and the amounts of circulating estrogens (estrogens exert a specific proliferative effect) are low; thus no mammary proliferation occurs. The secretory mammary activity is induced and maintained by prolactin and other metabolic hormones. Accordingly, during lactation no significant mammary proliferation takes place, and thereby protection against mastopathia or mammary tumor formation (cancer) appears to exist. It is believed that due to this period of mammary proliferative arrest (no mitoses are observed in secretory glandular cells) the disposition for atypical cellular changes is prevented, decreased, and/or delayed (Kaiser, 1969).

The risk of thromboembolic disease is the highest postpartum, and medicamentous suppression of lactation may increase this risk by several times. It seems that the incidence of puerperal thromboembolism is increased when high estrogen doses, such as a course of 80–120 mg of diethylstilbestrol, are given for suppression of lactation. When a low dose of a natural long-acting estrogen (estradiol valerate) in combination with a high dose of a long-acting androgen (testosterone enanthate), as commercially available in the ideal composition of Deladumone, is used for the same purpose, the risk of thromboembolism or menstrual disorders seems insignificant (Vorherr, 1973).

Because natural estrogens and probably more so synthetic nonsteroid estrogens are potentially carcinogenic, the latter should not be prescribed at all, and the administration of natural estrogens requires a critical medical evaluation. In women with a family or personal history of breast or genital cancer, however, any type of estrogen medication should be avoided; if estrogen medication is required such patients need close clinical observation.

Recently, warnings have been voiced against breast feeding by mothers with a family history of breast cancer or by those whose breast milk may contain viruslike particles (Dmochowski, 1972; Feller and Chopra, 1971; Moore *et al.*, 1971). In about 50–60% of lactating mothers with a personal or family history of breast cancer such viruslike particles have been observed in breast milk. It is well known that the transmission of mammary cancer via a milk virus (Bittner's virus) to the suckling offspring occurs in a certain strain of mice. It is not known whether such a viral transmission also takes place in certain nursing mothers possibly leading to development of breast cancer in their daughters during their later life. The demonstration of viruslike particles with biophysical properties of oncornavirions in breast milk and malignant tissue of breast cancer (Keydar *et al.*, 1973) is alarming and deserves further intensive studies. Accordingly, puerperas with a family history of breast cancer should not nurse their babies, and, perhaps, some day the milk of lactating mothers may be screened routinely for a possible content of viruslike particles. These newer views may caution us not to be too partial in recommending breast feeding; more so, it seems that the idea of nursing as a protective factor against breast cancer is not acceptable without reservations. Although the viral genesis of breast cancer in man has not yet been proven, the possibility exists that a virus is involved in this disease. On the other hand, relatives of breast cancer patients already run a two- to threefold higher risk of contracting mammary cancer; i.e., in such women the risk for breast cancer may amount to 15–20% (Anderson, 1972; Miller and Fraumeni, 1972). Furthermore, the familial aggregation of breast cancer occurs equally in female antecedents on both sides of the family on an obviously genetic basis. Therefore, if the cancerous disease were transmitted through breast milk, one would expect an excess of familial cases, only on the maternal side. But, this is not so, no relationship was found between breast feeding and mother–daughter occurrences of breast cancer (Tokuhata, 1969). Because this is one study only, for the time being women with a family or personal history of breast cancer should not nurse, until more research data are available clearly demonstrating that breast feeding under such conditions is of no potential danger to the breast-fed infant.

Because all in all the positive aspects of breast feeding are obvious, it is hoped that, in the future, mothers of newborn infants will be more encouraged, advised, and guided by physicians and nurses to breast feed if they wish to, as long as there are no contraindications such as drug intake, heavy smoking, mastitis, nephritis, tuberculosis, typhoid fever, serious infections or debilitating diseases, profuse postpartum hemorrhage, poor nutrition, epilepsy, or neurosis. *Note:* In women with a rhesus constellation and maternal serum antibody titers (fetal erythroblastosis), breast feeding is not contraindicated because the rhesus antibodies in the breast milk are not absorbed in the infant's intestinal tract and thus cannot evoke hemolytic disease in the baby.

CHAPTER VI

Lactation and Reproductive Function

A. RETURN OF FERTILITY POSTPARTUM

Usually 3 to 6 months pass postpartum before the normal female reproductive functions are restored, depending on whether or not the puerpera is nursing.

Menstruation and ovulation return more slowly in lactating women than in nonlactating puerperas (Table 40 and Fig. 43). According to an older study of 530 lactating and nonmenstruating patients, conception occurred in only 1.2% during the first 6 months after childbirth; whereas of 615 nonlactating women, 59.5% conceived within the first 6 months postpartum. These results are similar to those obtained in a recent investigation that showed that within 9 months after delivery 74% of nonlactating Egyptian women became pregnant, whereas the pregnancy rate was only 8% among lactating mothers during the same time span (Bonte and van Balen, 1969). These data are different from those of Peckham (1934) who reported that 12% of nursing mothers became pregnant within 6 months postpartum and 23% within 9 months postpartum. In 3–5% of lactating women postpartum amenorrhea may be followed by an amenorrhea of pregnancy, indicating that ovulation in the first postpartum cycle is possible. In another report, 33% of lactating women were found menstruating before the end of the third postnatal month, whereas 91% of the nonlactating mothers experienced menstruation within the same period of time (Zilliacus, 1967). Despite continued nursing for 9 months, two-thirds of the mothers were found menstruating at the end of that period. This is

TABLE 40

Return of Menstruation and Ovulation Postpartum

Postpartal women	Menstruation	Ovulation
Nonlactating	8 Weeks (average return); 65% of patients resume menstruation within 12 weeks postpartum	13 Weeks (average return); 40% of patients resume ovulation within 12 weeks postpartum
Lactating	12 Weeks (average return); 45% of patients resume menstruation within 12 weeks postpartum	18 Weeks (average return); 25% of patients resume ovulation within 12 weeks postpartum

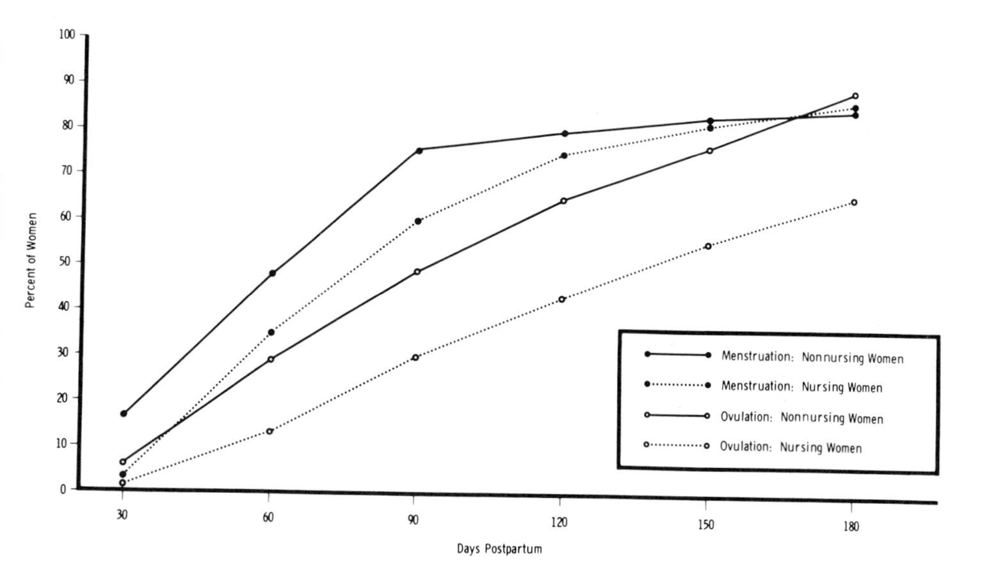

Fig. 43. Return of menstruation and ovulation postpartum. During the first 30 days after delivery, lactating and nonlactating puerperas are infertile. As early as 6 weeks postpartum, however, 5% of nursing mothers and 15% of nonnursing women have regained fertility as evidenced by ovulation. About 70% of nonlactating mothers menstruate and 40–50% ovulate 3 months after delivery; for lactating mothers it takes 4–5 months to reach respective percentage values. Four months portpartum fertility has returned in 65% of nonnursing women; whereas, it takes 6 months before the same percentage of lactating mothers ovulate. Despite the significantly decreased fertility of nursing women during the first 6 months postpartum, nursing as a contraceptive method is by far less reliable than oral or intrauterine contraception and therefore cannot be recommended. For more details and bibliography, see Vorherr (1973). (From Vorherr, 1973, by kind permission of the publisher.)

in agreement with a previous report (Peckham, 1934), where in 71% of mothers menstruation returned before cessation of lactation; about 30% of the women became pregnant within 1 year after delivery, 40% of whom were still lactating at conception. According to other data, 65% of the women menstruated prior to cessation of lactation (Sharman, 1966). Parity and fertile age are most likely without influence on the menstrual pattern postpartum. It has been suggested earlier that the tendency to menstruate and ovulate is greater in multiparas than in primiparas, but this is not uniformly agreed upon (Vorherr, 1973).

Generally, menstruation is not resumed before 4 to 5 weeks postpartum, and the first menstruation after delivery is usually unovulatory as confirmed by measurements of basal body temperature, vaginal smear, consistency of cervical mucus, endometrial biopsy, and urinary estrogen and pregnanediol excretion. Menstruations during the late puerperium or the postpuerperal period are usually preceded by ovulation. Regardless of whether the mother is nursing or not, usually no ovulation takes place before the fourth to the fifth week postpartum. Return of menstruation and ovulation postpartum is depicted in Table 41 and in Fig. 43. Menstruation returns on the average within 8 weeks postpartum in nonlactating women and within 12 weeks in lactating women. Similar time differences are observed between nonlactating and lactating women with regard to return of ovulation (Table 40) (Sharman, 1951; Cronin, 1968).

About 40% of lactating women menstruate during the first 2–3 months postpartum, and approximately 70–80% will resume menstruation within 4–6 months after delivery. Thus, over half of the nursing women may conceive during the phase of lactation, whereas the rest seem to be "protected."

The reason for the so called "lactation amenorrhea" and relative infertility has to be sought in an imperfect balance of hypothalamic–anterior pituitary function regarding gonadotropin secretion. Whenever the hypothalamo-anterior pituitary axis favors secretion and release of prolactin from the acidophilic cells of the anterior pituitary, secretion of FSH and LH seems to be diminished. In rats, the activity of tuberoinfundibular dopamine neurons has been found to increase during suckling, resulting in prolactin release from the anterior pituitary and in blockade of gonadotropin secretion. Also during pregnancy the activity of tuberoinfundibular dopamine neurons was increased, which may also account for the low circulating gonadotropin levels in gestation (Fuxe *et al.*, 1969; Jaffe *et al.*, 1969). It therefore appears that these hypothalamic dopamine neurons, sensitized by estrogens, participate in the regulation of gonadotropin secretion (blockade of FSH-LH release) and synthesis and release of prolactin

TABLE 41

Reproductive Function Postpartum [a]

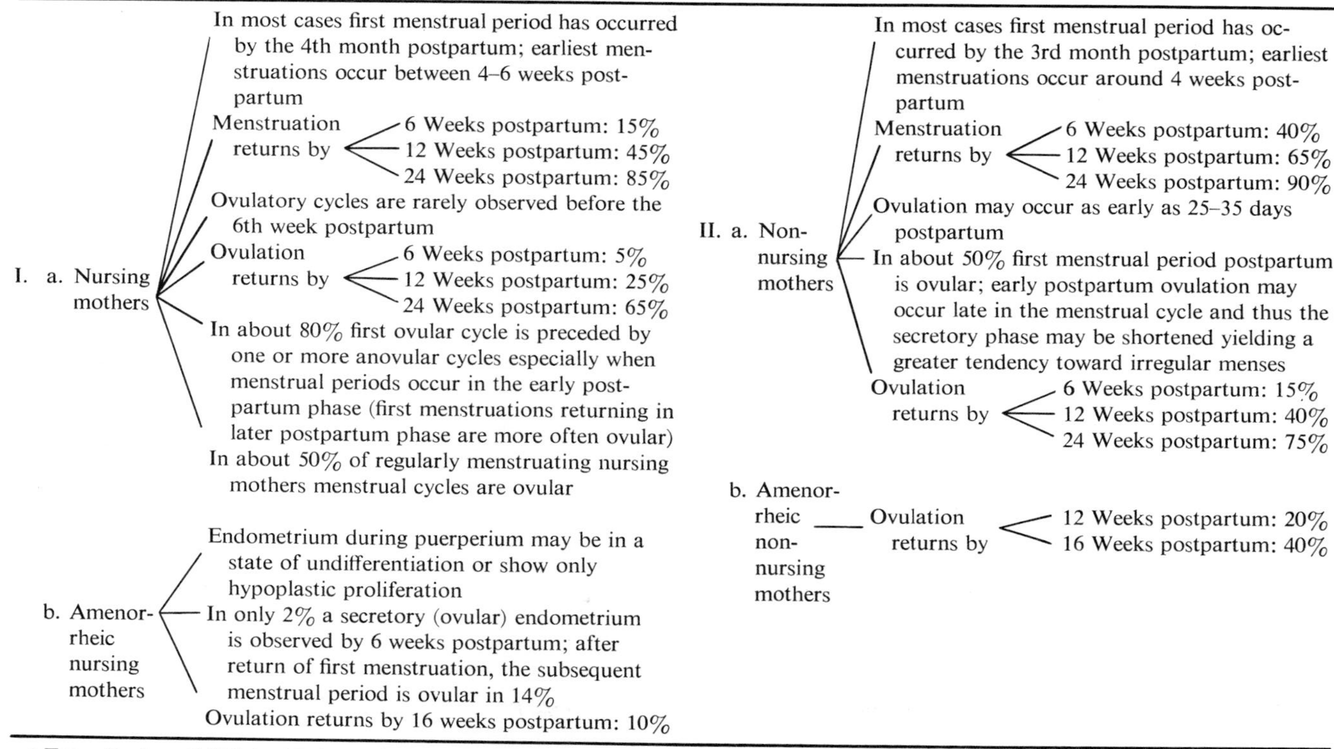

I. a. Nursing mothers	In most cases first menstrual period has occurred by the 4th month postpartum; earliest menstruations occur between 4–6 weeks postpartum
	Menstruation returns by: 6 Weeks postpartum: 15%; 12 Weeks postpartum: 45%; 24 Weeks postpartum: 85%
	Ovulatory cycles are rarely observed before the 6th week postpartum
	Ovulation returns by: 6 Weeks postpartum: 5%; 12 Weeks postpartum: 25%; 24 Weeks postpartum: 65%
	In about 80% first ovular cycle is preceded by one or more anovular cycles especially when menstrual periods occur in the early postpartum phase (first menstruations returning in later postpartum phase are more often ovular)
	In about 50% of regularly menstruating nursing mothers menstrual cycles are ovular
b. Amenorrheic nursing mothers	Endometrium during puerperium may be in a state of undifferentiation or show only hypoplastic proliferation
	In only 2% a secretory (ovular) endometrium is observed by 6 weeks postpartum; after return of first menstruation, the subsequent menstrual period is ovular in 14%
	Ovulation returns by 16 weeks postpartum: 10%
II. a. Non-nursing mothers	In most cases first menstrual period has occurred by the 3rd month postpartum; earliest menstruations occur around 4 weeks postpartum
	Menstruation returns by: 6 Weeks postpartum: 40%; 12 Weeks postpartum: 65%; 24 Weeks postpartum: 90%
	Ovulation may occur as early as 25–35 days postpartum
	In about 50% first menstrual period postpartum is ovular; early postpartum ovulation may occur late in the menstrual cycle and thus the secretory phase may be shortened yielding a greater tendency toward irregular menses
	Ovulation returns by: 6 Weeks postpartum: 15%; 12 Weeks postpartum: 40%; 24 Weeks postpartum: 75%
b. Amenorrheic non-nursing mothers	Ovulation returns by: 12 Weeks postpartum: 20%; 16 Weeks postpartum: 40%

[a] From Vorherr (1973) by kind permission of the publisher.

during pregnancy and lactation. Whether such a mechanism also exists in humans is not clear because 20 days postpartum normal levels of serum FSH and LH were already measurable (Jaffe *et al.,* 1969). Furthermore, in another study on rats, dopamine, as well as the pineal principle melatonin and its precursor serotonin, was administered into the third ventricle of the brain where it promoted increased release of FSH-LH and inhibited secretion of prolactin (Kamberi *et al.,* 1971). Thus, this subject still remains to be settled, especially since it has been shown that in rodents melatonin as well as melatonin-free bovine pineal extracts displayed antigonadotropin activity (Benson *et al.,* 1971).

A certain postnatal ovarian resistance to circulating FSH and LH may exist as a relic of the conditions of pregnancy in which high sex steroid plasma levels (negative feedback for pituitary gonadotropins) and large amounts of circulatory HCG (blunting ovarian responsiveness to gonadotropin stimulation) have led to ovarian quiescence. In one study (Źarate *et al.,* 1972) the administration of a total dose of menopausal gonadotropin of 2250 units from day 6 to day 11 after delivery could not raise the urinary output on estrogens, pregnanediol, and pregnanetriol; also the endometrium remained inactive, indicating an ovarian refractoriness postpartum. This ovarian quiescence of pregnancy seems to extend into the postpartum period and is probably supported by the "antigonadotropic" action of prolactin secreted increasingly during the period of nursing. Increased release of pituitary ACTH has also been observed in animals and women during lactation, leading to enhanced production of adrenal steroids that by means of a negative hypothalamic feedback mechanism may contribute to a decreased pituitary gonadotropin secretion and thus to ovarian inactivity. Because in postpartum women a urinary hyperexcretion of a "gonadotropin-inhibiting factor" has been observed, it is likely that such a factor present in the systemic circulation can contribute to the amenorrhea of lactation. Also, a direct inhibitory action of prolactin on the ovaries has been suggested, explaining the ovarian quiescence of postpartum lactating women. The finding that gonadotropin secretion is decreased only during the first 2 to 3 postnatal weeks, and around the sixth week is similar to that of normal fertile women, supports the assumption of some ovarian refractoriness postpartum (Keller, 1968; Jaffe *et al.,* 1969). During the first 2 weeks postpartum only low levels of FSH (half the amount found during the follicular phase) are measurable in plasma and urine. Pituitary gonadotropin secretion beginning after 2 to 3 weeks postpartum is not surprising when one considers the complete cessation of the function of the corpus luteum of pregnancy 4 to 5 days postpartum (Spellacy and Buhi, 1969); with postnatal disappearance of placental and luteal sex steroids from the circulation, the hypothalamo-anterior pituitary gonadotropic functions are then induced. Nevertheless in lactating mothers, the hypothalamoadenohypophysial mechanisms favoring prolactin synthesis and secretion may greatly counteract and delay the resumption of pituitary FSH-LH secretion. It cannot be answered yet as to what extent other pregnancy-induced adenohypophysial structural changes, which may be partly responsible for the puerperal pituitary hypofunction, lead to amenorrhea, genital hypoplasia, and loss in libido, symptoms as observed in cases with Sheehan's or Simmond's disease. Also, it has not yet been clarified whether, as earlier suggested, unknown hormonelike substances are released from lactating mammary tissues thus inhibiting ovarian function. Furthermore, uterine changes postpartum, with myometrial and endometrial involution, may play a direct role in reduced fertility. Inadequate endometrial regeneration, due to remnants of decidua and secundines or local endometritis, may delay the process of intrauterine wound healing, which normally takes 2–3 weeks for the endometrium around the placental seat and 3–7 weeks at the placental site. Because hypothalamo-pituitary-ovarian-uterine changes are much less pronounced in a pregnancy of only 2–4 months duration (interrupted by spontaneous or therapeutic abortion), the resumption of reproductive function postabortum is greatly accelerated; here the endometrium is regenerated within 7 days and ovulation may occur as early as 7–10 days after abortion. In 50–60% of cases, ovulation returns within 2–3 weeks postabortum (Table 42) and menstruation within 4 weeks (Fig. 44) (Vorherr, 1973).

In postpartum mothers the period of lactational amenorrhea provides some sort of contraception during the first 3 postnatal months. Women who do not breast feed (earliest return of menstruation is 25 days and of ovulation is 25–35 days) have approximately a 5% chance to regain fertility before the end of the "conventional" sterile period (6 weeks postpartum), whereas lactating mothers (earliest return of menstruation is 30 days and of ovulation is 35 days) rarely conceive during that time. Nevertheless, about 20–30% of lactating women can conceive between weeks 6 and 12 postpartum. Because more than half of the nursing mothers became pregnant during a period of 9 months lactation, some of whom remained amenorrheic in between pregnancies, nursing as a contraceptive method is by far inferior to the pill or the efficacy of an intrauterine device. An estimate as to the conception rates in nursing and nonnursing mothers is presented in Table 43.

Lactation certainly exerts an inhibitory effect on reproductive function. On the average a nursing period of 20 weeks will postpone the average return of menstruation to 12 weeks and ovulation to 18 weeks (Table 40). (Zilliacus, 1967; Rhodes, 1971; Brambilla and Sirtori, 1971; Kamal *et al.*, 1969a; Vorherr, 1973).

TABLE 42

Reproductive Function Postabortum [a]

1. Postabortum (including ectopic pregnancies): In about 60% ovulation occurs within 2–3 weeks; a luteal endometrial pattern may be observed as early as 7 days after abortion

2. Average return of ovulation after abortions occurring between
   - 8–15 Weeks of gestation: within 2–3 weeks postabortum
   - 16–20 Weeks of gestation: within 4–6 weeks postabortum

3. Average return of menstruation after abortions occurring between
   - 8–15 Weeks of gestation: within 4–5 weeks postabortum
   - 16–20 Weeks of gestation: within 6–7 weeks postabortum

4. In 75–90% the first menstruation postabortum is ovulatory and continues to be ovular in subsequent cycles. In the remaining cases, menstruation is anovular followed by ovular menstruation a month later. *Note*: in 5–15% of healthy nonpostpartum women, anovulatory cycles are observed.

5. Conclusion: Reproductive function postabortum is resumed promptly and contraception, if desired, should be instituted immediately.

[a] From Vorherr (1973).

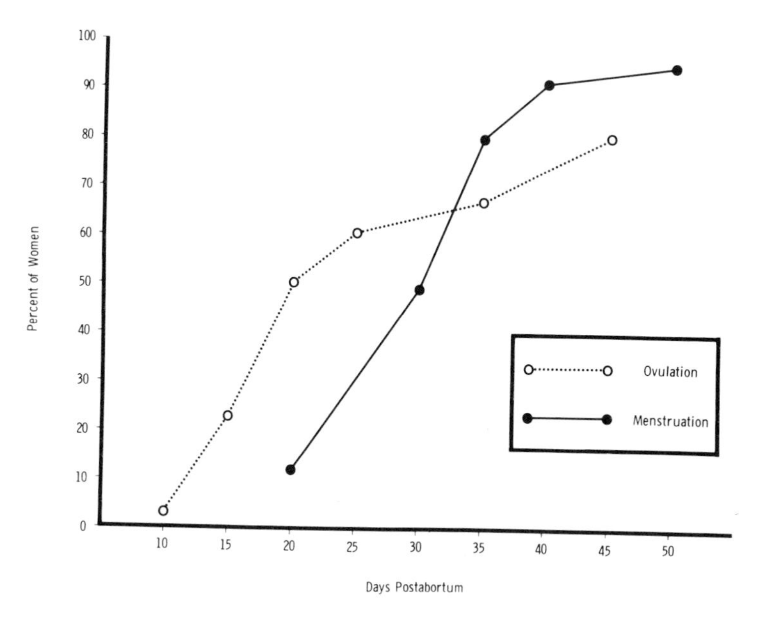

Fig. 44. Return of ovulation and menstruation postabortum. The reproductive functions return promptly after abortion in contrast to the situation postpartum, especially when the puerpera is nursing her baby (see Fig. 43). Postabortum, ovulation precedes menstruation; whereas, postpartum early menstrual bleedings are usually anovular. Ovulation occurs in over half of the abortion patients within the first 2–3 weeks, and accordingly menstruation is observed around 4–5 weeks postabortum. The reproductive functions are usually restored 5–6 weeks after abortion. *Note:* Among a given population of women, about 5–10% are temporarily or permanently amenorrheic during their fertile phase of life, and around 5–15% experience temporarily anovulatory cycles. This explains why percentage values of patients seldom exceed ninety in relation to return of ovulation and menstruation. (From Vorherr, 1973, see for more details and bibliography.)

B. CONTRACEPTION POSTPARTUM: EFFECT OF ORAL CONTRACEPTIVES ON LACTATING MOTHER AND BREAST-FED INFANT

Because many puerperas do not wish to breast feed and those who nurse their infants are inadequately protected against pregnancy, proper counseling and timely institution of contraception in these women is of great importance. Whereas after abortion contraception must be provided immediately, after delivery contraception may be started immediately or at least before the fifth week postpartum, whether the puerpera is nursing or

TABLE 43

Fertility of Postpartum Women [a]

1. Parity and fertile age are without influence on the menstrual pattern postpartum
2. Definite infertility exists during the first 4 weeks postpartum; first postpartum ovulation rarely occurs before 6 weeks after delivery

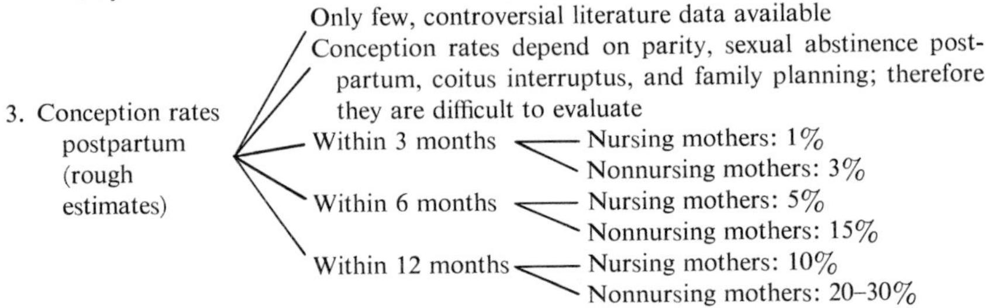

3. Conception rates postpartum (rough estimates)
   - Only few, controversial literature data available
   - Conception rates depend on parity, sexual abstinence postpartum, coitus interruptus, and family planning; therefore they are difficult to evaluate
   - Within 3 months
     - Nursing mothers: 1%
     - Nonnursing mothers: 3%
   - Within 6 months
     - Nursing mothers: 5%
     - Nonnursing mothers: 15%
   - Within 12 months
     - Nursing mothers: 10%
     - Nonnursing mothers: 20–30%
4. Postpartum, ovular cycles may be followed by anovulatory ones. The earlier menstruation returns postdelivery the higher the incidence of irregularities
5. In general regular postpartum menstrual cycles are observed
   - In nursing mothers: 70%
   - In nonnursing mothers: 90%
6. Irregular postpartum menstrual cycles become usually regular within 3 subsequent months

[a] From Vorherr (1973).

not. Administration of oral contraceptives or insertion of an intrauterine device (IUD) are the most effective and preferred methods of contraception. Oral contraceptives containing small progestin doses, not exceeding 0.5–2.5 mg of a 19-nortestosterone derivative and 50 μg of ethinylestradiol or 100 μg of mestranol, are recommended most (Loestrin, Norlestrin 1 and 2.5, Ovral, Demulen, Ortho-Novum 1 and 2.5, Norinyl 1 and 2.5). Because these agents inhibit ovulation successfully, they are as safe in preventing pregnancy as the higher dosages of progestin–estrogen combinations (Vorherr, 1973). In recent years various 19-norprogestin preparations without any estrogen have been evaluated for that purpose. The pregnancy rate among women taking "minidoses" of such an oral progestational agent seems to be similar to the rate for women using intrauterine devices and amounts to 1–4%. However, with oral progestins alone, the incidence of breakthrough bleeding and intermenstrual spotting is rather high; it is 10–30% during the first 3–6 cycles. Thereafter the rate diminishes to 3–5%. Combinations containing 1–2.5 mg of a 19-norprogestogen and 50–100 μg of ethinylestradiol or mestranol cause bleeding irregularities in only 5–10% during the first cycles, and later breakthrough bleeding or spotting is observed in only 2–3% of patients. Whereas oral progestogen–estrogen combinations usually inhibit ovulation through central diminution of gonadotropin secretion, ovulation is usually preserved with minidoses of 19-norprogestogens, but the corpus luteum function may be insufficient. The contraceptive effect of such small progestogen doses is due to alterations of the physicochemical properties of cervical mucus (thick mucus, inhibition of Spinnbarkeit and ferning with interference of transcervical sperm migration and sperm capacitation) and the endometrium (tendency for glandular hypoplasis and atrophy). Such a "minipill" containing progestin only (0.35 mg norethindrone) has recently been approved by the FDA and the product is marketed in the United States under the trade names, Micronor and Nor-QD. Most recently a "mini-minipill" containing 0.075 mg of Norgestrel (Ovrette) has also been released by the FDA. The conventional progestin–estrogen combinations presently applied are preferable to the "minipill" and to the IUD because of their higher efficacy (failure rate is less than 0.1% per hundred women years). Also, the expulsion rate of IUD's is rather high (10–30%) when inserted immediately postabortum or postpartum. About 90% of postabortum or postpartum women desire contraception; approximately 10–15% of the postpartum patients request tubal sterilization (pregnancy rate $<0.1\%$ with Irving method); about 30% of women take oral contraceptives and approximately 15% have an IUD inserted. Other means of contraception (condom, diaphragm with spermicidal jelly, vaginal foam, etc.) are less efficient (pregnancy rates 20%) and are applied only when above mentioned means are not feasible (Vorherr, 1973).

Although oral combination contraceptives are the most efficient means available in preventing pregnancy, some potential hazards threatening mother and breast-fed infant have to be taken into consideration when prescribing the pill (Hill, 1970). The potential adverse effects of oral contraceptives on milk yield of the lactating mother, uterine involution, and growth and development of the breast-fed infant, have been subject to extensive discussions. Moreover, the interrelationship of medication with oral estrogen or progestin–estrogen combinations and the occurrence of maternal thromboembolism, hypertension, breast and genital cancer has been debated thoroughly in recent years and is discussed in the light of literature data and newer developments by Vorherr (1973).

Data regarding the effects of oral contraceptives on mother (milk yield, genital involution) and infant's growth and development are presented in Table 44. That table shows that many data are poorly evaluated in the literature, and the results of various authors contradict each other. In general, it appears that preparations containing not more than 2.5 mg of a 19-norprogestogen and 50 μg of ethinylestradiol or 100 μg of mestranol (ethinylestradiol is twice as potent an estrogen as mestranol) present no

TABLE 44 **Effect of Contraceptive Steroids on Postpartum Lactation**

References	Drugs used and postpartum duration of administration	Milk yield	Comments
Curtis *et al.* (1963)	Enovid 2.5, 5 Norlutin 10 (for 3 months)	Not altered Not altered	No placebo controls; in one male infant whose mother received Enovid 5, gynecomastia developed
Rice-Wray (1963)	Anovlar 4 Ortho-Novum 10 (duration not reported)	Decreases (45% with Anovlar and 58% with Ortho-Novum)	No simple or placebo controls; no data given regarding time of treatment postpartum; methods of data evaluation not reported
Rice-Wray *et al.* (1963)	Ortho-Novum 2, 5, 10 (for over one year)	Decreased	No placebo controls; only reported that lactation was diminished or stopped in 1 out of 3 cases with 10 mg of norethindrone
Ferin *et al.* (1964)	Ethyl-estrenol 5 (for 4 weeks)	Not altered	Placebo controls
Pincus (1965)	Enovid 2.5 Enovid 5, 10, 20 (duration not reported)	Not altered Decreased by 40–80%	Textbook information; no details regarding data evaluation
Semm (1966)	Lyndiol 2.5 (for 1 to 3 weeks)	Not altered or slight increase	Placebo controls; period of treatment too short
Kaern (1967)	Norinyl 1 + 50 (for 6 days)	Decreased (complementary feeding was necessary in 12% of infants, as compared to 3.5% in the placebo group)	Placebo controls: period of treatment too short; "measurable diminution in quantity of milk" (no amounts reported)
Frank *et al.* (1969)	Ovulen (for 18 months)	"Nursing function occasionally impaired"	No placebo controls
Kamal *et al.* (1969[b])	Lyndiol 1, 2.5 Lynestrenol 0.5 Deladroxate 150/10 (for 24 weeks)	Not altered Not altered Slightly decreased	No change in infants' growth curves in either group

Kates *et al.* (1969)	C-Quens (during lactation)	No interference with lactation	No placebo controls; in some patients "increase in requirement for supplementary feeding;" no change in infants' growth and development
Kora (1969)	Ovulen (for 20 weeks)	Decreased by 30% (infants' weight gain 26% less)	No placebo controls
Gambrell (1970)	Ortho-Novum 1, 2	Not altered	No placebo controls; "patient's acceptance was excellent"
	Ovral, Ovulen	Not altered	
	Enovid 2.5, Oracon	Not altered	
	Gestest (for 6 weeks)	Not altered	
Miller and Hughes (1970)	ORF (1 mg norethindrone and 80 μg mestranol) (for 6–10 weeks)	Decreased (infants' weight gain 23% less)	Placebo controls; authors contradict themselves when comparing their results with those of Semm (1966)
Borglin and Sandholm (1971)	Lyndiol 2.5	Decreased by 70%	Placebo controls; period of treatment too short
	Ethinyl estradiol (50 μg)	Not altered	
	Mestranol (80μg) (for 7 days)	Not altered	
Karim *et al.* (1971)	Medroxyprogesterone acetate (150 mg)	Not altered	No placebo controls
	Norethisterone enanthate (200 mg) (for up to 18 months)	Not altered	No placebo controls
Koetsawang *et al.* (1972)	Ovulen (for 10 weeks)	Decreased by 32–55% (infants' weight gain 27% less)	No placebo controls; despite 55% decrease in milk volume after 10 weeks Ovulen treatment, additional feeding was needed in only 5% of cases
	C-Quens (for 10 weeks)	Decrease by 32% (infants' weight gain 10% less)	
Ramadan *et al* (1972)	Ovosiston (for 4 months)	Not altered	No placebo controls

hazard to mother and infant. The inhibition of milk yield with larger doses of oral combination contraceptives containing estrogen and 5–20 mg of a progestogen per tablet is due to the latter, because the estrogen dose is usually kept constant or lowered in such preparations. Oral contraceptives in small doses do not alter the milk yield, whether they may stimulate it (Karim *et al.*, 1971) remains to be confirmed. The most effective inhibiting factor for milk production is cessation of the suckling stimulus. This is demonstrated best by puerperas who had received a lactation-suppressing dose of Deladumone OB but later changed their minds; they could still nurse their babies successfully. In postpartum nonlactating women receiving an oral estrogen–progestin combination, no breast engorgement or pains were observed while the lochial flow decreased (Frank *et al.*, 1969).

When administration of Ovulen was begun 3 days after delivery (Frank *et al.*, 1969) (Table 44), all puerperas experienced cyclic withdrawal bleeding after the second treatment cycle. Breakthrough bleeding was 14% in the first treatment cycle declining to 3% in the second and subsequent cycles. During pill intake the endometrium displayed an atrophic appearance, but after discontinuance of the pill a prompt return to normal proliferative and secretory endometrium occurred (Frank *et al.*, 1969) in most women.

Despite the many articles written about the effect of oral contraceptives on milk yield and the infant's growth, there exists no definite report on negative pill effects on the mother. Also in the breast-fed infants of mothers using the pill, no adverse effects in regard to bone growth, genital development, or impaired fertility later in adulthood can be substantiated. Because in two different studies (Laumas *et al.*, 1967; van der Molen *et al.*, 1969) it was shown that about 0.05–1% of radioactive 19-norprogestogen taken by the mother was excreted into the breast milk, contraceptive steroids have been related to a possible acceleration of the infant's bone maturation (Laumas *et al.*, 1967). In one study (Breibart *et al.*, 1963) very high progestin doses (30–50 mg/day for weeks and months) were administered to the gravida because of threatened abortion. Here progestin was passed on to the fetus via the placenta and definitely affected the bone maturation, but this observation was misquoted in another article (Laumas *et al.*, 1967) stating that the high progestogen doses had been given postpartum and transmitted by breast milk to the breast-fed infant and thus caused accelerated bone maturation. Whether indeed sex steroids as contained in present pill preparations are transferred via breast milk into the system of the breast-fed infant in amounts sufficient to cause accelerated bone maturation or any other adverse effects appears to be rather doubtful. Since in newborn female rats the injection of a rather high testosterone dose (1 mg per a 6 g newborn rat) results in permanent sterility later in adulthood, it may be argued that 19-norprogestogens transferred to the female breast-fed infant may potentially cause a similar hypothalamic disturbance possibly resulting in sterility. In the first place, only small amounts of progestin reach the infant's circulation, and its androgenic potency is only a fraction of that of testosterone. Secondly, during the initial trials in Puerto Rico, in the midfifties, rather large amounts of progestogens (up to 20 mg per pill) were applied. So far, however, in the breast-fed infants of such mothers, neither accelerated bone maturation during infancy nor impairment of ovarian function and fertility in their reproductive age have been reported. Thirdly, it would not be correct to relate animal data obtained with mammoth doses of testosterone to the weak and potentially negligible androgenic effect of present pill-progestogen doses given to women for fertility control. In our animal experiments under rather physiologic conditions, lactating mother rats were given oral estrogen–progestin doses comparable on a weight basis to those given in the pill. Neither these "physiologic" estrogen–progestin doses nor those of up to fivefold of the corresponding human oral contraceptive doses lead to impairment of the suckling pups' weight gain, maturation, and reproductive function. Usually no change of milk constituents after intake of contraceptive steroids is observed (Vorherr, 1973).

CHAPTER VII

Suppression of Lactation

Obstetricians and pediatricians generally recommend breast feeding as ideal for mother and infant, but most puerperas do not wish to nurse their babies. Breast feeding is definitely contraindicated in women suffering from mastitis, tuberculosis, or any other serious disease. Also, puerperas taking certain potentially hazardous drugs (Table 24) that pass into the breast milk and may harm the newborn should not nurse. Most recently, in samples of human breast milk viruslike particles measuring approximately 50–100 mμ have been observed (Feller and Chopra, 1971; Dmochowski, 1972). These particles, which can be found by electron microscopic survey in about 10–20% of lactating women, resemble B or C-type RNA oncogenic tumor viruses, known to cause leukemia and breast cancer in mice. In human milk, such viruslike particles may appear in 3 different types: (1) herpeslike particles; (2) small viruslike particles; and (3) B-type viruslike particles. Such particles were seen in milk samples and in breast cancer biopsy specimens. They possess RNA-dependent DNA polymerase activity, and milk lacking these viruslike particles is also lacking DNA polymerase activity. In milk samples from 43 normal women and 4 women who had active or treated breast cancer, viruslike particles were observed in 9 (20.9%) and in 2 (50%), respectively (Feller and Chopra, 1971). In another study, viruslike particles were detected in milk of 5% of women without a history of breast disease and in 60% of patients with a family history of breast cancer (Moore *et al.*, 1971). As already indicated in Chapter V, Section B, women with a personal or family history of breast cancer should not nurse their baby.

Also, patients with a spontaneous or therapeutic abortion past 16 weeks of pregnancy, as well as those with a stillbirth, may require suppression of lactation, which can be accomplished by both clinical and medicamentous means.

A. CLINICAL SUPPRESSION OF LACTATION

The clinical measures for inhibition of lactation are restriction of fluid intake, binding the breasts or wearing a well-fitting brassiere, and avoidance of breast manipulation; in cases with painful breast engorgement the application of an ice pack or the intake of analgesics (aspirin and codeine, propoxyphene) may become necessary. Under this regimen and without any further medication, lactation usually ceases within one week, and some physicians believe that these clinical measures are as effective as the medicamentous inhibition of lactation with sex steroids. Nevertheless, the majority of these puerperas do have breast pains, engorgement, and discomfort caused by milk leakage. Accordingly, the support or replacement of these clinical measures with an effective medicamentous procedure is very much desired (Vorherr, 1972e).

B. MEDICAMENTOUS SUPPRESSION OF LACTATION

Prevention of lactation with estrogens had already been accomplished in the late thirties; in the late fifties, and since then, the i.m. administration of an androgen–estrogen combination has become the method of choice (Table 45). Also, orally active progestins or androgens alone and in combination with estrogens have been used for inhibition of lactation. When a long-acting preparation containing 360 mg of testosterone enanthate and 16 mg of estradiol valerate in sesame oil (Deladumone OB) is given as a single dose i.m. during the last stages of labor or immediately after delivery, the onset of lactation is prevented in most patients. This suppressive sex steroid effect on lactation has been thought to be due to inhibition of adenohypophysial release of prolactin by stimulating the hypothalamic prolactin-inhibiting factor to continued function. However, because Deladumone failed to alter HPr serum levels, a peripheral action of sex hormones antagonizing the prolactin effect on the mammary secretory epithelium has to be considered (del Pozo *et al.,* 1972; Meites, 1973; Tyson *et al.,* 1973) (for more details regarding mechanisms involved with the onset of lactation see Fig. 25). In postpartum women, high estrogen doses

TABLE 45

Suppression of Postpartum Lactation [a]

References	Agent(s)	Dosage and administration	Efficacy (degree and criteria)	Adverse effects
Lo Presto and Caypinar (1959)	Deladumone (long acting combination containing 90 mg testosterone and 4 mg estradiol/ml)	2 ml i.m. after first stage of labor	86% Successful (no breast pain or breast fullness)	None (no uterine subinvolution or change in character of lochia; restoration of normal ovarian function)
	Deladumone	2 ml i.m. within 2 hr after delivery	43% Successful	
	Deladumone	3 ml i.m. after first stage of labor	91.5% Successful	
	Deladumone	3 ml i.m. during third stage of labor	74% Successful	
Watrous *et al.* (1959)[b]	Deladumone	2 ml i.m. within 40 min after delivery	50% Successful (no or mild breast engorgement, no pain or lactation); 30% satisfactory (moderate engorgement, mild breast pain which did not require treatment)	No undesirable androgenic effects or edema formation; slight local pain reaction after drug injection in 3.8% of patients; delayed postpartum hemorrhage in one patient
	Deladumone	3 ml i.m. within 40 min after delivery	86% Successful; 14% satisfactory	
	Deladumone	4 ml i.m. within 40 min after delivery	89% Successful; 11% satisfactory	
	Tylosterone (0.25 mg diethylstilbestrol and 5 mg methyltestosterone)	3 Tablets PO daily for 5 consecutive days	70% Successful; 22% satisfactory	Breast engorgement in 5% of patients; rebound lactation in about 4%; tibial edema in some patients

Bare *et al.* (1963)	Depo-Testadiol (long-acting combination containing 50 mg testosterone and 2 mg estradiol/ml)	2 ml i.m. near end of first stage of labor	49% Successful (no breast pain, engorgement or lactation; no necessity for additional clinical measures)	None (no masculinization, withdrawal bleeding, delay in onset of menstruation, or rebound lactation)
	Deladumone OB (long-acting combination containing 180 mg testosterone and 8 mg estradiol/ml)	2 ml i.m. near end of first stage of labor	79% Successful	
	Placebo	2 ml i.m. near end of first stage of labor	46% Successful	
Kantor *et al.* (1963)	Lactostat (long-acting combination containing 150 mg testosterone and 10.5 mg estradiol/ml)	2 ml i.m. in late first or during second or third stage of labor	64% Excellent (no reaction in breast); 19% satisfactory (transient mild breast engorgement or pain); 17% failure (breast engorgement and pain)	None (no local or general reactions after drug injection; no delay in onset of menstruation)
	Placebo	2 ml i.m. in late first or during second or third stage of labor	13% Satisfactory; 87% failure	
MacDonald and O'Driscoll (1965)	Diethylstilbestrol	10 mg PO daily on postpartum days 1–5	95% Successful (no breast discomfort)	Not considered
	Placebo	1 Tablet PO b.i.d. on postpartum days 1–5	90% Successful	
Cole and Pitts (1966)	Mixogen (short-acting and longer-acting combination containing 100 mg testosterone and 5 mg estradiol/ml)	2 ml i.m. (given immediately after delivery in some patients; in most patients, lactation had been established for 3–8 days before drug was given)	100% Successful (no breast pain, engorgement, or rebound lactation; milk leakage stopped within 3–8 days after injection)	None (no endometrial withdrawal bleeding, change in quantity and quality of lochia, or androgenic effects)
Gillibrand and Huntingford (1968)	Quinestrol	4 mg PO immediately after delivery	32% Successful (no lactation, breast engorgement, or pain)	Not reported, investigators suggested that administration of oral estrogen-

TABLE 45 *(continued)*

References	Agent(s)	Dosage and administration	Efficacy (degree and criteria)	Adverse effects
Gillibrand and Huntingford (1968) —*continued*	Enovid 10 mg (10 mg norethynodrel and 0.15 mg mestranol per dose)	30 mg PO on postpartum days 1–4; 20 mg on days 5–7; 10 mg on days 8–10	83% Successful	progestin combination in which estrogen dose is low might lower the risk of thromboembolism
Turnbull (1968)	Diethylstilbestrol	20 mg PO daily for 9 consecutive days	84% Successful (no breast pain or engorgement)	"Rebound" lactation in 75% of patients required further treatment; prolonged red lochial discharge from 32 to 50 days postpartum; thromboembolism in 0.59%
	Usual clinical ("natural") means (breast binder, fluid restriction, ice pack, analgesics)	—	68% Successful	None (no rebound lactation; red lochial discharge cleared by 28th day postpartum; thromboembolism in 0.23%)
McGlone (1969)	Quinestrol	4 mg PO within 6 hr after delivery	77–86% Successful (soft breasts; no breast discomfort, uterine subinvolution, or withdrawal bleeding); 14–18% partially successful (mild breast engorgement and discomfort)	None (no uterine subinvolution or endometrial withdrawal bleeding)
	Placebo	1 Tablet PO within 6 hr after delivery	18–27% Successful; 32–50% partially successful	
Watson (1969)	Diethylstilbestrol	10 mg PO t.i.d. on postpartum days 1 and 2; 5 mg t.i.d. on days 3–5	79% Successful (no lactation); 21% moderate (slight lactation which did not require treatment)	Altered menstrual cycle in 10–15%
	Quinestrol	4 mg PO immediately after delivery	42% Successful; 52% moderate	

Morris *et al.* (1970)	Deladumone OB	2 ml i.m. immediately after placental expulsion	78% Successful (no breast discomfort, engorgement or lactation); 18% partially successful (mild to moderate breast discomfort, engorgement and lactation)	None (no change in lochia, or involution of breasts and uterus; no delay in onset of menstruation; no androgenicity)
	Placebo	2 ml i.m. immediately after placental explusion	20% Successful; 50% partially successful	None
	Tace (chlorotrianisene)	144 mg PO within 6 hr after delivery and 72 mg b.i.d. for one day	63% Successful; 33% partially successful	None (no change in amount or color of lochia)
	Placebo	2 Capsules PO within 6 hr after delivery and 1 capsule b.i.d. for one day	42% Successful; 46% partially successful	None
Cruttenden (1971)	Quinestrol	4 mg PO immediately after delivery	76% Successful (breast consistency soft, no discomfort); 18% satisfactory (mild breast engorgement, no discomfort)	Not reported
	Placebo	1 Tablet immediately after delivery	43% Successful; 23% satisfactory	Not reported
Mann (1971)	Quinestrol	4 mg PO immediately after delivery	93% Successful (no lactation or only minimal milk leakage, no breast engorgement)	Rebound lactation in 7%
	Diethylstilbestrol	15 mg PO daily for 7 consecutive days	92% Successful	Rebound lactation in 40%

[a] From Vorherr (1972e).
[b] Untreated controls: 7% successful, 24% satisfactory.

lead to peripheral inhibition of secretory prolactin action, but the estrogens fail to inhibit HPr secretion before and during the infant's suckling (Tyson *et al.,* 1973). The same inhibitory mechanisms apply when female or male sex steroids are given orally; here usually the treatment is initiated immediately after delivery and continued for 1 week with decreasing drug dosages (see Table 45). These orally active sex steroid preparations are ethinyl estradiol, diethylstilbestrol, synthetic progestins, and methyltestosterone. In recent years a synthetic long-acting estrogen, quinestrol, given in a single oral dose of 4 mg has also been used for suppression of lactation; this agent is well reabsorbed in the gut, stored in the body fat, and then slowly released over the following 6 to 14 weeks.

For any given drug or combination of drugs utilized for suppression of lactation, the clinical effectiveness has to be established and weighed against potentially dangerous side effects. Accordingly, estrogens and progestins, administered early postpartum to inhibit lactation, may interfere with proper uterine involution and lead to endometrial hyperstimulation with dysfunctional uterine bleeding. Furthermore, in animals high doses of female sex steroids have been found to cause endometritis, myometritis (Pyometra), salpingitis, and fatal peritonitis. Also natural estrogens or, perhaps more so, synthetic nonsteroid estrogenic substances (diethylstilbestrol) may induce neoplastic disease or accelerate the development of a preexisting latent malignancy (for more details see below).

1. Agents Used for Suppression of Lactation and Dose Regimens Applied—Drugs of Choice

In Table 45 various agents and dose regimens applied for suppression of postpartum lactation are compiled, and the respective therapeutic effectiveness of these sex steroids is outlined. The table shows that large discrepancies exist among different investigators evaluating the same drug or agents with a similar action. Thus, the high "success rates" achieved with placebos and the striking differences in studies from various groups of investigators using the same drug and similar dosages are rather confusing. Also the criteria for evaluation of the inhibitory drug effect on lactation vary from one investigator to another. Some investigators measure the inhibitory drug response by the degree that treatment relieves breast pains, while others use the prevention of breast engorgement and of milk leakage as main subjective and objective criteria for successful suppression of lactation.

Apart from these rather controversial and partially inconclusive reports concerning the effectiveness of sex steroid administration for inhibition of lactation, it seems that at present the intramuscular injection of a long-acting testosterone–estrogen combination (Deladumone OB) toward the end of labor or immediately postpartum is the most preferred method; it represents a simple way to achieve adequate suppression of lactation in about 80% of cases. Such a drug combination exerts a sustained effect over 2 to 4 weeks, and the relatively low amounts of estrogen released rarely interfere with myometrial involution or resumption of postpartum menstruation. Thus, it has become a routine method for many physicians to administer this androgen–estrogen combination during the second or third stage of labor or immediately thereafter, and the clinical measures for inhibition of lactation are considered to be an adjunct to medicamentous suppression of lactation. According to some investigators (Lo Presto and Caypinar, 1959) (see Table 45), the best results are achieved when Deladumone is given after the first stage of labor, and the success rates are lower when the drug is administered during the third stage of labor and thereafter. But, these findings are not confirmed by other investigators (Watrous *et al.*, 1959; Cole and Pitts, 1966) (see Table 45) who report comparable and even higher success rates when the long-acting estrogen–androgen combination is given immediately postpartum. Nevertheless, the efficiency and simplicity of this procedure is established, and the oral intake of sex steroids for suppression of lactation has been largely abandoned lately. In one study (Cole and Pitts, 1966) most of the patients were treated with Deladumone when lactation had already been established for 3 to 8 days postpartum, and even then the hormonal suppressive effect on lactation was impressive. These results, however, have not yet been confirmed; it seems rather that once prolactin-induced lactation is established, it becomes more difficult to suppress it. Because no disadvantages regarding progress of labor or adverse effects on the fetus are noticed when the injection is given early in labor, it may indeed be a good idea to inject the hormone combination at this stage. The stimulus of suckling or any breast manipulation may greatly antagonize the inhibitory drug effect. This is demonstrated best in postpartum women to whom a suppressive dose of Deladumone had been given immediately postpartum and who thereafter changed their mind and wanted to nurse their baby; they were still able to breast-feed successfully. This indicates that the stimulus of suckling can always override the inhibitory effect exerted by sex steroids. It also reveals that the stimulus of suckling with removal of milk from the breast is a major factor in the maintenance of lactation. The potentially virilizing effect of testosterone contained in such drugs is negligible and is partially antagonized by the admixed estrogen. At the same time the very minor danger of estrogen-induced endometrial proliferation and hyperplasia as well as of myometrial subinvolution per se is further decreased by the admixed testosterone, which probably antagonizes to some extent the stimulative effect of estrogen on myometrium and endometrium. Nevertheless, the possibility exists that after medicamentous suppression of lactation with an androgen–estrogen compound fertility may be impaired in the early months postpartum. However, in most women the ovarian function, the involution of the uterus, and the resumption of normal menstruation is not affected by Deladumone treatment. It also is important that the androgen dose chosen is high enough, as shown by Lo Presto and Caypinar (Table 45), at least 180 mg of a depot testosterone appear to be necessary for suppression of lactation (Bare *et al.,* 1963; Watson, 1969; MacDonald and O'Driscoll, 1965; McGlone, 1969; Cole and Pitts, 1966; Kantor *et al.,* 1963; Gillibrand and Huntingford, 1968; Mann, 1971; Cruttenden, 1971; Watrous *et al.,* 1959; Lo Presto and Caypinar, 1959).

2. Potentially Harmful Drug Effects Encountered during Medicamentous Suppression of Lactation

Some precautions are necessary in medicamentous suppression of lactation. Thus, any estrogen or estrogen-containing drug combination is contraindicated in puerperas with (1) a family history of breast or genital cancer (there seems to be a connection between estrogens and development of endometrial cancer, Vorherr, 1973) or a history of suspected or established mammary or genital malignancy; (2) thromboembolic disorders, cerebral apoplexy, or a past history of these conditions; (3) impaired liver function; (4) undiagnosed vaginal bleeding; (5) Hodgkin's disease and malignant melanoma; (6) elective surgery; and (7) hypertension.

In addition, an alarming connection has been established between diethylstilbestrol treatment of pregnant women and the occurrence of vaginal adenocarcinomas in their daughters 15–25 years later. In the fifties pregnant women with threatened abortion were treated with high doses (5–135 mg/day) of nonsteroid synthetic estrogen (diethylstilbestrol predominantly) for a prolonged period of time (weeks, months). Now, 15–25 years later, some of their daughters (over 120 cases are known so far) developed vaginal adenocarcinoma (almost half of them have died from the disease); it appears that the time of latency until clinical diagnosis of vaginal cancer is inversely related to the diethylstilbestrol dosage applied during the respective pregnancy (Greenwald *et al.,* 1971; Herbst *et al.*, 1971; Folkman, 1971). The carcinogenic or co-carcinogenic agent, diethylstilbestrol, was passed on to these unfortunate descendants during fetal life, and this is the first known incident where a carcinogenic agent was transmitted from mother to fetus. No report exists that those diethylstilbestrol treated mothers or their male offspring developed mammary or genital cancer or any other malignancy in a repetitive pattern. It seems retrospectively mandatory to evaluate and follow mothers who received diethylstilbestrol during pregnancy as well as all female and male offspring and their descendants. It seems that fetal tissues are more susceptible than adult tissues to malignant induction by estrogenic substances. Furthermore, synthetic nonsteroid estrogens may act primarily as initiators of cellular malignancy, whereas the ovarian sex steroids secreted at puberty in these girls most likely act as promoters of carcinomatous growth through the burst of mitotic activity they exert.

Because the liver is the main metabolizing organ for sex steroids, the administration of sex steroids is contraindicated for all patients with liver dysfunction or disease. Precautions have to be taken regarding sex-steroid-induced water and electrolyte retention, which may become hazardous to patients with cardiac or renal disease or to those with epilepsy and migraine. In some sensitive patients receiving an androgen–estrogen compound, virilization, as manifested by hirsutism, deepening of the voice, and clitoris hypertrophy, has been observed. If these masculinizing effects occur, they regress spontaneously with the exception of the deepened voice. Additional estrogen treatment will accelerate the disappearance of the symptoms of androgenicity. In some puerperas undue stimulation of endometrium and breast tissues has been reported after Deladumone administration. These undesired effects will subside spontaneously as the estrogen effect of the medication wears off; they also may be alleviated faster by an additional injection of testosterone. After the administration of the oily Deladumone solution, pains at the site of the injection and the formation of sterile abscesses have been observed in very rare instances. In general, it can be stated that the injection of an androgen–estrogen depot is without major side effects, and, at present, it is the most simple and efficient medicamentous procedure for suppression of lactation.

There is no doubt, however, that estrogens or synthetic nonsteroid estrogenic compounds given in relatively high doses for suppression of lactation may interfere with uterine involution and may cause endometrial dysfunctional bleeding even though it is seldom reported (Table 45). It is also known that estrogens are a contributing factor for thromboembolism, the incidence of which is already high during the puerperium (Table 46) (Tindall, 1968). Thus, the use of relatively high estrogen doses will in-

TABLE 46

Factors Responsible for Increased Puerperal Thromboembolism [a]

Genuine factors [b]	Predisposing factors
1. Increased levels of blood-clotting factors I, II, V, VII, VIII, IX, X	1. Administration of high estrogen doses for suppression of lactation
2. Formation of thromboplastin substances in placenta, decidua, and amniotic membranes	2. Increasing age and parity; obesity
3. Decreased erythrocyte-plasma suspensibility	3. Anesthesia and surgery; physical inactivity postsurgery (venous stasis of lower extremities)
4. Increased platelet number (postpartum thrombocytosis) and enhanced platelet adhesiveness	4. Mechanical damage to venous endothelium (difficult forceps deliveries)
5. Increased levels of fibrinolysis inhibitors (antiactivators, antiplasmin)	5. Previous history of thromboembolism
	6. Patients with heart disease
	7. Patients with blood groups A, B, AB
	8. Patients with pancreatitis, pancreatic carcinoma, and ulcerative colitis
	9. Hypothermia

[a] The data presented in this table are derived from McDevitt and Smith (1969); Hume *et al.* (1970); and Turnbull *et al.* (1971).
[b] The state of blood hypercoagulability during puerperium is due to these factors.

crease the risk of thromboembolic disease and mortality. Out of the reported 44 patients with postpartum thromboembolism, the disease had led to pulmonary emboli in 8 women and caused death of one mother; all these women had been treated with stilbestrol for suppression of lactation (Turnbull, 1968). In this study it was also observed that with increasing maternal age (women over 25 years) the incidence of thromboembolism was 1/1000 in women who breast-fed compared to 9.3/1000 in those treated with stilbestrol for suppression of lactation. Furthermore, it was shown that stilbestrol caused an 87% rise in plasma levels of factor IX, whereas in puerperas in whom lactation was "naturally" suppressed only a 43% increase in factor IX plasma levels was observed. This state of factor IX-induced hypercoagulability is probably due to increased formation of factor IX in the liver as a consequence of stilbestrol treatment (Kantor *et al.,* 1963). Also, estrogens may lead to enhanced platelet function and acceleration of blood coagulation, negative aspects that are not observed with progestins. Estrogens may produce a certain state of blood hypercoagulability, but there seem to be other factors (decreased circulation, endothelial damage, inflammation, decreased antifibrinolytic activity) that are more important in the development of thromboembolic disease than estrogens. In another study, however, oral estrogen was found to increase spontaneous fibrinolytic activity of blood toward hypocoagulability, an effect also ascribed to progestins (Vorherr, 1973). Furthermore, in many patients (up to 70%) the withdrawal of an oral 1 week estrogen treatment results in engorgement of the breast with subsequent lactation as observed normally postpartum; this is called "rebound lactation." Such a "rebound lactation" can be easily demonstrated in puerperal rats. Here, lactation can be inhibited by repeated injections of relatively high doses of female sex steroids to such an extent that the suckling litter does not receive enough milk and dies within 1–2 weeks. When the treatment is discontinued and the mother rat is given pups from another lactating rat, she can nurse the adopted litter, indicating a typical "rebound lactation" (H. Vorherr, unpublished observations, 1971).

3. Interference of Other Drugs with Medicamentous Suppression of Lactation

Patients selected for suppression of lactation should never take tranquilizer drugs (phenothiazines, meprobamate, reserpine), which may bring about and sustain lactation by increased adenohypophysial prolactin release caused by the hypothalamic prolactin-inhibiting factor being freed from its restraining function. Theophylline has also been reported to stimulate prolactin release in laboratory animals. Thus, coffee and tea should be temporarily excluded from the diet of women undergoing suppression of lactation to eliminate the possibility of xanthine-induced prolactin secretion. Because fluid intake greatly lowers pituitary prolactin secretion and serum prolactin levels (Buckman *et al.,* 1973), in contrast to former beliefs, such women should be encouraged to drink plenty of water.

4. Stimulation of Ovarian Function to Counteract Lactation

In previous investigations it was found that an antiestrogenic compound, clomiphene, can act as a stimulant for secretion of a hypothalamic gonadotropin-releasing factor to bring about secretion of adenohypophysial FSH and LH. Whenever the hypothalamoadenohypophysial axis is operative for gonadotropin secretion and ovarian sex steroid production, the prolactin effect on mammary epithelium is counteracted by estrogen and progesterone. Under this condition the anterior pituitary prolactin secretion is not increased above the normal levels observed in nonpuerperal women. Thus the evaluation of clomiphene for suppression of postpartum lactation (via stimulation of pituitary gonadotropin secretion) was theoretically well founded, but the clinical trial failed to provide encouraging results (Miyamoto *et al.,* 1963). Also the rather frequent clomiphene side effects (flushing, nausea and vomiting, ovarian cysts, increased fertility, and multiple pregnancies) are a major impediment to its use in suppression of lactation.

5. Potential Future Drugs for Suppression of Lactation

Because about 20% of patients fail to respond to Deladumone or experience undesired side effects, the ideal suppressant is yet to be found. It remains to be investigated in the future whether other agents inhibiting prolactin release such as ergocornine (in rats) or L-dopa and bromergocryptine (in humans) may be clinically useful and superior to currently applied measures for the suppression of lactation (Malarkey *et al.,* 1971; Lu *et al.,* 1971). In rats ergocornine depressed pituitary prolactin secretion and prolactin serum levels by increasing hypothalamic PIF function and/or by exerting a direct inhibitory effect on pituitary lactotrophs (Wuttke *et al.,* 1971). Most recently, suppression of postpartum lactation was attempted with bromergocryptine (5 mg b.i.d. orally for 6 days, followed by 5 mg/day for 3 days) and was successful in about 95% of patients judged by absence of mammary engorgement; a rebound lactation occurred in 25% of cases. No blood pressure changes and no thromboembolism were observed (Varga *et al.,* 1972a,b). In another study using 2.5 mg of bromergocryptine b.i.d. for 2 weeks followed by 2.5 mg once per day for one week, lactation was successfully suppressed in 100% of patients treated, i.e., neither milk flow, breast congestion, nor breast pain were observed (Rolland and Schellekens, 1973). In their study placebo treatment was 100% ineffective, whereas usually success rates of 30–80% have been reported with placebo administration (Table 45). Bromergocryptine was found effective in depressing elevated serum prolactin levels, thus inhibiting puerperal lactation. In patients with galactorrhea and prolactin serum levels of 60–80 ng/ml, a single oral dose of 2.5 or 4 mg of bromergocryptine lowered serum prolactin to normal (10 ng/ml) or below within 2 hours after administration; the drug effect persisted for more than 12 hours (del Pozo *et al.,* 1972). Thus it appears that bromergocryptine is suited for suppression of postpartum lactation by direct inhibition of adenohypophysial prolactin secretion. It should be noted that other ergot alkaloids (methylergonovine, for instance) can also suppress lactation, and these oxytocics are therefore contraindicated in nursing puerperas. Whether bromergocryptine is indeed superior to Deladumone for suppression of postpartum lactation remains to be proved. Whereas in the first reports no side effects due to bromergocryptine were reported, it appears that nausea and vomiting occur rather frequently in patients taking the drug (Lutterbeck *et al.,* 1971). In one study, "trouble-free" suppression of lactation was achieved in 95% of puerperas by oral administration of pyridoxine, 200 mg t.i.d. for 6 consecutive days (Foukas, 1973). It is thought that pyridoxine is taken up by hypothalamic neurones and serves locally in the form of pyridoxal phosphate as a coenzyme of dopa decarboxylase, which converts dopa into dopamine. Through dopamine PIF secretion becomes enhanced, and thereby prolactin release is depressed. These results need to be confirmed by future studies. Also medication with L-dopa may inhibit prolactin secretion and thus inhibit puerperal lactation (for more details see Chapter VIII, Section C).

At present the intramuscular injection of a long-acting estrogen–androgen combination toward the end of labor or immediately after delivery appears to be the most efficient and least hazardous procedure for medicamentous suppression of lactation. The amount of natural estrogen contained in such a depot injection is small, and the hormone is absorbed within 2–3 weeks from the injection site. It appears, therefore, very unlikely that estrogen-induced thromboembolic phenomena or other serious complications are encountered with such a drug regimen.

C. BREAST INVOLUTION POSTLACTATION

Lactation is usually terminated around the fourth to the sixth postnatal month when secretory mammary activities decline, and thus the caloric and mainly the protein supply for the growing infant become inadequate. In order to avoid unnecessary breast engorgement and pains, lactation should not be stopped abruptly but rather should be continued on a reduced scale for a week or two or more. Thereafter, milk should be removed only by pumping off when breast engorgement and pains cannot be satisfactorily relieved by analgesic drugs and breast binding.

After cessation of lactation the total mammary involutional process encompasses a period of 3 months; the degree of breast involution is time related and depends on the individual's constitution. Furthermore, the functional and morphological postlactational changes depend on whether lactation was abruptly discontinued or whether it was gradually diminished over a longer period of time. Thus mammary involution is more intense and rapid in women who stop nursing abruptly, whereas mothers who extend the weaning over a longer period of time, nursing once a day or every second day, will experience a slower mammary involution. Postlactational involution encompasses all breast tissues, and, after the regression of pregnancy and lactation-induced changes, the breasts are less dense, less nodular-glandular, and less protuberant than in the prepregnant state. Breast feeding does not seem to influence to a significant extent the degree of postpartum mammary involution. However, postlactational mammary involution differs greatly from postmenopausal breast involution; whereas in the former only a minor reduction of mammary alveoli is observed, in the latter there is predominant regression of the parenchymal lobular-alveolar structures. Thus the postpartum breast involution is more confined to a certain "inactivity atrophy" of glandular-alveolar cells without causing substantial epithelial losses (Dabelow, 1957). This inactivity of alveolar epithelial cells is demonstrated by a decrease of nuclear DNA and RNA content as the involutional process advances toward the prepregnant state. Because mammary connective tissue and fat tend to remain increased postpartum, the breasts usually remain slightly larger after termination of lactation, especially in cases of frequent pregnancies.

The mammary glandular involution begins with cessation of nursing, when milk is accumulating in alveoli and smaller milk ducts, causing distention and mechanical atrophy of the epithelial structures; also rupture of dilated alveolar walls with formation of larger hollow spaces may occur. Due to alveolar distention, pressures on mammary capillaries are exerted resulting in their partial or complete occlusion. The alveolar epithelium becomes compressed by the overfilled alveoli, and flat epithelial cells are situated near the membrana propria. Because alveolar oxygen and nutrient supply is reduced due to this condition, mammary milk secretion is greatly depressed, and alveolar secretory cells disintegrate and desquamate. With progressing epithelial degeneration and necrosis, the cellular nuclei change from their vesiclelike structure into a pyknotic form, and the alveoli lose most of their capillary network. Through disintegration of the membrana propria, the cellular borders disappear, and the respective epithelial cells desquamate into a formless cellular detritus containing fat droplets. This induces phagocytic and reabsorptive processes reducing the number and size of glandular elements. Through phagocytosis of the degenerated and necrotic alveolar material, the mammary lobular-acinar structures become smaller and fewer, the alveolar lumina narrow, and the ductular mammary system begins to govern the picture. These autolytic and regressional processes of cellular necrobiosis and necrosis begin at the periphery of the breast progressing toward its center (Table 47). Reduction of lobular-glandular elements is followed by new formation of connective tissue

TABLE 47

Biology and Biochemistry of Mammary Involution Postlactation

1. Autophagic processes: Intraplasmic degradation of secretory material with development of large autophagic cytoplasmic vacuoles (sign for sublethal cellular injury)
2. Heterophagic processes: Invasion of mammary tissue areas with lethal cellular damage by macrophages and other phagocytic cells for removal of autolytic and necrotic cells
3. Induction of intracellular lysosomal enzymes (cathepsin, acid phosphatase, nuclease, β-glucuronidase) by autophagic and heterophagic processes
4. Mammary involutional process (consecutive) changes)
   - Absence of secretory stimulation by prolactin (nonnursing or cessation of lactation)
   - Failure of milk removal causes mammary alveoli to become maximally dilated and milk filled leading to increased intraalveolar hydrostatic pressure with flattening of cells and obstruction of capillary blood supply
   - Intracellular vacuole formation
   - Partial cellular autolysis and necrosis
   - Appearance of phagocytic cells ridding the mammary tissue of necrotic cells and eliminating stagnated intraalveolar milk
   - Autophagic and heterophagic involutional processes involve lysosomal enzymes, accumulated at the site of the Golgi apparatus, where they are probably produced; lysosomal enzymes eliminate useless intracellular secretory or non-secretory material by lytic degranulation processes
   - Invasion and deposition of connective tissue and fat between the involuted mammary alveolar structures
   - Duration of the total involutional process 2–3 months

septa. Concomitant with these involutional parenchymal changes is the appearance of phagocytic cells (histiocytes, lymphocytes, monocytes, neutrophilic leukocytes) and fibroblasts, surrounding, infiltrating, removing, and replacing the degenerative tissues. These mammary round cell infiltrates disappear after complete regression of glandular elements. Because of obstruction and involution of alveolar capillaries, an anastomotic shifting of blood to fat tissue occurs, whereby vascular mammary stasis is circumvented. Under this condition of enhanced blood supply the mass of fat tissue increases. Toward the end of the phagocytic and reparative processes, the remaining epithelial alveolar cells increase in height; their basal nuclei detach from the membrana propria and move apically. Thereby the alveolar lumen gradually narrows and it may, in some instances, temporarily disappear because of the new formation of high, thick, centrally nucleated alveolar cells. This increase in cellular height is partially due to enhanced intracellular lipid accumulation. Through the involutional postlactational process, the alveolar lining changes from a secretory to a nonsecretory 2-layered alveolar stem-cell type epithelium. The alveolar epithelial "steatosis" (intracellular lipid accumulation) leads to dislocation of cytoplasmic organelles and to reduction of ergastoplasm and granular protein at the Golgi apparatus. It also appears that at this stage of "steatosis" some fibrous degeneration of the glandular epithelium occurs. At the time of parenchymal regression the deposition of mammary fat increases, and the connective tissue becomes denser. The involution of the mammary gland postpartum is greatly accelerated in nonnursing puerperas.

Although with discontinuation of nursing the prolactin stimulation for milk synthesis and secretion ceases, facilitating the process of mammary involution, the event of postlactational mammary involution seems governed more by local mechanical factors (alveolar distention, tissue hypoxia, phagocytosis) than by hormonal ones.

Colostrum secretion is observed in premenstrual women, in gravidas, and in puerperas during the first week postpartum as well as in the early phase of postlactational involution. The postlactational intraalveolar colostrum is a product derived from secretory and degenerative processes, and it is removed by phagocytic histiogenic migratory cells during the process of involution. The disquamated and phagocytic autolysed alveolar material is transported away via mammary lymphatics or venous capillaries into the systemic circulation for further metabolism and disposal (Zilliacus, 1967; Dabelow, 1957; Bässler and Schäfer, 1969; Newton, 1961; Lieser and Bässler, 1969).

Also, with cessation of lactation, the pituitary prolactin cell hyperplasia regresses, and, in the absence of suckling stimuli, the increased prolactin secretion subsides and a baseline secretion of prolactin yielding plasma levels of about 10 ng/ml of prolactin is assumed. The process of adenohypophysial prolactin cell involution has been studied on lactating rats from which the suckling young had been removed for several days. A progressive involution of the cellular organelles involved in protein synthesis with sequestration of endoplasmic reticulum was observed. The incorporation of ribosomes into autophagic vacuoles and of secretory granules into multivesicular lytic bodies containing high acid phosphatase activity seems to indicate an autolysosomic process leading to digestion of endogenous materials. Thus in prolactin secretory cells, lysosomes may be considered regulators of the secretory process by providing a mechanism that takes care of overproduction of secretory products during lactation and that brings about involution of hypertrophic pituitary prolactin cells postlactation (Smith and Farquhar, 1966). With involution of the anterior pituitary prolactin cells, the hypothalamoadenohypophysial reproductive functions are fully resumed leading to FSH-LH-releasing-hormone-induced gonadotropin secretion and resulting in ovarian stimulation, ovulation, and menstruation.

D. MENOPAUSAL MAMMARY INVOLUTION

With declining ovarian function in the climacteric woman (premenopause) and thus diminished blood levels of sex steroids, senile mammary involution begins. At any stage of premenopausal mammary involution the glandular epithelium regresses first, and then along with it parts of the connective tissue disappear. As mammary parenchymal regression advances, mammary fat deposition in individually different quantities may be increasing. In these cases enhanced fat deposition is observed first in the periphery of the breast, progressing toward the center of the mamma. Usually new formation of fat is observed around glandular epithelium and increases with progressive reduction of epithelial elements. Two phases of senile mammary involution have been described:

1. Preclimacteric—climacteric phase of mammary involution between 35 to 45 years of age. During this stage only a moderate decrease in glandular mammary epithelium with round cell infiltration and decomposition of acinar and lobular tissues occurs.

2. Menopausal—postmenopausal (senile) phase of mammary involution between 45 to 75 years of age. Here a drastic reduction of glandular tissue occurs, and an increase in fat deposition and relative predominance of connective tissue are observed. At this stage the round cell infiltration

disappears, and hyalinization of connective tissue and loss of lobular and alveolar structures are most apparent. The end results of senile mammary involution are the remaining mammary ducts, which may show cystic degeneration (Fig. 45). Thus, at the stage of pronounced senile involution only fat, connective tissue, and mammary ducts, from which the mammary evolution began, remain. In advanced glandular parenchymal involution, the stromal connective tissue mass is relatively increased, framing the

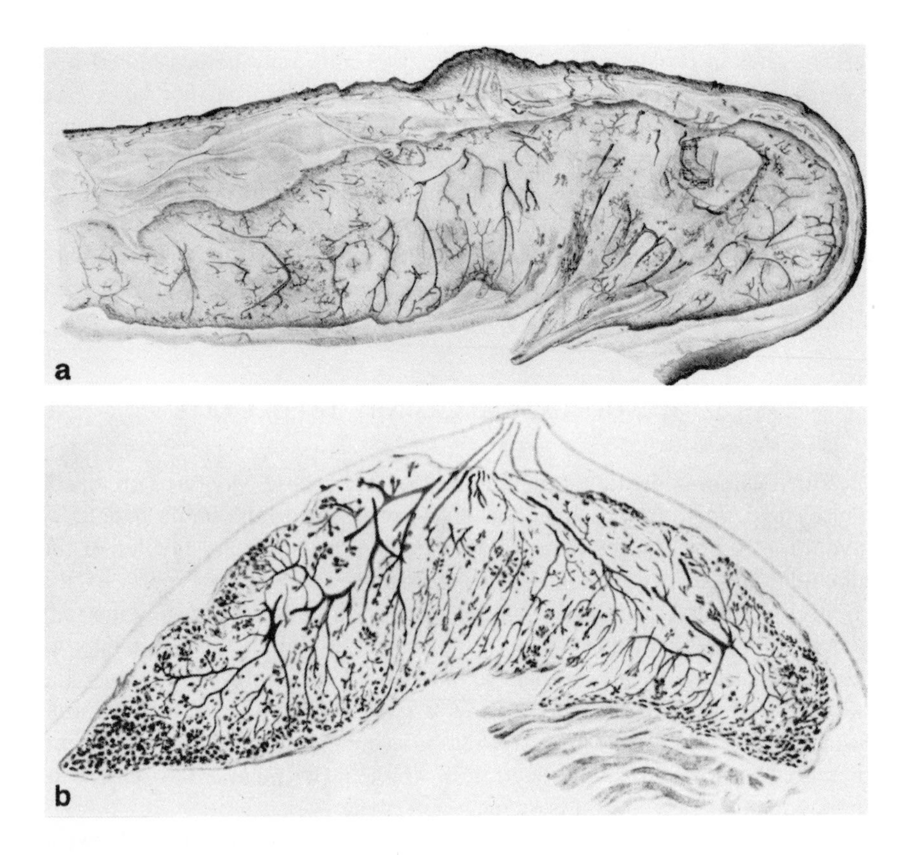

Fig. 45. Mammary structure of a menopausal and a fertile woman (from Dabelow, 1957). (a) Section through the breast of a menopausal woman. This thick, Alum Carmine stained tissue slice shows a well-preserved connective mammary tissue apparatus that is richly intermingled with fat. The ductular structures are pronounced; whereas the lobular-alveolar parts are almost completely involuted. (b) Section through the breast of a fertile woman. For the sake of comparison to the menopausal breast structure, a tissue slice from a nulliparous normally cycling woman with a normal ductular-lobular-alveolar configuration is presented.

sparse epithelial elements. Thus, in nonobese postmenopausal women the breasts may have shrunk to a small discoid mass through atrophy of the glandular body leading to regression of mammary lobi and lobuli. Also a decrease in vascular parenchymal branching occurs, whereas the vascularization of fat remains unchanged. In obese individuals, no reduction in the size of the breast may take place because the regressed mammary glandular tissue is abundantly replaced by fat tissue.

Histologic investigations revealed that at the beginning of the premenopausal mammary involution the alveoli collapse to form small acinar berries in which the 2-layer lining is maintained with the basal membrane intact. At that involutional stage the mammary duct lumina are round, lined by a 2-layer low epithelium, and round cells are dispersed within the mesenchyma signifying tissue resorption. With progressing involution, the glandular epithelial order and lining become disrupted; parenchymal cells degenerate and are phagotized and transported away. Remaining at the end stage of senile mammary involution are only small islands of epithelial parenchyma embedded in relatively hard fibrous tissue. The connective tissue differentiation into intralobulary, perilobulary, and interlobulary parts, as observed during the fertile period of life, no longer exists. Also, at the terminal phase of mammary involution, the round cell infiltration disappears, and a senile resting stage is reached in which round cells are no longer present. Nevertheless, in the senium elimination of parenchymal epithelium continues at a reduced rate by way of fatty degeneration (Dabelow, 1957).

CHAPTER VIII

Inappropriate Lactation: Galactorrhea

Since the first description of 2 patients with persistent postpartum galactorrhea by Chiari *et al.* (1855; cited in Speert, 1958) and that of 28 additional patients by Frommel (1881), many instances of galactorrhea and amenorrhea have been reported in the literature (see Table 48 and Fig. 46). Abnormal lactation has been observed under various conditions in nulliparous and parous women, and some galactorrhea syndromes have been named after the physicians who first reported them (Table 48). These eponyms for various conditions of galactorrhea seem to be appropriate because they are of historical, diagnostic, and prognostic interest. Today, the multifaceted aspects of galactorrhea are far less confusing because of the availability of specific prolactin assays, allowing accurate measurement of serum or plasma prolactin and thus placing the diagnosis of galactorrhea on a sound pathophysiologic basis (Tables 49 and 50 and Fig. 47). Galactorrhea is associated with increased pituitary prolactin secretion and deficient gonadotropin production (oligo- or amenorrhea); in addition, abnormal secretion patterns of thyroid hormone, growth hormone, and ACTH are observed (see Table 48). Increased prolactin blood levels do not necessarily lead to galactorrhea, for, although prolactin serum concentration is enhanced in most women after surgical stress or in gravidas, no milk secretion is observed. Thus, it appears that besides prolactin the amount of sex steroids in the circulation is of importance, blocking prolactin effect on the mammary alveolar cells (Tyson *et al.,* 1973). Galactorrhea in the male is extremely rare, but it has been seen in men with chromophobe adenoma, angiosarcoma of the pituitary, chorioepithelioma of the testis, pineal tumors, hypernephroma, adrenocortical tumor, and in some cases without any known pathologic condition (Relkin, 1965). In most cases of gynecomastia (mammary development in the male), however, no increase in serum prolactin is observed (Turkington, 1972d; Volpé *et al.*, 1972).

Galactorrhea is most frequently a consequence of modern pharmacotherapy and manual breast hyperstimulation (breast stimulation and/or coitus may bring about an up to eightfold increase in prolactin serum levels; Stearns *et al.*, 1972). Galactorrhea should not be considered a disease but a symptom of an underlying disorder that may be of a serious nature. Because, in the absence of suckling, at 14–21 days after parturition lactation ceases (Haskins *et al.*, 1964), any milk flow continuing beyond 3–6 months after cessation of breast stimulation by the breast-fed infant has to be considered abnormal or inappropriate lactation. Any condition of inappropriate milk secretion in association with amenorrhea or oligomenorrhea is termed "galactorrhea." The incidence of galactorrhea is about 1–2% (Pernoll, 1971; Greenblatt and Gambrill, 1972). In about 10–30% of regularly menstuating women, predominantly in parous women, some breast secretion may occur (Wenner, 1967; Friedman and Goldfien, 1969b; Shevach and Spellacy, 1971) and can be accepted as normal. Age, degree of parity, and phase of menstrual cycle do not seem to influence the incidence of breast secretions in normal menstruating women (Friedman and Goldfien, 1969b). On the other hand, spontaneous secondary amenorrhea is observed in 5–7% of women of fertile age (Friedman and Goldfien, 1969b), and pill-induced temporary secondary amenorrhea occurs in about 1%. In most of these amenorrhoic women no abnormal lactation is observed.

Although development of galactorrhea may be the first indication of a serious underlying disease, there is no evidence that galactorrhea patients show a higher incidence of breast cancer due to the continued stimulation of mammary epithelium by prolactin. In rat experiments dimethyl benzathracene-induced mammary carcinoma was found to be a prolactin-dependent tumor (Pearson, 1969). It is not known whether prolactin hypersecretion may induce mammary carcinomatous growth in humans; blood prolactin levels are usually not increased in breast cancer patients (Dickey and Minton, 1972). Prolactin, however, may contribute to accelerated growth of mammary carcinoma, as indicated by the fact that in pregnant women whose prolactin serum levels are severalfold increased, the course of malignant breast disease is more fluminant. Moreover, it has been observed that L-dopa treatment (depression of prolactin secretion) in patients with advanced breast cancer (bone metastases) may bring about

TABLE 48

Galactorrhea Syndromes

Disorder	Pathophysiology	Therapy [a]
1. Forbes-Albright syndrome; pituitary chromophobe adenoma	Galactorrhea, amenorrhea, hypothyroidism, headache, failing vision, diplopia; facultative acromegaly; autonomous and/or prolactin-inhibiting factor (PIF) dependent prolactin-secreting tumor; reduced secretion of PIF, FSH-LH, and potentially of TSH, ACTH, HGH.	Hypophysectomy and/or tumorectomy; hormone substitution; suppression of prolactin hypersecretion of remaining or recurrent tumor by bromergocryptine (B-Er) or L-dopa (L-Do).
2. Sheehan syndrome; postpartum pituitary infarction	Galactorrhea, amenorrhea (utero-ovarian atrophy), poor nutrition, decreased thyroid and adrenal function; deficient secretion of pituitary hormones with maintained function of lactotrophs; abnormal short-loop feedback mechanism of prolactin may depress hypothalamic PIF function causing prolactin hypersecretion.	Hormonal substitution; possibly induction of ovulation (clomiphene and/or gonadotropins); eventually trial with B-Er or L-Do.
3. Other organic brain lesions	Galactorrhea, clinical symptoms of respective brain disease; craniopharyngioma, metastatic carcinoma, ectopic pinealoma, pituitary angiosarcoma, basophilic adenoma, trauma, and infection causing depression in PIF function with prolactin hypersecretion.	Causal; eventually administration of B-Er or L-Do.
4. Chiari-Frommel syndrome; idiopathic, persistent postpartum galactorrhea	Galactorrhea occurs mostly after first delivery; in up to 50% of patients later a pituitary tumor may be diagnosed; low serum FSH-LH and sometimes low TSH (amenorrhea, secondary hypothyroidism); failure of PIF secretion and of cyclic FSH-LH release: abnormal feedback mechanism by FSH-LH and prolactin may lead to diminished secretion of hypothalamic-releasing hormones for FSH-LH (TSH) and for PIF (prolactin hypersecretion).	Spontaneous remissions in almost half of patients; eventual induction of ovulation (pregnancy) but high recurrence rates of galactorrhea; hormonal substitution; eventually B-Er or L-Do.

5. Ahumada-del Castillo-Argonz syndrome; idiopathic galactorrhea in nulliparous women	Prolactin hypersecretion and gonadotropin hyposecretion in nonpuerperal patients; failure of PIF and FSH-LH releasing hormone secretion leading to prolactin hypersecretion (galactorrhea) and gonadotropin hyposecretion (secondary amenorrhea mostly).	Spontaneous remissions rare; sex hormone substitution; eventual induction of ovulation (pregnancy) success rate low; trial with B-Er or L-Do.
6. Van Wyk-Grumbach syndrome; primary juvenile hypothyroidism	Galactorrhea and menorrhagia in prepuberal hypothyroid girls; enlargement of sella turcica but no pituitary tumor; hypersecretion of FSH-LH, (hyperestrenism) TSH, MSH (hyperpigmentation); breast development, absence of axillary or pubic hair; hypertrophy of labia minora; low serum TH levels; when serum TH is low, hypothalamic PIF secretion may be decreased, and production of prolactin, FSH-LH, and MSH is increased; abnormal feedback by FSH-LH and MSH with increased secretion of respective hormones.	Thyroid hormone replacement usually stops galactorrhea; skin pigmentation and enlargement of sella disappear; development of true menarche and puberty; eventually trial with B-Er or L-Do.
7. Primary adult hypothyroidism	Galactorrhea mostly in connection with previous pregnancy; hypersecretion of prolactin and of TSH; hyposecretion of gonadotropins and of TH; ACTH and HGH secretion may be diminished; decreased TH serum levels and/or abnormal feedback of TSH may bring about depression of secretion of PIF (increased prolactin secretion) and of FSH-LH (hypoestrogenism; amenorrhea) and possibly ACTH and HGH releasing hormones; low serum TH leads to increase in pituitary TSH secretion.	Thyroid hormone replacement mostly successful: cessation of galactorrhea, recurrence of menstruation and ovulation; eventually B-Er or L-Do.
8. Zondek-Bromberg-Rozin hyperhormonotropic syndrome	Galactorrhea, hyperthyroidism and hyperestrogenic bleeding in postpuerperal women; excess of hypothalamic TSH-RH (TSH-releasing hormone) induces hypersecretion of FSH-LH (hyperestrenism) and of TSH (TH); abnormal feedback by TSH and FSH may depress PIF activity and allow prolactin hypersecretion.	Antithyroid drugs; possibly partial thyroidectomy; progestins; clomiphene and/or gonadotropins; eventually B-Er or L-Do.

TABLE 48 *(continued)*

Disorder	Pathophysiology	Therapy [a]
9. Selective hyperpituitarism in postmenopausal women	Galactorrhea, occasionally observed in postmenopausal women with ovarian failure (estrogen deficiency) and increased FSH-LH production leading to depression of PIF function and prolactin hypersecretion.	Breast examination; possibly small doses of an estrogen–androgen combination and/or B-Er or L-Do.
10. Drug intake	Meprobamate, reserpine, phenothiazines, prednisone, α-methyldopa and sex steroids (pill) may cause galactorrhea and amenorrhea; abnormal feedback (sex steroids) or depression of PIF function (tranquilizers) leading to prolactin hypersecretion and FSH-LH hyposecretion.	Discontinuance of drug intake: high spontaneous remission rates; possibly induction of ovulation (good results), trial with B-Er or L-Do.
11. Psychosis, anxiety states	Galactorrhea in patients with schizophrenia and states of anxiety and endogenous depression; hypothalamic dysfunction of PIF activity with prolactin hypersecretion.	Treatment of respective mental disorder; possibly B-Er or L-Do.
12. Hyperstimulation of breast and chest	Prolonged suckling, frequent breast manipulation, cystic breast disease, thoracic burns, surgery and herpes zoster may elicit galactorrhea through a neurogenic afferent reflex pathway causing depression of PIF secretion with increased pituitary prolactin release.	Avoidance of undue neurogenic stimulation; spontaneous remissions after breast and chest surgery, etc.; perhaps B-Er or L-Do.
13. Utero-ovarian disorders	Galactorrhea in connection with benign and malignant utero-ovarian tumors, hysterectomy and castration; under these conditions ovarian sex steroid secretion is affected (too low,	Causal; spontaneous remissions 2–3 weeks after surgery (hysterectomy); possibly sex hormone

	too high, too much fluctuation) causing hypothalamic dysfunction in regard to FSH-LH and PIF activity with prolactin hypersecretion.	substitution and/or B-Er or L-Do.
14. Adrenocortical disorders	Adrenocortical carcinoma or hyperplasia in connection with dehydroepiandrosterone (DHA) and cortisol hypersecretion or their enhanced fluctuation may impair ovarian estrogen or pituitary FSH-LH and hypothalamic PIF secretion resulting in amenorrhea and galactorrhea.	Causal, and eventually supportive steroid treatment.
15. Liver disease; poor nutrition and health; diabetes mellitus	Especially in patients with liver problems; decreased metabolism of sex hormones resulting in hypothalamic dysfunction in regard to secretion of FSH-LH releasing hormone and of PIF; also hypothalamic PIF dysfunction in diabetes mellitus.	Causal; trial with B-Er or L-Do.
16. Ectopic production of lactogenic hormone	Hypernephroma, choriocarcinoma, bronchogenic carcinoma or pineal psammosarcoma may secrete prolactin or prolactin-like lactogenic substances causing galactorrhea.	Causal; trial with B-Er or L-Do in cases with secretion of remaining or recurrent tumor.
17. End-organ hypersensitivity	Sparse breast secretion in normally menstruating women, probably due to increased sensitivity of mammary secretory epithelium to normal prolactin serum levels.	Breast examination and reassurance; trial with an oral contraceptive with a high progestin content or giving a progestin only; perhaps B-Er or L-Do.

[a] B-Er is preferred over L-Do as a choice of treatment; both are still investigative agents in this respect.

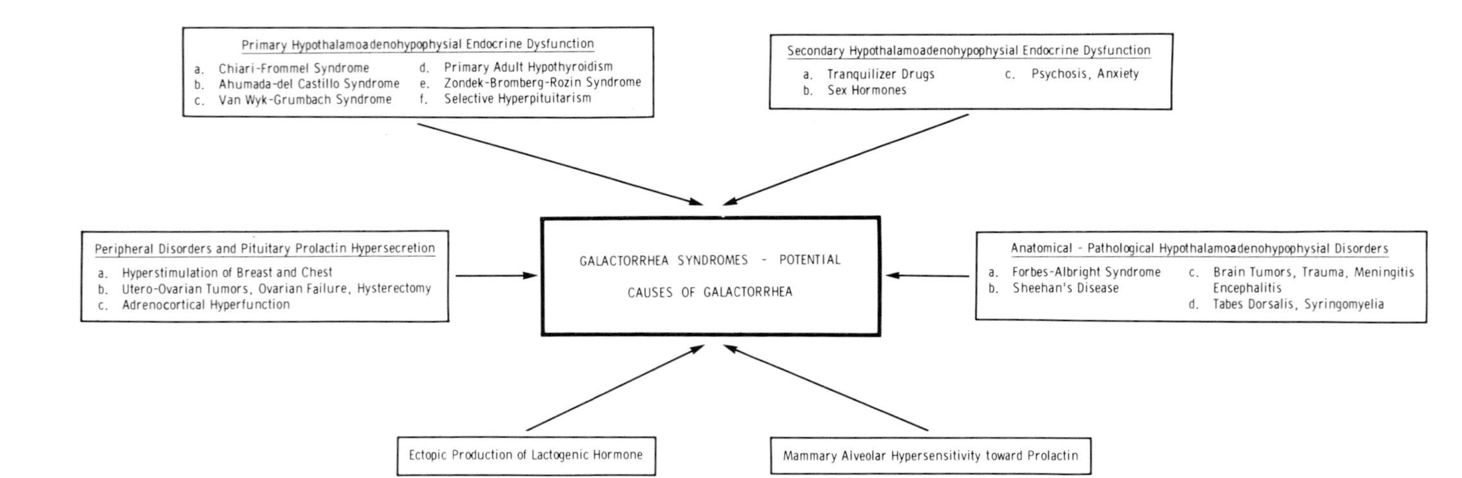

Fig. 46. The various groups of galactorrhea (abnormal lactation) are primary and secondary hypothalamoadenohypophysial endocrine dysfunctions; anatomical–pathological hypothalamoadenohypophysial disorders; peripheral disorders leading to prolactin hypersecretion; ectopic production of prolactin or prolactin-like substances; and mammary alveolar hypersensitivity toward normal levels of circulating prolactin. In idiopathic or functional galactorrhea no cause or pathology can be identified, as is the case in the group of primary hypothalamoadenohypophysial endocrine dysfunction. Exogenous galactorrhea induced by drugs, hyperstimulation of breast and chest, and surgery (e.g., hysterectomy) may be distinguished from endogenous galactorrhea, which comprises the rest of above listed galactorrhea syndromes or causes. Modern drug therapy and manual breast stimulation are the most frequent causes for galactorrhea, which should be considered a symptom of an underlying disorder rather than a disease in itself.

TABLE 49

Fundamentals of Prolactin Hypersecretion

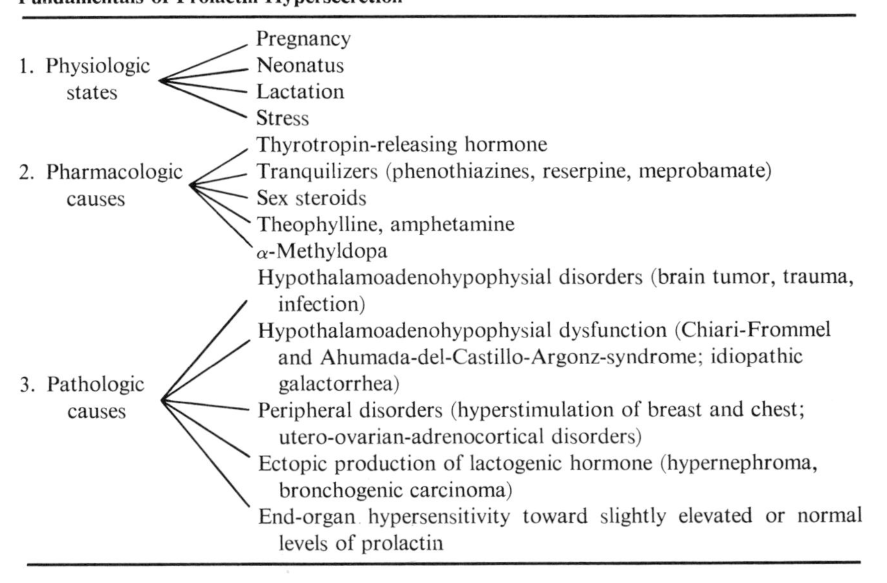

1. Physiologic states
   - Pregnancy
   - Neonatus
   - Lactation
   - Stress
2. Pharmacologic causes
   - Thyrotropin-releasing hormone
   - Tranquilizers (phenothiazines, reserpine, meprobamate)
   - Sex steroids
   - Theophylline, amphetamine
   - α-Methyldopa
3. Pathologic causes
   - Hypothalamoadenohypophysial disorders (brain tumor, trauma, infection)
   - Hypothalamoadenohypophysial dysfunction (Chiari-Frommel and Ahumada-del-Castillo-Argonz-syndrome; idiopathic galactorrhea)
   - Peripheral disorders (hyperstimulation of breast and chest; utero-ovarian-adrenocortical disorders)
   - Ectopic production of lactogenic hormone (hypernephroma, bronchogenic carcinoma)
   - End-organ hypersensitivity toward slightly elevated or normal levels of prolactin

TABLE 50

Pathology of Prolactin Secretion

1. No hereditary prolactin deficiency known; congenital isolated growth hormone deficiency exists (ateliotic dwarfs)
2. Hypersecretion of prolactin in patients with chromophobe adenoma and other morphologic or functional hypothalamoadenohypophysial disturbances resulting in galactorrhea
3. Hyposecretion or hypersecretion of prolactin in some patients with Sheehan's syndrome; hyposecretion in puerperas who fail to lactate postpartum despite normal mammary development and oxytocin treatment
4. Basic underlying disorders of prolactin hypersecretion
   - Hypothalamus: hypothalamic prolactin-inhibiting factor (PIF) dysfunction with chronic excessive drive for adenohypophysial prolactin synthesis may lead to development of pituitary tumor; prolactin hypersecretion may be accompanied by growth hormone and gonadotropin hyposecretion and conversely
   - Anterior pituitary: autonomous adenohypophysial disorder (chromophobe adenoma) leading to increased synthesis and release of prolactin

clinical remission of the disease (relief of bone pain, reduction in tumor mass). L-Dopa treatment (2 g/day) lowered basal prolactin serum levels of 26 ng/ml and 37 ng/ml by about 50% in ⅔ of breast cancer patients; this was accompanied by signs of clinical remission (Dickey and Minton,

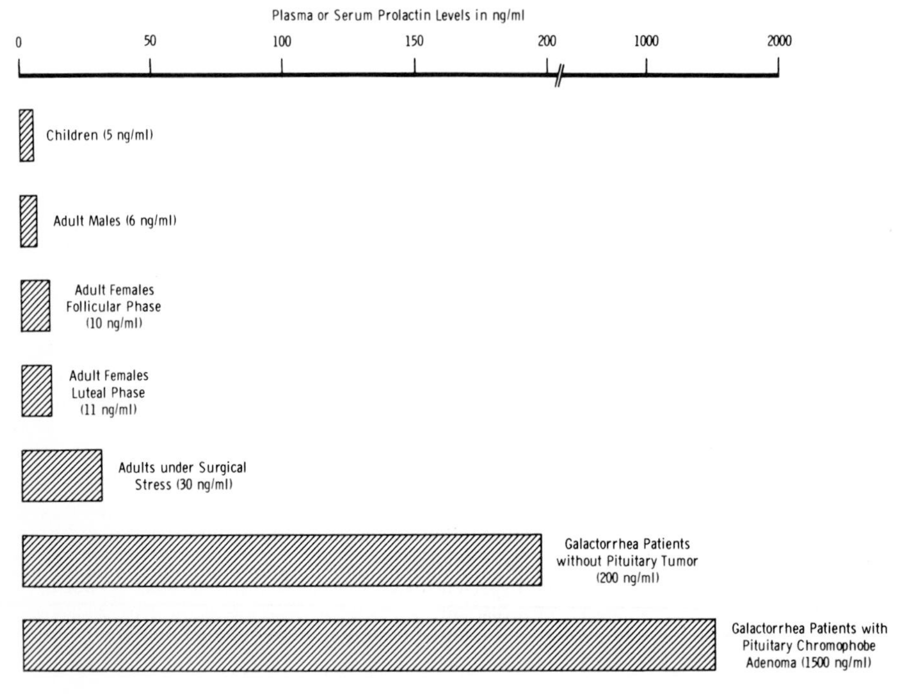

Fig. 47. Prolactin blood levels in children, adults, patients under surgical stress, and in patients with galactorrhea. Prolactin plasma or serum levels in children appear slightly lower than those in adults (for HGH the reverse applies). In adult females no significant difference or ovulation peak in prolactin levels is observed during the proliferative and secretory phase of the menstrual cycle. Under surgical stress, prolactin concentration in blood increases to 20–40 ng/ml. Prolactin hypersecretion is significant in galactorrhea patients without pituitary tumor; the prolactin blood levels, however, are highest in patients with galactorrhea caused by pituitary chromophobe adenomas. From Friesen (1972) and various other sources as indicated in the text.

1972). It appears that serum prolactin is not increased in breast cancer patients; but in about 5–10% of breast cancer patients tumor growth is prolactin-dependent, i.e., prolactin may promote cancerous growth in such patients (Hobbs *et al.*, 1973).

Galactorrhea is diagnosed by spontaneous outflow or by manual expression of milk. The term "galactorrhea" is applicable for excretion of a few drops of milky fluid containing fat globules as well as for daily amounts of 20–50 ml of milk. In general, abnormal lactation can be divided into idiopathic (functional) galactorrhea without any demonstrable pathologic disorder and galctorrhea due to endogenous (pituitary tumor, etc.) and exogenous (drugs, breast manipulation) causes. The classification of galactorrhea is presented in Fig. 46.

Any galactorrhea patient with a one-sided breast secretion that contains fat globules will also have milk secretory epithelium somewhere in the other breast, even when no spontaneous secretion or expulsion of milk occurs; deep-seated secretory alveoli may not produce enough milk to reach the larger milk ducts and milk sinuses and thus the outside. Breast biopsy specimens from galactorrhea patients reveal mammary tissues that are similar in appearance to those of an involuting breast following normal lactation. The mammary lobuli vary in size, some containing partly dilated alveoli with secretory epithelium others only small nonsecretory alveoli (Levine *et al.,* 1962).

The composition of "galactorrhea milk" appears to be similar to breast milk obtained during normal lactation (Table 51). Some investigators (Levine *et al.,* 1962; Brown *et al.,* 1965) reported much higher values of fat, protein, and lactose in galactorrhea milk, a fact that might be explained by the small, concentrated milk volume in these patients. In our studies of a patient's galactorrhea milk, which was produced in copious amounts by breast hyperstimulation, it was found that the milk constituents did not differ from puerperal milk.

TABLE 51

Composition of Normal Milk and "Galactorrhea Milk" [a]

Milk components and properties	Normal milk	"Galactorrhea milk"
Components		
Fat	3.7 gm %	3–8 gm %
Lactose	7.0 gm %	3–5 gm %
Total protein	1.2 gm %	2–7 gm %
Sodium	15 mg %	70 mg %
Potassium	50 mg %	5 mg %
Calcium	35 mg %	38 mg %
Chlorine	45 mg %	50 mg %
Phosphorus	15 mg %	2 mg %
Ash	20 mg %	40–70 mg %
Properties		
Specific gravity	1030–1033	1031
Milk pH	6.8–7.3	7.3
Daily volume	400–800 ml	1–120 ml

[a] Data derived from various sources as indicated in the text.

A. GALACTORRHEA SYNDROMES AND POTENTIAL CAUSES FOR GALACTORRHEA

Table 48 presents the various galactorrhea syndromes and their possible causes, as described in the literature. In the following section, the various types of galactorrhea, their pathophysiology, clinical manifestations, and therapy are discussed.

1. Forbes-Albright Syndrome

The Forbes-Albright syndrome (Forbes *et al.,* 1954) occurs in nulliparous or parous women with pituitary tumor (chromophobe adenoma); there are about 40 cases reported in the literature (Hughes *et al.,* 1972). Approximately 70–80% of all pituitary tumors are chromophobe adenomas (most frequently occurring at age 40 to 50). Because 20–30% of chromophobe adenomas are prolactin secreting (100–150 mg/day), yielding prolactin levels of 500–2000 ng/ml serum (Friesen *et al.,* 1972), such chromophobes may be considered as prelactotrophic cells or even more likely as degranulated secretory lactotrophs. Pituitary gonadotropin secretion and urinary estrogen excretion are reduced or may be undetectable (Dignam *et al.,* 1969; Gates *et al.,* 1973) prior to eventual decline of HGH (growth hormone), TSH (thyroid-stimulating hormone), and ACTH activities (Nabarro, 1972); in some of these patients adrenocortical function may be increased (17-ketosteroids slightly above average) (Relkin, 1965; Young *et al.,* 1967).

Depressed hypothalamic prolactin-inhibiting factor (PIF) activity may be the cause for increased prolactin secretion by the pituitary tumor; also autonomous prolactin hypersecretion by the lactotrophic tumor cells has to be considered (Peake *et al.,* 1969; Turkington, 1972a, c).

In patients with the Forbes-Albright syndrome, galactorrhea, sometimes with breast engorgement, and amenorrhea are the principal symptoms. Most patients are amenorrheic; some have irregular menses. Estrogen deficiency with failure to achieve withdrawal bleeding is observed. Other manifestations are hypothyroidism, headache, failing vision, diplopia (Hekmatpanah, 1969; Rushworth, 1971), facultative acromegaly, obesity, occasionally acne, hirsutism, and seborrhea (Nasr *et al.,* 1972; Relkin, 1965). In patients with eosinophilic (acidophilic) adenoma (15% of pituitary tumors, most frequently occurring at age 30 to 40) and increased somatotrophic activity (20% of all pituitary tumors are associated with acromegaly) (Friesen *et al.,* 1972), yielding HGH serum levels of 10–600 ng/ml, serum prolactin is usually not increased and may even be decreased (Turkington, 1972a). Conversely, the HGH content of a prolactin-secreting chromophobe adenoma was found to be greatly decreased to 0.02–0.1 μg/mg weight (Peake *et al.,* 1969) compared with the HGH values of a normal pituitary of about 8 μg/mg; this HGH value is similar to the prolactin content of a normal adult pituitary (Pasteels *et al.,* 1972). Recently, however, an eosinophilic adenoma in a 4½ year old girl secreting both HGH (200–400 ng/ml plasma) and HPr (50–250 ng/ml plasma) was reported (Guyda *et al.,* 1973). In this case, besides excessive growth and acromegaly, neither puberal development nor galactorrhea were observed. Basophil pituitary adenomas are relatively rare (about 5%). Usually they produce Cushing's disease through ACTH hypersecretion; rarely is an excess of TSH or prolactin observed.

Treatment of the cause of the syndrome is required. Two patients with early diagnosed chromophobe adenomas, yielding prolactin levels of 100–130 ng/ml, were cured of galactorrhea when serum prolactin was lowered to normal (2–15 ng/ml) by transfrontal craniotomy, hypophysectomy, and tumorectomy (Canfield and Bates, 1965; Nasr *et al.,* 1972; Turkington, 1972a). As nonsurgical therapy, inhibition of prolactin hypersecretion by bromergocryptine (B-Er)* or as an alternative (second choice) L-dopa (L-Do)* has been tried with some degree of success (Table 54) (Besser *et al.,* 1972; Turkington, 1972a), especially in cases of incomplete pituitary tumor surgery or recurrent tumors as manifested by increased serum prolactin levels. Other ergot derivatives (ergotamine tartrate) are also effective in lowering serum prolactin. Substitutional therapy with TH (thyroid hormone) and estrogens may also be advisable in such cases.

2. Sheehan's Disease

Sheehan's syndrome (postpartum hemorrhage leading to pituitary thrombotic infarction and necrosis) occurs in 0.01–0.02% of postpartum women (Sheehan and Davis, 1968; Drury, 1969). Pituitary hormone secretion including prolactin is usually deficient, and such patients may fail to lactate postpartum (Turkington, 1972d; Martin *et al.,* 1970). In some instances, however, inappropriate lactation is observed in Sheehan patients (Levine *et al.,* 1962). This is due to the maintained function of pituitary lactotrophs (compensatory activity of hypothalamoadenohypophysial function) (Richardson, 1970; Relkin, 1965) in cases in which hyposecretion

---

* Bromergocryptine and L-dopa are still investigative drugs for suppression of postpartum lactation and treatment of galactorrhea.

of TSH, ACTH, FSH-LH, and HGH is noted. It can be assumed that some patients who were reported as having Chiari-Frommel syndrome with accompanying malnutrition and progeria in fact had Sheehan's syndrome with relatively mild symptoms.

In Sheehan patients, ischemic necrosis of parts of the anterior pituitary gland is usually observed, causing deficiency of secretion of various adenohypophysial hormones. With intact prolactin secretion, an abnormal short-loop negative feedback mechanism of prolactin may inhibit secretion of hypothalamic-releasing and inhibiting factors, leading to continued prolactin secretion and deficiency in adenohypophysial output of trophic hormones. Conversely, reduced secretion of some pituitary hormones may stimulate secretion of other remaining pituitary hormones such as prolactin (Relkin, 1965). These patients suffer from oligo- or amenorrhea and utero-ovarian atrophy due to reduced FSH-LH secretion (decreased ovarian estrogens), and they show decreased thyroid and adrenal function. Here hormonal substitution is indicated. Some patients were cured of galactorrhea and amenorrhea by administration of TH (Ross and Nusynowitz, 1968). B-Er and L-Do may also be tried (Table 54). If pregnancy is desired, a trial with clomiphene and/or gonadotropins is indicated.

3. Organic Brain Lesions

Hypothalamoadenohypophysial disorders may be caused by tumors, specifically hypothalamic tumors such as craniopharyngioma, metastatic carcinoma, ectopic pinealoma, basophilic adenoma, pituitary angiosarcoma, as well as by trauma, pituitary stalk section, meningitis, encephalitis, and sarcoidosis (hypothalamic granulomas) (Relkin, 1965; Richardson, 1970; Krieger, 1971; Turkington 1972d). Most recently a case of histiocytosis-x disease with a posterior hypothalamic lesion resulting in diabetes insipidus, diabetes mellitus, and galactorrhea has been reported (Gates *et al.,* 1973).

In such conditions disturbance of PIF function may lead to prolactin hypersecretion, but galactorrhea is not always observed. Only some of these patients suffer from galactorrhea, which usually subsides when the underlying cause is treated. In refractory cases B-Er or L-Do may be administered.

4. Chiari-Frommel Syndrome

In patients with Chiari-Frommel syndrome (persistent postpartum lactation) the normal postpartum lactation is extended over months or years without any breast stimulation (Frommel, 1881; Speert, 1958; Haskins *et al.,* 1964; Canfield and Bates, 1965; Relkin, 1965; Thompson and Kempers, 1965; Pernoll, 1971). About 40% of Chiari-Frommel patients have irregular menses prior to pregnancy. Galactorrhea usually appears after the first delivery regardless of whether the mother breast-feeds or not (Rankin *et al.,* 1969). Serum FSH-LH levels are usually diminished; occasionally they are normal (Dignam *et al.,* 1969). As a rule, ACTH secretion is normal, i.e., 17-hydroxysteroids, plasma cortisol, and 17-ketosteroids are not increased. Growth hormone secretion in Chiari-Frommel patients is also normal (Benjamin *et al.,* 1969). About 25% of these patients suffer from secondary hypothyroidism (decreased serum levels of TSH and TH); some display psychosis or malnutrition (Rankin *et al.,* 1969; Relkin, 1965). Approximately 100 cases of Chiari-Frommel syndrome are reported in the literature (Hughes *et al.,* 1972); in up to 50% of patients a pituitary tumor of the Forbes-Albright type may be diagnosed in the course of the disease. In the absence of pituitary chromophobe adenoma, the Chiari-Frommel syndrome may occur as part of the Zollinger-Elison complex with familial recurrent peptic ulceration and hyperplastic adenomatous and carcinomatous changes in the pituitary, thyroid, parathyroid, pancreas, and adrenal glands (Briggs and Powell, 1969).

The underlying pathophysiology of the Chiari-Frommel syndrome may be explained by failure of PIF secretion and of cyclic FSH-LH release resulting in prolactin hypersecretion and gonadotropin hyposecretion. An abnormal short-loop negative feedback of prolactin on PIF secretion probably causes continued prolactin hypersecretion (Table 48) and decreased FSH-LH release; in some cases FSH-LH levels may be normal (Dignam *et al.,* 1969).

Clinical manifestations of the Chiari-Frommel syndrome are galactorrhea with possible breast engorgement, oligo- or amenorrhea, and in some cases hypothyroidism, obesity, and uterine and ovarian failure (small uterus and ovaries, in isolated cases ovaries are polycystic) (Zárate *et al.,* 1968). On rare occasions hirsutism or acne are observed.

The treatment of choice, if pregnancy is desired, is induction of ovulation with gonadotropins and/or clomiphene; the reported results are good. Substitution with sex hormones and eventually administration of TH, B-Er, or L-Do have also been effective (Figs. 48 and 49). Spontaneous remission and restoration of menses occurred within 3 to 5 years in 41% of patients who then conceived spontaneously (Rankin *et al.,* 1969). The recurrence rate of galactorrhea following gonadotropin and/or clomiphene induced pregnancy is very high (Rankin *et al.,* 1969).

5. Ahumada-del Castillo-Argonz Syndrome

The Ahumada-del Castillo-Argonz syndrome occurs in nulliparous women in the absence of a pituitary tumor; about 70 cases are reported in

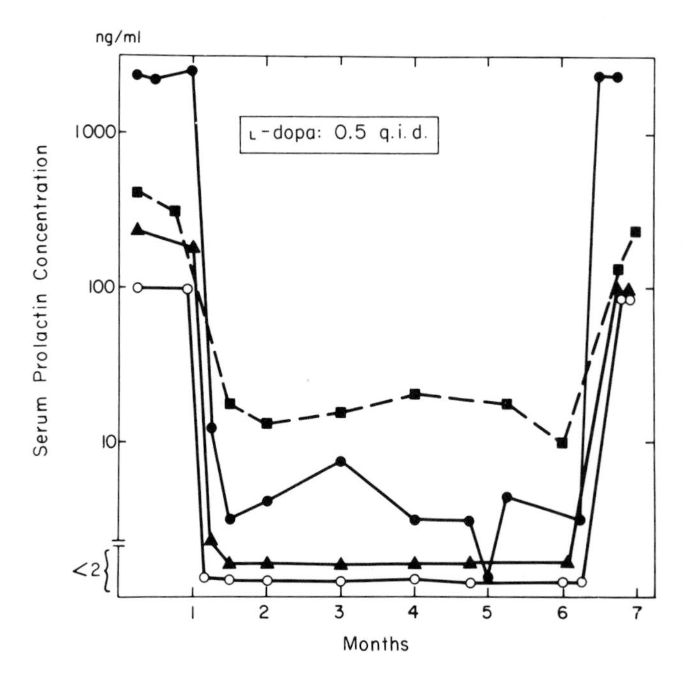

Fig. 48. L-Dopa treatment of idiopathic galactorrhea. In four patients with idiopathic galactorrhea, L-dopa (2 g/day, orally) lowered elevated serum prolactin levels (left side of picture) to normal and cured galactorrhea. L-Dopa treatment of galactorrhea in women of fertile age induced increased urinary gonadotropin excretion and, at the same time, return of normal menstruation. The drug activates hypothalamic prolactin-inhibiting factor function and simultaneously induces increased secretion of hypothalamic FSH-LH releasing hormone. L-Dopa appears to offer a treatment modality for patients with galactorrhea. When L-dopa administration was terminated after 5 months of treatment, serum prolactin levels increased again and galactorrhea recurred (right side of picture), which must be considered a draw-back of the therapy. From Turkington (1972c) by kind permission of the author and publisher.

the literature (Ahumada and del Castillo, 1932; Argonz and del Castillo, 1953; Hughes *et al.,* 1972). Here, pituitary growth hormone and thyroid hormone secretion are normal (Benjamin *et al.,* 1969); ACTH secretion and urinary 17-hydroxy- and 17-ketosteroid levels usually are also normal. In some cases urinary 17-ketosteroid excretion reaches the upper limit of normalcy, i.e., 15 mg/24 hours. Urinary gonadotropin levels are very low or undetectable in such patients. The pathophysiology of this syndrome is similar to that of the Chiari-Frommel disease characterized by prolactin hypersecretion and gonadotropin hyposecretion. The clinical symptoms are galactorrhea, with possible breast engorgement, and amenorrhea. Occa-

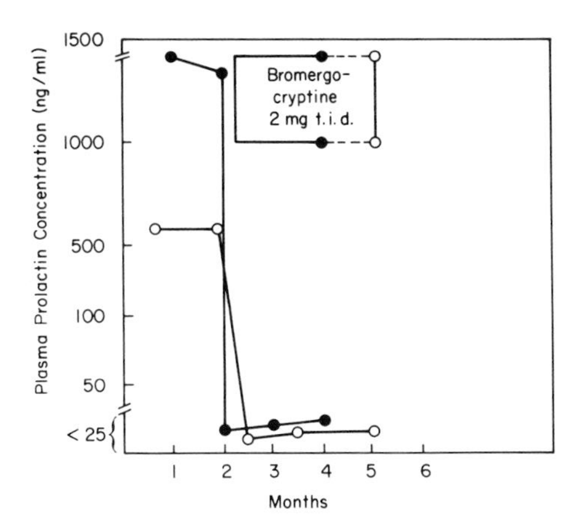

Fig. 49. Bromergocryptine treatment of idiopathic galactorrhea. In two patients with idiopathic galactorrhea (closed circles: patient with Ahumada-del Castillo-Argonz syndrome; open circles: patient with galactorrhea after oral contraception) bromergocryptine (6 mg/day, orally) lowered increased plasma prolactin levels to normal or below and cured galactorrhea. Normal menstruation returned in both patients after 2–3 weeks of bromergocryptine treatment and remained regular after the drug was discontinued, indicating that bromergocryptine offers an effective treatment modality for patients with galactorrhea. This ergot derivative directly inhibits prolactin secretion by adenohypophysial lactotrophs. From Besser *et al.* (1972) by kind permission of the authors and the publisher.

sionally amenorrhea is primary, but mostly it is of a secondary nature. Estrogen deficiency evidenced by a small uterus, inactive endometrium, hypoestrogenic vaginal smear, hot flashes, nervousness, and cystic ovaries is common (Richardson, 1970). This syndrome has also been observed in emotionally disturbed patients.

The treatment of choice, if pregnancy is desired, is induction of ovulation with clomiphene and/or gonadotropins, but the results are unsatisfactory, i.e., it is difficult to achieve pregnancy in such patients. Other therapeutic modalities are substitution of sex hormones or administration of B-Er, or L-Do to suppress increased prolactin secretion. Spontaneous remissions occur only rarely (Rankin *et al.,* 1969).

6. Van Wyk-Grumbach Syndrome

This syndrome of primary juvenile hypothyroidism presents itself with galactorrhea and menorrhagia in prepuberal girls (Van Wyk and Grumbach, 1960). It is probably caused by hypersecretion of FSH-LH (hyperestrinism, ovarian cysts, ovulatory failure) and of TSH and melanocyte-stimulating hormone (MSH); thyroid hormone secretion is greatly reduced as indicated by low serum concentration of TH and low PBI serum values, i.e., primary hypothyroidism. In connection with high serum levels of TSH, hypothalamic PIF secretion is diminished, and pituitary FSH-LH secretion is increased leading to precocious menstruation (Richardson, 1970) and galactorrhea. Abnormal short and long-loop negative feedback mechanism of FSH-LH and MSH may also be responsible for diminution of PIF secretion thus contributing to prolactin hypersecretion. Increased levels of thyrotropin-releasing hormone (TRH) also bring about enhanced prolactin secretion and its release into the circulation. It is also possible that the sensitivity of pituitary lactotrophs toward TRH is increased, resulting in prolactin hypersecretion (Turkington, 1972d).

The clinical manifestations are galactorrhea, hypothyroidism presenting myxedema, dry skin, brittle hair, constipation, precocious menstruation associated with menorrhagia, breast development, absence of pubic or axillary hair, hyperpigmentation of the skin, hypertrophy of labia minora, enlargement and demineralization of sella turcica. No pituitary tumor can be detected; the bone age in these patients is retarded.

This syndrome is successfully treated with thyroid hormone replacement. Thyroid therapy makes the patient euthyroid. Galactorrhea and the associated symptoms disappear, and true menarche develops along with normal puberty. In refractory cases B-Er, or L-Do may eventually be administered to lower increased serum prolactin to a normal level and thus cure galactorrhea.

7. Primary Adult Hypothyroidism

About 14 cases of this galactorrhea syndrome have so far been reported, all but one in connection with previous pregnancy (Futterweit and Goodsell, 1970; Kinch *et al.,* 1969; Ross and Nusynowitz, 1968; Bayliss and van't Hoff, 1969). Prolactin hypersecretion (2000 ng/ml in one case) (Edwards *et al.,* 1971) has been recorded along with normal or low urinary gonadotropin excretion; estrogen deficiency may be present. TSH secretion is above normal while thyroid hormone secretion is subnormal. In rare instances, decrease of serum thyroxine may lead to increased FSH-LH, MSH, and ACTH secretions in connection with galactorrhea. Usually decrease of TH secretion causes depression of the production of PIF and FSH-LH releasing hormone as well as stimulation of thyrotropin-releasing hormone (TRH), which may result in selective hyperpituitarism with prolactin and TSH hypersecretion (Kinch *et al.,* 1969; Richardson, 1970). Also, an abnormally short- and long-loop negative feedback mechanism of TSH may lead to hyposecretion of PIF and FSH-

LH releasing hormone and thereby to prolactin hypersecretion and to FSH-LH hyposecretion, respectively.

Manifestations of the syndrome are hypothyroidism with dry skin and hair, sparse pubic and axillary hair, myxedema, and oligo- or amenorrhea. In the absence of galactorrhea hypothyroid patients usually suffer from menorrhagia.

The most successful treatment is thyroxine replacement (0.1–0.2 mg/day) restoring menstruation and fertility and achieving cessation of galactorrhea within 4 to 6 weeks. In refractory cases B-Er, or L-Do may be administered.

8. Zondek-Bromberg-Rozin Syndrome

This hyperhormonotrophic syndrome appears rarely in postpuerperal women. It is associated with hypersecretion of prolactin, FSH-LH (hyperestrogenism), TSH, and TH (hyperthyroidism); HGH secretion may be decreased. Galactorrhea has also been reported in patients with hyperthyroidism and hypoestrogenism; here excessive thyroxine secretion may lead to inhibition of FSH-LH secretion (or may be triggered by decreased FSH-LH activity), which in turn results in decreased output of ovarian estrogens and thus in oligo- and amenorrhea and galactorrhea due to prolactin hypersecretion (Zondek *et al.,* 1951; Relkin, 1965).

Excess of hypothalamic TRH brings about increased adenohypophysial secretion of TSH, FSH-LH, and prolactin; in addition abnormal short- and long-loop negative feedback mechanisms of FSH-LH and TSH probably also lead to prolactin hypersecretion by depressing PIF activity.

The clinical picture shows hyperthyroidism with poor nutritional status, hyperestrogenic bleeding, slightly enlarged uterus, endometrial hyperplasia, and profuse galactorrhea (50–100 ml milk/day) related to a previous pregnancy. Hypoglycemia, anemia, hyperpigmentation of breasts, and sterility have also been observed.

Antithyroid drugs and possibly partial thyroidectomy are recommended as therapy, as well as progestins. If pregnancy is desired, clomiphene and/or gonadotropins may be applied to induce ovulation. In some resistant cases a trial with B-Er, or L-Do is indicated.

9. Selective Hyperpituitarism

Postmenopausal ovarian failure (refractoriness to FSH-LH) induces increased pituitary secretion of FSH-LH, and in rare instances this may be connected with galactorrhea. An abnormal short-loop negative feedback mechanism of FSH-LH may lead to a depression of PIF function and thus to prolactin hypersecretion. Also, the reduced blood levels of sex hormones in postmenopausal women may facilitate the action of prolactin on mammary alveolar cells for milk secretion. The syndrome appears as galactorrhea accompanied by postmenopausal estrogen deficiency. Small doses of an estrogen–androgen combination most likely will relieve the problem.

10. Drug-Induced Galactorrhea

A number of drugs may cause galactorrhea in puerperal and nonpuerperal women; they are meprobamate, reserpine, phenothiazines, phenothiazine derivatives (tricyclic antidepressants), α-methyldopa (Pernoll, 1971; Turkington, 1972e; Arrata and Howard, 1972; Shearman and Turtle, 1970). Drug-induced galactorrhea often occurs in connection with gonadotropin and growth hormone hyposecretion, especially with phenothiazines (Benjamin *et al.,* 1969).

Monthly temporary withdrawal of sex hormones (pill) or discontinuing the use of oral contraceptives may lead to galactorrhea, either sparse or cyclic; usually serum and urinary gonadotropin levels are low, leading to amenorrhea. In some patients FSH-LH levels are normal (Dignam *et al.,* 1969). Estrogen deficiency may be pronounced, and then the endometrium becomes atrophic. Adrenocortical steroid secretion is usually normal.

Depletion of hypothalamic catecholamine stores by tranquilizers and α-methyldopa results in prolactin hypersecretion (50–1000 ng/ml serum), due to depression of PIF function, and in gonadotropin-hyposecretion; after diazepam intake no changes in prolactin serum levels occurred (Turkington, 1972e). Prolonged suppression of hypothalamo-pituitary gonadotropic and PIF function is possible when contraceptive steroids combine with specific hypothalamic receptors and thus depress secretion of FSH-LH releasing hormone and PIF (Arrata and Howard, 1972). A temporary, 7 day, monthly withdrawal of an oral combination contraceptive may bring about decreased PIF function with increased prolactin secretion and/or effect on the mammary epithelium. After permanent discontinuance of oral contraceptives the state of hypothalamoadenohypophysial dysfunction (decreased secretion of FSH-LH releasing hormone and of PIF) may become evident, resulting in galactorrhea. After the intake of sex hormones is discontinued, an abnormal short- and/or long-loop negative feedback regarding FSH-LH, in conjunction with their own secretion, and regarding PIF function, may be operative causing amenorrhea and galactorrhea. In addition, prolactin may be able to inhibit secretion of hypothalamic-releasing hormones for FSH-LH, HGH, TSH, and PIF secretion through a short-loop feedback (Arrata and Howard, 1972).

Clinical manifestations of the syndrome are amenorrhea and galactorrhea. Treatment consists primarily of discontinuation of drug intake. Ovulation, if pregnancy is desired, should be induced with clomiphene citrate and/or gonadotropins leading to conception in about 50% of the patients (Friedman and Goldfien, 1969a). In case drug withdrawal does not bring about relief of symptoms, administration of B-Er, or L-Do appears promising. The rate of spontaneous remission of drug-induced secondary amenorrhea and galactorrhea is high; up to 95% menstruate within 12 to 18 months once the use of drugs is discontinued (Friedman and Goldfien, 1969a).

11. Psychosis and State of Anxiety

Amenorrhea and galactorrhea can be observed in women suffering from schizophrenia and endogenous depression; some patients with Ahumada-Argonz-del Castillo syndrome also fall into this category.

Psychological and emotional factors play a role in prolactin hypersecretion through impairment of hypothalamic PIF function. In these patients, treatment of the underlying disease will bring about remission of galactorrhea; in some patients administration of B-Er, or L-Do may be of value.

12. Hyperstimulation of Breast and Chest

Peripheral neurogenic stimuli such as prolonged suckling, intensive and frequent breast manipulation, thoracic burns, herpes zoster, thoracic surgery (thoracotomy, lung lobectomy, mastectomy), and cystic breast disease may cause galactorrhea (Relkin, 1965; Crist *et al.,* 1971; Kolodny *et al.,* 1972). According to some studies manual breast and nipple stimulation caused a rise of serum prolactin from the baseline level of 5–6 ng/ml to 108 ng/ml within 5 minutes of self-stimulation or stimulation by the husband. Fifteen minutes later serum prolactin levels were still 37 ng/ml (Kolodny *et al.,* 1972). Interestingly, self-stimulation in men caused no change in prolactin serum levels, but after stimulation by the wife an increase of serum prolactin from 6 ng/ml to 25 ng/ml was observed. Accordingly, lactation can be induced in nonpregnant, nulliparous farm animals by milking, as well as in women by suckling; in ruminants the galactorrhea response could be enhanced by estrogen and estrogen-progesterone administration (Cowie, 1972). Most recently it was reported that oral estrogen treatment (5 mg/day) and nipple stimulation for 14 consecutive days produced galactorrhea (prolactin hypersecretion and FSH-LH hyposecretion) in healthy fertile women. Whereas during estrogen treatment nipple stimulation failed to increase serum prolactin further, with administration of thyrotropin-releasing hormone an additional rise of serum prolactin levels was observed. As in postpartum lactation, 3 days after estrogen withdrawal, breast engorgement and milk flow occurred and this "galactorrhea milk's" fat content was almost threefold higher than that of normal postpartum milk (Khojandi and Tyson, 1973).

By means of manual nipple and breast stimulation or through suckling, a neurogenic pathway allows impulses to travel via sensory nerves and their accompanying sympathetic fibers, such as medial and lateral branches of the thoracic intercoastal nerves (II–VI); the impulses then reach the brain via the spinal cord causing depression of hypothalamic PIF secretion, and consequently prolactin hypersecretion and lactation occur.

Any type of the above mentioned factors in regard to breast and chest stimulation can be the cause of galactorrhea. Therefore, therapy requires avoiding neurogenic hyperstimulation of the breast; in cases with chest and breast disease and surgery, temporary administration of B-Er, or L-Do may be indicated.

13. Uteroovarian Disorders

Uteroovarian tumors such as carcinomas, fibromas, cystomas, and dermoids, with or without endocrine function, may be the underlying cause of galactorrhea. Abnormal lactation can also be seen in connection with surgery and stress related to these disorders. Castration, partial ovarian resection, precocious and physiologic menopause, and depression of ovarian function by androgens can be linked to galactorrhea. In all these conditions sex hormone secretion is affected; it may be above normal, subnormal, or there may be too much fluctuation of sex hormone secretion. All these conditions may bring about changes in prolactin secretion or its effectiveness on the mammary epithelium.

Sex hormone hypersecretion may cause a negative long-loop feedback mechanism of secretion of hypothalamic-releasing (FSH-LH releasing) hormone and inhibiting (PIF) factors resulting in decreased adenohypophysial FSH-LH secretion and prolactin hypersecretion. On the other hand, sex hormone hyposecretion or abrupt sex steroid withdrawal as achieved by castration may result in FSH-LH hypersecretion with negative short-loop feedback mechanism in regard to PIF secretion resulting in enhanced prolactin secretion. Following hysterectomy with or without ovariectomy, in 2–10% of patients, the operative stress may cause diminution in PIF function and thus evoke prolactin hypersecretion. Hysterectomy also leads to the withdrawal of a possible uterine growth-promoting substance that is thought to potentiate the effect of estrogen on various target organs; this, in turn, may diminish the PIF function leading to rebound release of prolactin (Sheld, 1968; Sheld and Charme, 1969); also diminished estrogen action on the mammary epithelium may facilitate prolactin-induced milk secretion.

Therapy requires treatment of the utero-ovarian disorder. It may be necessary to substitute estrogens and administer B-Er, or L-Do. Posthysterectomy galactorrhea usually subsides spontaneously within 2–3 weeks after surgery.

14. Adrenocortical Disorders

Adrenal carcinoma, adrenal hyperplasia in connection with hypersecretion and/or enhanced fluctuation of levels of cortisol, estrogens, and androgens may cause galactorrhea. Galactorrhea may also be due to primary adrenocortical insufficiency (Addison's disease) with selective hypersecretion of prolactin (about 30 ng/ml serum) and ACTH (Refetoff *et al.*, 1972).

Adrenocortical steroids such as dehydroepiandrosterone (DHA) may affect the ovarian function directly by stimulating estrogen secretion or indirectly by affecting pituitary FSH-LH secretion (Barlow, 1964). Small quantities of DHA may stimulate release of FSH-LH whereas larger amounts may inhibit it. Adrenocortical disease requires causal treatment; cortisone therapy may be indicated in order to achieve normalization of ACTH secretion and prolactin serum levels, which will promote cessation of galactorrhea.

15. Liver Disease, Hepatic Insufficiency, Poor Nutrition and Health, and Diabetes Mellitus

Patients suffering from hepatic disease, malnutrition, poor health, or diabetes mellitus sometimes experience galactorrhea, especially those patients with liver disease (Richardson, 1970).

Decreased metabolism of sex hormones and/or prolactin appears to be responsible for galactorrhea. Increased levels of sex steroids or fluctuation of serum levels of sex hormones due to impaired liver function are liable to bring about a negative long- and short-loop feedback mechanism of FSH-LH leading to diminished PIF (increased prolactin release) and FSH-LH secretion. Hypothalamic dysfunction in diabetes mellitus patients may result in depressed PIF activity and increased release of pituitary prolactin.

Primarily, it is necessary to treat the underlying disease; administration of B-Er, or L-Do may be tried.

16. Ectopic Production of Lactogenic Hormone

Patients with nonendocrine tumors may be afflicted with galactorrhea. Hypernephroma, choriocarcinoma, bronchogenic carcinoma, or pineal psammosarcoma can be responsible for ectopic production of prolactin or prolactin-like lactogenic substances; pineal tumors may also impair PIF function (Pernoll, 1971; Turkington, 1972d).

The tumors responsible for galactorrhea require treatment. If active tumor tissue is still present after therapy, B-Er, or L-Do may be administered and are also recommended for patients with recurrent disease and galactorrhea.

17. End-Organ Hypersensitivity

Galactorrhea occurs sometimes without known origin and with normal or only slightly increased prolactin serum levels. In these cases, it is possible that fluctuation of the level of sex hormones, e.g., premenstrual decrease in estrogens and progesterone, may bring about sparse, cyclic breast secretion due to minor increases in pituitary prolactin release or to increased sensitivity of mammary epithelium toward prolactin. Measurement of prolactin serum concentrations may not reveal such minor changes. In some patients oligo- or amenorrhea may be intermittently observed, and here the occurrence of breast secretion can be explained by increased sensitivity of mammary epithelium to normal serum levels of prolactin.

When galactorrhea occurs in patients with normal prolactin levels, the breasts should be carefully examined. In case of a normal finding the patient needs reassurance. In cases with oligo- or amenorrhea, if the secretion is irritating, an oral contraceptive should be administered temporarily. Administration of an oral progestin (Norlutate) may also bring relief. Eventually B-Er, or L-Do may be tried.

B. CAUSAL EVALUATION OF THE GALACTORRHEA PATIENT

The diagnostic procedures that are used to determine the cause of galactorrhea are outlined in Table 52. The patient's case history often reveals drug intake or undue breast manipulation eliciting galactorrhea. Events such as early menopause and previous surgery may also play a role. An eye field examination of galactorrhea patients is important because a prolactin-secreting pituitary tumor may exist: about 25% of all intracranial tumors are chromophobe adenomas and one-third of them are prolactin secreting. Also, measurement of serum prolactin and growth hormone levels are essential for differential diagnosis. In patients with a prolactin-secreting pituitary tumor, prolactin blood levels are usually much higher (500–5000 ng/ml) than those observed in patients with idiopathic galactorrhea in whom prolactin serum levels usually amount to 100–300 ng/ml on the average (Forsyth *et al.*, 1971; Besser *et al.*, 1972; Turkington, 1972a, b,c) (Tables 28 and 54). Another differential diagnostic procedure is the prolactin-serum-water-load test (Buckman *et al.*, 1973). After an oral water load (20 ml/kg/30 minutes) in normal persons and patients with idiopathic galactorrhea, serum prolactin levels fall within 2 hours to about 10% of baseline levels; while in patients with prolactin-secreting pituitary tumors, the water load fails to suppress prolactin levels to less than 50% of pretest values. Moreover, any patient with Cushing's syndrome, even in the absence of sellar enlargement, should be suspected to have a pituitary

TABLE 52

Diagnostic Procedures for Galactorrhea Patients

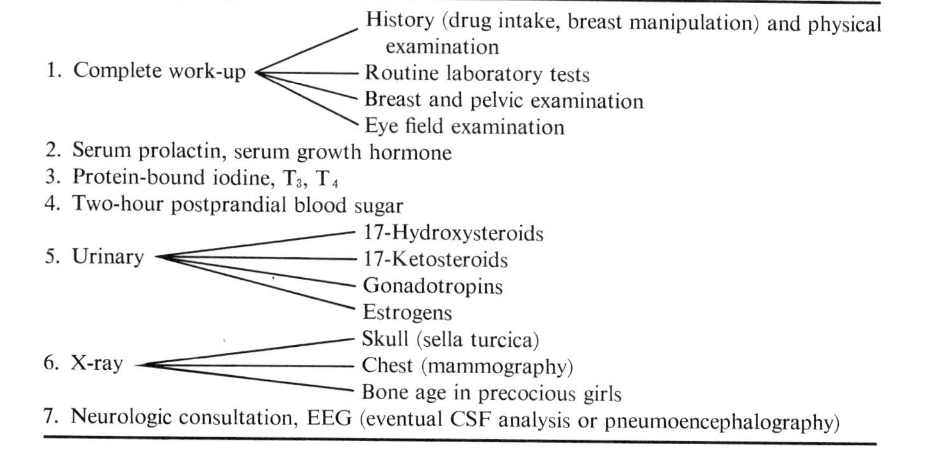

1. Complete work-up
   - History (drug intake, breast manipulation) and physical examination
   - Routine laboratory tests
   - Breast and pelvic examination
   - Eye field examination
2. Serum prolactin, serum growth hormone
3. Protein-bound iodine, T_3, T_4
4. Two-hour postprandial blood sugar
5. Urinary
   - 17-Hydroxysteroids
   - 17-Ketosteroids
   - Gonadotropins
   - Estrogens
6. X-ray
   - Skull (sella turcica)
   - Chest (mammography)
   - Bone age in precocious girls
7. Neurologic consultation, EEG (eventual CSF analysis or pneumoencephalography)

tumor (Young *et al.*, 1967). Detailed evaluation of hypothalamoadenohypophysial function and of the activity of various interrelated endocrine glands by blood and/or urine hormonal analyses and other techniques may be considered in patients with galactorrhea (Table 52). In some galactorrhea patients CSF analysis, carotid arteriography, or pneumoencephalography may become necessary; in other selected cases a metyrapone or an adrenocortical ACTH stimulation test may be indicated. In some instances the measurement of blood androgens may be preferred as the more accurate procedure to estimate urinary 17-ketosteroids. Also, besides measurements of protein-bound iodine, evaluation of serum TSH, T_3, and T_4 levels may be of diagnostic significance. The different steps of diagnostic procedures and parameters to be evaluated depend on the individual patient's history, physical findings, and diagnostic needs.

C. MANAGEMENT OF THE GALACTORRHEA PATIENT

The various types of management of the galactorrhea patient are outlined in Table 53; in Table 54 recent treatment results of galactorrhea by new drugs, L-dopa and bromergocryptine, are compiled. It seems that bromergocryptine and L-dopa are very promising drugs for the treatment of idiopathic galactorrhea; even in some cases with pituitary chromophobe adenoma, L-dopa has been found to lower serum prolactin levels effectively (Malarkey *et al.*, 1971; Edmonds *et al.*, 1972; Turkington, 1972c). It has therefore been suggested that some prolactin-secreting chromophobe adenomas may not function autonomously but may be under hypothalamic influence because through L-dopa treatment the serum prolactin concentra-

TABLE 53

Management of the Galactorrhea Patient

- I. Causal therapy
  - A. Surgery, irradiation or chemotherapy of pituitary, brain, adrenal or ovarian tumors
  - B. Elimination of drugs, treatment of psychosis, avoidance of breast manipulation
  - C. Hormonal substitutional treatment of hypothyroidism, treatment of hyperthyroidism and hypercorticism
- II. Therapy of idiopathic galactorrhea
  - A. Amenorrhea treatment
    1. No desire for pregnancy: substitutional therapy with oral estrogens and progestins
    2. Desire for pregnancy: clomiphene, gonadotropins
  - B. Galactorrhea treatment
    1. Progestins
    2. Bromergocryptine or L-dopa

tion could be lowered. L-Dopa increases hypothalamic catecholamine stores, promoting increased release of PIF, FSH-LH, and HGH releasing hormones, resulting in enhanced secretion of FSH-LH and HGH and in decreased secretion of prolactin. Since some pituitary prolactin-secreting tumors respond to L-dopa treatment, it appears that in such cases tumor growth may be induced or maintained through hypothalamic hypofunction of PIF. Nevertheless, a direct effect of L-dopa on anterior pituitary lactotrophs cannot be fully excluded and could also explain the depression of serum prolactin levels in patients with chromophobe adenoma. Whereas it is believed that L-dopa acts on the hypothalamus reducing PIF function, bromergocryptine probably reduces prolactin secretion by a direct inhibitory effect on adenohypophysial lactotrophs; again, as with L-dopa, an additional depressive effect on hypothalamic PIF cannot be disregarded. Although both L-dopa and bromergocryptine can effectively lower serum prolactin to normal or below normal levels in patients with idiopathic galactorrhea, bromergocryptine seems to be the more promising drug because galactorrhea did not recur after its discontinuance; when L-dopa was discontinued, galactorrhea recurred (Table 54). Whereas L-dopa was capable of lowering increased prolactin serum levels to normal, galactorrhea could not be cured with a 3 day L-dopa treatment regimen (Malarkey *et al.,* 1971) (Table 54). After a dose of 2.5 mg of bromergocryptine, increased serum prolactin levels of 60–80 ng/ml in patients with idiopathic galactorrhea were lowered within 2 hours to 5–10 ng/ml (del Pozo *et al.,* 1972), indicating the efficiency of this compound in depressing serum prolactin. Under both L-dopa and bromergocryptine treatment, urinary gonadotropin titers rise to normal within a few weeks, and regular menses return (Table 54 and Figs. 48 and 49). Under treatment with bromergocryptine, galactorrhea disappeared, and ovulatory menstruation returned without relapse after cessation of therapy (Table 54) (Besser *et al.,* 1972; del Pozo *et al.,* 1972).

Patients with minimal galactorrhea who do not want a pregnancy need only to be reassured and advised to have regular checkups over a prolonged period of time because in many cases with idiopathic galactorrhea a pituitary tumor may be diagnosed later. Some galactorrhea patients may benefit from a low estrogen dose (Hughes *et al.,* 1972). Also, progestins have their place in the treatment of idiopathic galactorrhea, especially since progestin hazards such as carcinogenicity, thromboembolism, and hypertension are minimal or nonexisting (Vorherr, 1973). In all patients whose menstrual function cannot be restored, galactorrhea will persist (Thompson and Kempers, 1965). In this respect, bromergocryptine and L-dopa are of great therapeutic interest because they may lead to normali-

TABLE 54

Treatment of Galactorrhea with Bromergocryptine and L-Dopa

References	Serum prolactin levels				Comments
		Galactorrhea patients			
	Normal women	Before treatment	Therapeutic agent and dose regimen	During treatment	
Lutterbeck *et al.* (1971)	Prolactin levels not measured	Prolactin levels not measured (pituitary tumor, and idiopathic galactorrhea)	Bromergocryptine 1–3 mg t.i.d. orally for 6 months	Prolactin levels not measured	After 1 to 2 months of treatment galactorrhea ceased in ⅔ of patients
Malarkey *et al.* (1971)	3–7 ng/ml	144–2800 ng/ml (pituitary tumors)	L-Dopa 0.5 g q.i.d., orally for 3 days	28–800 ng/ml	Sustained suppression could not be obtained with this dose regimen; no improvement of galactorrhea within the short period of treatment
		14–60 ng/ml (idiopathic galactorrhea)	L-Dopa 0.5 g q.i.d., orally for 3 days	2–12 ng/ml	
Besser *et al.* (1972)	<25 ng/ml	100–1500 ng/ml (pituitary tumor, idiopathic and pill-induced galactorrhea)	Bromergocryptine 3–9 mg/day, orally for 4–10 months	<25 ng/ml	In all cases plasma prolactin levels were suppressed to <25 ng/ml; return of regular menses within 3–6 weeks after treatment, in one male return of potency; only in ⅕ patients galactorrhea recurred after cessation of treatment

del Pozo *et al.* (1972)	10–15 ng/ml	60–80 ng/ml (idiopathic galactorrhea)	Bromergocryptine 2.5 mg t.i.d. orally for 2–4 months	5–10 ng/ml	Galactorrhea ceased during treatment; return of menses and ovulation
Edmonds *et al.* (1972)	4–25 ng/ml	40–50 ng/ml (status after surgical removal of pituitary tumor, radiotherapy, and pituitary stalk section)	L-Dopa 0.25 g t.i.d., orally for 8 weeks, then 0.5 g t.i.d.	20–30 ng/ml	Galactorrhea ceased within 48 hours of treatment and remained suppressed for 8 weeks then recurred and could not be suppressed even with double the daily dose of L-dopa
Turkington (1972c)	<2 ng/ml	110–240 ng/ml (pituitary tumors)	L-Dopa 0.5 g q.i.d., orally for 3–6 months	100–120 ng/ml	Within 2–3 days of treatment prolactin serum levels returned to normal in most patients with idiopathic galactorrhea; with cessation of galactorrhea, resumption of normal gonadotropin secretion and ovarian function (menstruation) occurred; withdrawal of L-dopa was associated with recurrence of galactorrhea
		10–1130 ng/ml (idiopathic galactorrhea)	L-Dopa 0.5 g q.i.d., orally for 3–6 months	<2–56 ng/ml	

zation of pituitary-ovarian function and thus bring about ovulation and menstruation. At present bromergocryptine appears to be the drug of choice.

In conclusion, it may be stated that (1) causal treatment of galactorrhea should be always attempted; (2) substitutional therapy with small estrogen doses may be of value; (3) progestins (oral, parenteral) may be of benefit in various conditions, especially in cases with precocious puberty; and (4) bromergocryptine and L-dopa treatment of galactorrhea has already demonstrated its effectiveness and relative safety. Bromergocryptine (derived from ergotoxine) as all other ergot derivatives (ergotamine, ergonovine) may cause nausea, vomiting, and headache. With L-dopa, albeit not reported yet in galactorrhea patients, side effects such as dizziness, nausea, vomiting, anorexia, and hypotension have to be considered. Further clinical evaluation and experience with bromergocryptine and L-dopa are needed to decide on their efficacy, indications, and clinical use for certain types of galactorrhea or for suppression of normal postpartum lactation.

CHAPTER IX

Conclusion and Future Horizons

Our knowledge of mammary morphology, blood, lymph, and nerve supply of the breast is quite satisfactory. Also, breast development from puberty to adolescence, under the influence of sex steroids and various other metabolic hormones, is rather well understood.

The menstrual cycle can be correlated not only with genital but also mammary changes. The exact mechanisms responsible for the cyclic alterations of mammary tissue are, however, not well known. Moreover, the factors determining the regulation of the degree of mammary proliferation and epithelial secretory activity, which depend on neural, hormonal, and humoral means, are not understood either, though it can be assumed that adenohypophysial prolactin and ovarian sex hormones and the consequence of their withdrawal at the time of menstruation play a major role in this respect. Rather unknown also are the processes bringing about regression of proliferated and secretory mammary structures during menstruation and a few days thereafter; this short period represents the time of reduced ovarian activity until renewed breast stimulation in a subsequent menstrual cycle takes place. The intracellular enzymatic-autolytic processes leading to this temporary breast involution at the time of menstruation are even more obscure.

During gestation a tremendous growth stimulation of the various breast tissues and of the anterior pituitary takes place, and, even abortion, after a period of no more than 16 weeks of pregnancy, is accompanied by lactation. The existence of a distinct pituitary lactogenic hormone (prolactin), which represents a major and probably sole specific stimulant for mammary alveolar milk synthesis and its secretion into the alveolar lumen, has been established almost beyond any doubt. But our understanding of the exact intracellular biochemical processes involved in the synthesis of milk fat, proteins, and lactose and factors influencing and regulating it, as well as the transmembranal release of these constituents together with ions and water into the alveolar lumen, is still in the embryonic stage.

The nutritional value of the various milk components is well known; therefore, artificial feeding formulas can be prepared with cow's milk and adjusted appropriately.

Drugs taken by the lactating mother may appear in her milk and cause adverse effects in the breast-fed infant. At the present time it is not clear which drugs are actually harmful to the suckling baby and what the critical amounts are. Literature reports on drug passage into milk provoking adverse effects in the infant are either scanty, or they lack proof of a definite cause-effect relationship. Therefore, more studies are required to elucidate whether, for instance, contraceptive pills taken by a nursing mother diminish the milk yield and harm the infant by causing virilization of the female and feminization of the male breast-fed infant or whether only those oral contraceptives containing high doses of estrogens and progestins evoke these undesirable effects.

The mechanical, nerval, and hormonal factors involved with the onset and maintenance of lactation are well recognized; also, the mechanisms bringing about pituitary prolactin (milk synthesis and secretion) and oxytocin (milk ejection) release have become known in recent years. Nevertheless, it remains to be investigated further whether the neural afferent reflex pathways for release of prolactin and oxytocin into the circulation as traced in animal experiments are the same in the human organism. Furthermore, it needs to be clarified whether the hypothalamic control for prolactin release is solely due to the function of a prolactin-inhibiting factor, or whether, at times, a stimulatory influence is also present or prevailing. Therefore, it awaits elucidation whether a stimulus for prolactin release not only frees the hypothalamic-inhibiting factor from its restraining function on transmembranal prolactin release but also causes secretion of a hypothalamic prolactin-releasing hormone(s) to bring about release of the lactogenic hormone from the pituitary prolactin cells. Furthermore, knowledge of secretion rates, plasma levels, and metabolism of prolactin and more so of oxytocin is far from complete, and extensive future studies are required to shed more light on these mechanisms observed under physiologic and pathophysiologic conditions.

The subject of breast-feeding versus bottle-feeding, considering the advantages and disadvantages of either procedure, is again in dispute. Nursing seems to decrease the incidence of maternal breast cancer, and for this reason breast-feeding is recommended. On the other hand, warnings have been voiced most recently that breast milk of some mothers may contain viruslike particles that may be passed on to the breast-fed infant and may cause breast cancer or other malignancies later in its life cycle. Most recently, viruslike particles have been observed by electron microscope, especially in women with a family history of breast cancer. Although breast-feeding benefits infant and mother, this subject certainly needs future well-controlled prospective studies in order to allow a well-founded conclusion. Also, retrospective investigations in women with breast cancer or with a family history of mammary carcinoma may be worthwhile in order to establish meaningful correlations between nursing and development of breast cancer later in the life of the nursed daughter.

Lactation definitely postpones the return of maternal reproductive functions. But the exact mechanisms underlying these changes have not yet been clarified. Breast feeding may bring about or maintain pregnancy-induced changes in hypothalamoadenohypophysial gonadotropic functions as well as in the ovary. That a hormonelike substance released from the lactating breast tissues plays a role in postpartum ovarian "quiescence," seems rather unlikely, but it warrants further investigations.

Suppression of lactation with a long-acting androgen–estrogen combination is presently considered the safest and most efficient method available; it is, however, not an optimal procedure. Therefore, future investigations with other pharmacologic agents, such as ergocornine, bromergocryptine, and L-dopa, suited for inhibition of pituitary prolactin release and thus suppression of lactation may provide an alternative. Conversely, agents that may bring about increased release of pituitary prolactin as, for instance, reported for purines (theophylline), chlorpromazine, and thyrotropin-releasing hormone, may play a role in the treatment of inadequate mammary secretory activity at the onset of or during lactation. Also highly purified human prolactin preparations or synthetic prolactin may be utilized some day to stimulate mammary milk secretory activity. The regulation of prolactin secretion is also crucial in the treatment of abnormal lactation (galactorrhea); also, prolactin appears to be involved in the maintenance of breast cancer growth in some women (see page 226).

The physical and biochemical processes occurring in cells of glandular tissues during postpartum mammary involution and after cessation of lactation, whereby the extant breast tissues developed during pregnancy and utilized during lactation become reduced, are not yet well known. Improved histiochemical and enzymatic methods of analysis could lead to a better understanding of this subject. The factors operative in premenopausal and postmenopausal parenchymal mammary involution are not well understood. Moreover, the causes responsible for changes of the normal ductular-alveolar mammary epithelium into cell(s) with abnormal biochemical behavior progressing into development of malignant cell(s) are not at all known. Here methods that allow recognition of cellular abnormalities are needed to detect the cause of abnormal control of nuclear DNA in dependence on normal and abnormal sex steroid milieu and to explore other unknown operative agents. Early detection and possibly correction of mammary growth processes deviated from the norm are crucial and desirable goals.

In this book an effort has been made to gather all the essential information on morphology, physiology, and interrelated endocrinology of the breast in one place. Although the vast literature data on these subjects have been greatly condensed, the text still provides adequate background information to be utilized in future investigations. Particularly, it is hoped that this book will stimulate further research in those areas in which our knowledge is far from complete.

Bibliography

Adamsons, K., Jr., Engel, S. L., van Dyke, H. B., Schmidt-Nielsen, B., and Schmidt-Nielsen, K. (1956). The distribution of oxytocin and vasopressin (antidiuretic hormone) in the neurohypophysis of the camel. *Endocrinology* **58,** 272–278.

Ahmed, A. B. J., George, B. C., and Dingman, J. F. (1966). Independent regulation of arginine vasopressin (AVP) and oxytocin (OT) secretion in man. *Clin. Res.* **14**, 277.

Ahumada, J. C., and del Castillo, E. B. (1932). Sobre un caso de galactorrea y amenorrea. *Bol. Soc. Obstet. Ginecol.* **11**, 64–78.

Aloj, S. M., Edelhoch, H., Handwerger, S., and Sherwood, L. M. (1972). Correlations in the structure and function of human placental lactogen and human growth hormone. II. The effects of disulfide bond modification on the conformation of human placental lactogen. *Endocrinology* **91,** 728–737.

Amoroso, E. C., and Porter, D. G. (1966). Anterior pituitary function in pregnancy. *In* "Pituitary Gland" (G. W. Harris and B. T. Donovan, eds.), Vol. II, pp. 364–411. Univ. of California Press, Berkeley.

Anderson, D. E. (1972). A genetic study of human breast cancer. *J. Nat. Cancer Inst.* **48,** 1029–1034.

Arena, J. M. (1970). Contamination of the ideal food. *Nutr. Today* **5,** No. 4, 2–8.

Argonz, J., and del Castillo, E. B. (1953). A syndrome characterized by estrogenic insufficiency, galactorrhea and decreased urinary gonadotropin. *J. Clin. Endocrinol. Metab.* **13**, 79–87.

Arias, I. M., and Gartner, L. M. (1964). Production of unconjugated hyperbilirubinaemia in full-term new-born infants following administration of pregnane-3 (alpha), 20 (beta)-diol. *Nature (London)* **203**, 1292–1293.

Armstrong, D. T. (1970). Reproduction. *Annu. Rev. Physiol.* **32**, 439–470.

Aroskar, J. P., Chan, W. Y., Stouffer, J. E., Schneider, C. H., Murti, V. V. S., and du Vigneaud, V. (1964). Renal excretion and tissue distribution of radioactivity after administration of tritium labeled oxytocin to rats. *Endocrinology* **74,** 226–232.

Arrata, W. S. M., and Howard, A. (1972). The amenorrhea-galactorrhea syndrome following oral contraceptives. *J. Reprod. Med.* **8**, 139–142.

Babson, S. G. (1971). Feeding the low-birth-weight infant. *J. Pediat.* **79,** 694–701.

Bach, S. J., and Messervy, A. (1969). Observations of the diffusible calcium fraction in the serum of the cow during oestrus and during parturition. *Vet. Rec.* **84,** No. 9, 210–213.

Bare, W. W., Zaleznik, E., and Levin, H. (1963). Double-blind evaluation of two androgen-estrogen preparations and a placebo for suppression of lactation. *Amer. J. Obstet. Gynecol.* **87**, 276–279.

Bargmann, W. (1968). Neurohypophysis. Structure and function. *In* "Handbuch der experimentellen Pharmakologie" (B. Berde, ed.), Vol. 23, pp. 1–39. Springer-Verlag, Berlin, Heidelberg, and New York.

Bargmann, W., and Scharrer, E. (1951). The site of origin of hormones of the posterior pituitary. *Amer. Sci.* **39**, 255–259.

Barlow, J. J. (1964). Adrenocortical influences on estrogen metabolism in normal females. *J. Clin. Endocrinol. Metab.* **24**, 586–596.

Bashore, R. A. (1967). Immunoassay of oxytocin. *Obstet. Gynecol.* **29**, 431.

Bässler, R. (1970). The morphology of hormone induced structural changes in the female breast. *Curr. Top. Pathol.* **53**, 1–89.

Bässler, R., and Schäfer, A. (1969). Elektronenmikroskopische Cytomorphologie der Gynäkomastie. *Virchows Arch., A* **348**, 356–373.

Baum, D. J. (1971). Nutritional value of human milk. *Obstet. Gynecol.* **37,** 126–130.

Bayliss, P. F. C., and van't Hoff, W. (1969). Amenorrhoea and galactorrhoea associated with hypothyroidism. *Lancet* **2,** 1399–1400.

Beck, P., Parker, M. L., and Daughaday. W. H. (1965). Radioimmunologic measurement of human placental lactogen in plasma by a double antibody method during normal and diabetic pregnancies. *J. Clin. Endocrinol. Metab.* **25**, 1457–1462.

Benjamin, F., Casper, D. J., and Kolodny, H. H. (1969). Immunoreactive human growth hormone in conditions associated with galactorrhea. *Obstet. Gynecol.* **34,** 34–39.

Benson, B., Matthews, M. J., and Rodin, A. E. (1971). A melatonin-free extract of bovine pineal with antigonadotropic activity. *Life Sci., Part I* **10**, 607–612.

Benson, R. C. (1968). The puerperium. *In* "Handbook of Obstetrics and Gynecology," pp. 198–215. Lange Med. Publ., Los Altos, California.

Berde, B. (1959). "Recent Progress in Oxytocin Research," pp. 29–32, 61, and 67. Thomas, Springfield, Illinois.

Berle, P., and Apostolakis, M. (1970). Demonstration of prolactin in the human plasma and its possible function in lactogenesis and galactopoesis. *Intl. J. Gynaecol. Obstet.* **8**, 240.

Berle, P., and Apostolakis, M. (1971). Prolaktin-Konzentrationen im menschlichen Plasma während Schwangerschaft und Wochenbett. *Acta Endocrinol.* (*Copenhagen)* **67**, 63–72.

Berle, P. Apostolakis, M., and Link, A. (1971). Nachweis laktotroper Aktivität im Plasma während des ovariellen Cyclus der Frau. *Arch. Gynaekol.* **210**, 124–130.

Besser, G. M., Parke, L., Edwards, C. R. W., Forsyth, I. A., and McNeilly, A. S. (1972). Galactorrhoea: Successful treatment with reduction of plasma prolactin levels by brom-ergocryptine. *Brit. Med. J.* **3**, 669–672.

Bisset, G. W., Clark, B. J., and Lewis, G. P. (1967). The mechanism of the inhibitory action of adrenaline on the mammary gland. *Brit. J. Pharmacol. Chemother.* **31**, 550–559.

Board, J. A. (1968). Plasma human growth hormone in the puerperium. *Amer. J. Obstet. Gynecol.* **100**, 1106–1109.

Boden, G., Lundy, L. E., and Owen, O. E. (1972). Influence of levodopa on serum levels of anterior pituitary hormones in man. *Neuroendocrinology* **10**, 309–315.

Bonte, M., and van Balen, H. (1969). Prolonged lactation and family spacing in Rwanda. *J. Biosoc. Sci.* **1**, 97–100.

Borglin, N.-E., and Sandholm, L-E. (1971). Effect of oral contraceptives on lactation. *Fert. Steril.* **22**, 39–41.

Bradbury, M., Vorherr, H., Hoghoughi, M., and Kleeman, C. (1968). ADH and oxytocin in cerebrospinal fluid. *Clin. Res.* **16**, 262.

Brambilla, F., and Sirtori, C. M. (1971). Gonadotropin-inhibiting factor in pregnancy, lactation, and menopause. *Amer. J. Obstet. Gynecol.* **109,** 599–603.

Breibart, S., Bongiovanni, A. M., and Eberlein, W. R. (1963). Progestins and skeletal maturation. *N. Engl. J. Med.* **268**, 255.

Brew, K. (1969). Secretion of α-lactalbumin into milk and its relevance to the organization and control of lactose synthetase. *Nature (London)* **222**, 671–672.

Briggs, R. Le, and Powell, J. R. (1969). Chiari-Frommel syndrome as a part of the Zollinger-Ellison multiple endocrine adenomatosis complex. *Calif. Med.* **111**, 92–96.

Brown, D. M., Jenness, R., and Ulstrom, R. A. (1965). A study of the composition of milk from a patient with hypothyroidism and galactorrhea. *J. Clin. Endocrinol. Metab.* **25**, 1225–1230.

Bryant, G. D., Siler, T. M., Greenwood, F. C., Pasteels, J. L., Robyn, C., and Hubinont, P. O. (1971). Radioimmunoassay of a human pituitary prolactin in plasma. *Hormones* **2**, 139–152.

Buckman, M. T., Kaminsky, N., Conway, M., and Peake, G. T. (1973). Water load —a new test for prolactin suppression. *Clin. Res.* **21**, 250.

Burford, G. D., Jones, C. W., and Pickering, B. T. (1971). Tentative identification of a vasopressin-neurophysin and an oxytocin-neurophysin in the rat. *Biochem. J.* **124**, 809–813.

Caldeyro-Barcia, R., and Méndez-Bauer, C. (1966). Oxytocin in labor and lactation. *Proc. Pan-Amer Congr. Endocrinol. 6th, 1966.* Int. Congr. Ser. No. 112, pp. 89–96.

Canfield, C. J., and Bates, R. W. (1965). Nonpuerperal galactorrhea. *N. Engl. J. Med.* **273**, 897–902.

Cantarow, A., and Schepartz, B. (1967). Milk. *In* "Biochemistry," 4th ed., pp. 793–796. Saunders, Philadelphia, Pennsylvania.

Catz, C. S., and Giacoia, G. P. (1972). Drugs and breast milk. *Pediat. Clin. N. Amer.* **19**, 151–166.

Chard, T., Boyd, N. R. H., Forsling, M. L., McNeilly, A. S., and Landon, J. (1970). The development of a radioimmunoassay for oxytocin: The extraction of oxytocin from plasma, and its measurement during parturition in human and goat blood. *J. Endocrinol.* **48**, 223–234.

Chrambach, A., Bridson, W. E., and Turkington, R. W. (1971). Human prolactin: Identification and physical characterization of the biologically active hormone by polyacrylamide gel electrophoresis. *Biochem. Biophys. Res. Commun.* **43**, 1296–1303.

Cobo, E., De Bernal, M. M., Gaitan, E., and Quintero, C. A. (1967). Neurohypophyseal hormone release in the human. II. Experimental study during lactation. *Amer. J. Obstet. Gynecol.* **97**, 519–529.

Coch, J. A., Brovetto, J., Cabot, H. M., Fielitz, C. A., and Caldeyro-Barcia, R. (1965). Oxytocin-equivalent activity in the plasma of women in labor and during the puerperium. *Amer. J. Obstet. Gynecol.* **91**, 10–17.

Coch, J. A., Fielitz, C., Brovetto, J., Cabot, H. M., Coda, H., and Fraga, A. (1968). Estimation of an oxytocin-like substance in highly purified extracts from the blood of puerperal women during suckling. *J. Endocrinol.* **40**, 137–144.

Cole, B. W., and Pitts, N. E. (1966). Suppression of lactation by a single injection of an androgen-oestrogen combination. *Practitioner* **196**, 139–143.

Corbin, A., and Milmore, J. E. (1971). Hypothalamic control of pituitary follicle-stimulating hormone synthesis. *Endocrinology* **89**, 426–431.

Cowie, A. T. (1972). Comparative physiology of lactation. *Proc. Roy. Soc. Med.* **65**, 1084–1085.

Crist, T., Hendricks, C. H., and Brenner, W. E. (1971). Lactation following lobectomy. *Amer. J. Obstet. Gynecol.* **110**, 738–739.

Cronin, T. J. (1968). Influence of lactation upon ovulation. *Lancet* **2**, 422–424.

Cross, B. A. (1961). Neural control of oxytocin secretion. *In* "Oxytocin" (R. Caldeyro-Barcia and H. Heller, eds.), pp. 24–47. Pergamon, Oxford.

Cruttenden, L. A. (1971). Inhibition of lactation. *Practitioner* **206**, 248.

Cummins, H. (1966). The skin and breasts. *In* "Morris' Human Anatomy" (B. J. Anson, ed.), 12th ed., pp. 125–131. McGraw-Hill (Blakiston), New York.

Curtis, E. M., Newsom, N. H., and Grant, R. P. (1963). Oral contraceptives in the immediate puerperum. *J. Med. Ass., Ga.* **52**, 425–428.

Dabelow, A. (1957). Die Milchdrüse. *In* "Handbuch der mikroskopischen Anatomie des Menschen." (W. Bargmann, ed.), Vol. 3, part 3, Haut und Sinnesorgane, pp. 277–485. Springer-Verlag, Berlin.

del Pozo, E., Brun del Re, R., Varga, L., and Friesen, H. (1972). The inhibition of prolactin secretion in man by CB-154 (2-Br-α-ergocryptine). *J. Clin. Endocrinol. Metab.* **35**, 768–771.

Desjardins, C., Paape, M. J., and Tucker, H. A. (1968). Contribution of pregnancy, fetuses, fetal placentas and deciduomas to mammary gland and uterine development. *Endocrinology* **83**, 907–910.

Dickey, R. P., and Minton, J. P. (1972). L-Dopa effect on prolactin, follicle-stimulating hormone, and luteinizing hormone in women with advanced breast cancer: A preliminary report. *Amer. J. Obstet. Gynecol.* **114,** 267–269.

Dignam, W. J., Parlow, A. F., and Daane, T. A. (1969). Serum FSH and LH measurements in the evaluation of menstrual disorders. *Amer. J. Obstet. Gynecol.* **105**, 679–695.

Dmochowski, L. (1972). Viruses and breast cancer. *Hosp. Pract.* **7**, 73–81.

Drury, M. I. (1969). Sheehan's disease. *Ob Gyn Digest* (August) pp. 32–39.

Dunn, J. D., Arimura, A., and Scheving, L. E. (1972). Effect of stress on circadian periodicity in serum LH and prolactin concentration. *Endocrinology* **90,** 29–33.

Eayrs, J. J., and Baddeley, R. M. (1956). Neural pathways in lactation. *J. Anat.* **90,** 161–171.

Edmonds, M., Friesen, H., and Volpé, R. (1972). The effect of levodopa on galactorrhea in the Forbes-Albright syndrome. *Can. Med. Ass. J.* **107**, 534–538.

Edwards, C. R. W., Forsyth, I. A., and Besser, G. M. (1971). Amenorrhoea, galactorrhoea, and primary hypothyroidism with high circulating levels of prolactin. *Brit. Med. J.* **3**, 462–464.

Ehara, Y., Siler, T., VandenBerg, G., Sinha, Y. N., and Yen, S. S. C. (1973). Circulating prolactin levels during the menstrual cycle: episodic release and diurnal variation. *Amer. J. Obstet. Gynecol.* **117**, 962–970.

El Tomi, A. E. F., Crystle, C. D., and Stevens, V. C. (1971). Effects of immunization with human placental lactogen on reproduction in female rabbits. *Amer. J. Obstet. Gynecol.* **109**, 74–77.

Erdheim, J., and Stumme, E. (1909). Über die Schwangerschaftsveränderung der Hypophyse. *Beitr. Pathol.* **46**, 1–132.

Ewer, R. W. (1968). Familial monotropic pituitary gonadotropin insufficiency. *J. Clin. Endocrinol. Metab.* **28**, 783–788.

Fabian, M., Forsling, M. L., Jones, J. J., and Pryor, J. S. (1969). The clearance and antidiuretic potency of neurohypophysial hormones in man, and their plasma binding and stability. *J. Physiol. (London)* **204**, 653–668.

Fekete, K. (1930). Beitraege zur Physiologie der Graviditaet. *Endokrinologie* **7,** 364–369.

Fell, L. R., Beck, C., Brown, J. M., Catt, K. J., Cumming, I. A., and Goding, J. R. (1972). Solid-phase radioimmunoassay of ovine prolactin in antibody-coated tubes. Prolactin secretion during estradiol treatment, at parturition, and during milking. *Endocrinology* **91,** 1329–1336.

Feller, W. F., and Chopra, H. C. (1971). Virus-like particles in human milk. *Cancer* **28**, 1425–1430.

Ferin, J., Charles, J., Rommelart, G., and Beuselinck, A. (1964). Ovarian inhibition during lactation. *Int. J. Fert.* **9**, 41–43.

Fisch, L., Sala, N. L., and Schwarcz, R. L. (1964). Effect of cervical dilatation upon uterine contractility in pregnant women and its relation to oxytocin secretion. *Amer. J. Obstet. Gynecol.* **90**, 108–114.

Fitzpatrick, R. J. (1961a). The estimation of small amounts of oxytocin in blood. *In* "Oxytocin" (R. Caldeyro-Barcia and H. Heller, eds.), pp. 358–379. Pergamon, Oxford.

Fitzpatrick, R. J. (1961b). Blood concentration of oxytocin in labour. *J. Endocrinol.* **22,** xix–xx.

Floderus, S. (1949). Changes in the human hypophysis in connection with pregnancy. *Acta Anat.* **8**, 329–346.

Folkman, J. (1971). Transplacental carcinogenesis by stilbestrol. *N. Engl. J. Med.* **285**, 404–405.

Forbes, A. P., Henneman, P. H., Griswold, G. C., and Albright, F. (1954). Syndrome characterized by galactorrhea, amenorrhea and low urinary FSH: Comparison with acromegaly and normal lactation. *J. Clin. Endocrinol. Metab.* **14,** 265–271.

Forsyth, I. A. (1970). The detection of lactogenic activity in human blood by bioassay. J. *Endocrinol.* **46,** (February), iv–v.

Forsyth, I. A., and Myres, R. P. (1971). Human prolactin. Evidence obtained by the bioassay of human plasma. *J. Endocrinol.* **51**, 157–168.

Forsyth, I. A., Besser, G. M., Edwards, C. R. W., Francis, L., and Myres, R. P. (1971). Plasma prolactin activity in inappropriate lactation. *Brit. Med. J.* **3,** 225–227.

Foukas, M. D. (1973). An antilactogenic effect of pyridoxine. *J. Obstet. Gynaecol. Brit. Commonw.* **80**, 718–720.

Frank, R., Alpern, W. M., and Eshbaugh, D. E. (1969). Oral contraception started early in the puerperium. A clinical study. *Amer. J. Obstet. Gynecol.* **103,** 112–120.

Frantz, A. G., Rabkin, M. T., and Friesen, H. (1965). Human placental lactogen in choriocarcinoma of the male. Measurement by radioimmunoassay. *J. Clin. Endocrinol. Metab.* **25,** 1136–1139.

Friedman, S., and Goldfien, A. (1969a). Amenorrhea and galactorrhea following oral contraceptive therapy. *J. Amer. Med. Ass.* **210**, 1888–1891.

Friedman, S., and Goldfien, A. (1969b). Breast secretions in normal women. *Amer. J. Obstet. Gynecol.* **104**, 846–849.

Friesen, H. G., and Guyda. H. (1971). Biosynthesis of monkey growth hormone and prolactin *in vitro. Endocrinology* **88**, 1353–1362.

Friesen, H. G. (1972). Prolactin: Its physiologic role and therapeutic potential. *Hosp. Pract.* **7**, (September), 123–130.

Friesen, H., Shome, B., Belanger, C., Hwang, P., Guyda, H., and Myers, R. (1971). The synthesis and secretion of human and monkey placental lactogen (HPL and MPL) and pituitary prolactin (HPr and MPr). *Excerpta Med. Int. Congr. Ser.* **244**, 224–238.

Friesen, H., Webster, B. R., Hwang, P., Guyda, H., Munro, R. E., and Read, L. (1972). Prolactin synthesis and secretion in a patient with the Forbes Albright syndrome. *J. Clin. Endocrinol. Metab.* **34**, 192–199.

Frommel, R. (1881). Ueber puerperale Atrophie des Uterus. *Z. Geburtsh. Gynaekol.* **7**, 305–313.

Futterweit, W., and Goodsell, C. H. (1970). Galactorrhea in primary hypothyroidism: Report of two cases and review of the literature. *Mt. Sinai J. Med., New York* **37**, 584–589.

Fuxe, K., Hökfelt, T., and Nilsson, O. (1969). Factors involved in the control of the activity of the tubero-infundibular dopamine neurons during pregnancy and lactation. *Neuroendocrinology* **5**, 257–270.

Gambrell, R. D. (1970). Immediate postpartum oral contraception. *Obstet. Gynecol.* **36**, 101–106.

Ganong, W. F. (1971). The gonads: Development and function of the reproductive system. *In* "Review of Medical Physiology," pp. 307–338. Lange Med. Publ. Los Altos, California.

Gardner, E., Gray, D. J., and O'Rahilly, R. (1969). Veins, lymphatic drainage, and breast. *In* "Anatomy," 3rd ed., pp. 112–114. Saunders, Philadelphia, Pennsylvania.

Gates, R. B., Friesen, H., and Samaan, N. A. (1973). Inappropriate lactation and amenorrhoea: Pathological and diagnostic considerations. *Acta Endocrinol. (Copenhagen)* **72**, 101–114.

Gautvik, K. M., Weintraub, B. D., Graeber, C. T., Maloof, F., Zuckerman, J. E., and Tashjian, A. H., Jr. (1973). Serum prolactin and TSH: Effects of nursing and pyroGlu-His-ProNH_2 administration in postpartum women. *J. Clin. Endocrinol, Metab.* **36**, 135–139.

Gillibrand, P. N., and Huntingford, P. J. (1968). Inhibition of lactation with combined oestrogen and progestogen. *Brit. Med. J.* **4**, 769.

Gilliland, P. F., and Prout, T. E. (1965). Immunologic studies of octapeptides. II. Production and detection of antibodies to oxytocin. *Metab., Clin. Exp.* **14**, (August), 918–923.

Ginsburg, M., and Heller, H. (1953). Antidiuretic activity in blood obtained from various parts of the cardiovascular system. *J. Endocrinol.* **9**, 274–282.

Ginsburg, M., and Ireland, M. (1966). The role of neurophysin in the transport and release of neurohypophysial hormones. *J. Endocrinol.* **35**, 289–298.

Goluboff, L. G., and Ezrin, C. (1969). Effect of pregnancy on the somatotroph and the prolactin cell of the human adenohypophysis. *J. Clin. Endocrinol. Metab.* **29**, 1533–1538.

Greenblatt, R. B., and Gambrill, R. D. (1972). The increasing recognition of galactorrhea. *Current Med. Digest* **39**, 1036–1037.

Greenwald, P., Barlow, J. J., Nasca, P. C., and Burnett, W. S. (1971). Vaginal cancer after maternal treatment with synthetic estrogens. *N. Engl. J. Med.* **285**, 390–392.

Greenwood, F. C., Hunter, W. M., and Klopper, A. (1964). Assay of human growth hormone in pregnancy at parturition and in lactation. *Brit. Med. J.* **1**, 22–24.

Grynfeltt, J. (1937). Étude du processus cytologique de la sécrétion mammaire. *Archives d'anatomie microscopique* **33**, 177–209.

Gürson, C. T., and Etili, L. (1968). Relation between endogenous lipoprotein lipase activity, free fatty acids, and glucose in plasma of women in labour and of their newborns. *Arch. Dis. Childhood* **43**, 679–683.

Guyda, H., Robert, F., Colle, E., and Hardy, J. (1973). Histologic, ultrastructural, and hormonal characterization of a pituitary tumor secreting both hGH and prolactin. *J. Clin. Endocrinol. Metab.* **36**, 531–547.

György, P. (1971). Biochemical aspects. *Amer. J. Clin. Nutr.* **24**, 970–975.

Haagensen, C. D. (1971). Anatomy of the mammary gland. *In* "Diseases of the Breast," 2nd ed., pp. 1–67. Saunders, Philadelphia, Pennsylvania.

Handwerger, S., Pang, E. C., Aloj, S. M., and Sherwood, L. M. (1972). Correlations in the structure and function of human placental lactogen and human growth hormone. I. Modification of the disulfide bonds. *Endocrinology* **91**, 721–727.

Hanson, F. W., Powell, J. E., and Stevens, V. C. (1970). Follicle-stimulating hormone levels during pregnancy. *Obstet. Gynecol.* **36**, 667–670.

Harley, J. M. G. (1969). The endocrine control of the breasts. *Practitioner* **203**, 153–157.

Harrison, V. C., and Peat, G. (1972). Significance of milk pH in newborn infants. *Brit. Med. J.* **4**, 515–518.

Haskins, A. L., Moszkowski, E. F., and Cohen, H. (1964). Chiari-Frommel syndrome. *Amer. J. Obstet. Gynecol.* **88**, 667–670.

Hawker, R. W., and Robertson, P. A. (1957). Oxytocin in human female blood. *Endocrinology* **60**, 652–657.

Hawker, R. W., Walmsley, C. F., Roberts, V. S., Blackshaw, J. K., and Downes, J. C. (1961). Oxytocic activity of blood in parturient and lactating women. *J. Clin. Endocrinol. Metab.* **21**, 985–995.

Hekmatpanah, J. (1969). Pituitary tumors. *Surg. Clin. N. Amer.* **49**, 163–178.

Herbst, A. L. Ulfelder, H., and Poskanzer, D. C. (1971). Adenocarcinoma of the vagina. Association of maternal stilbestrol therapy with tumor appearance in young women. *N. Engl. J. Med.* **284**, 878–881.

Herlant, M. (1964). The cells of the adenohypophysis and their functional significance. *Int. Rev. Cytol.* **17**, 299–382.

Herlant, M., and Pasteels, J. L. (1967). Histophysiology of human anterior pituitary. *Methods Achiev. Exp. Pathol.* **3**, 250–305.

Herlyn, U., Jantzen, K., Flaskamp, D., Hoffmann, H., and von Berswordt-Wallrabe, I. (1969). A modification of the pigeon crop sac assay for lactotrophic hormone determinations by means of the addition of prednisolone. *Acta. Endocrinol. (Copenhagen)* **60**, 555–560.

Hill, H. (1970). Oral contraception. *Practitioner* **205**, 5–12.

Hobbs, J. R., Salih, H., Flax, H., and Brander, W. (1973). Prolactin dependence in human breast cancer. *Proc. Roy. Soc. Med.* **66**, 866.

Hollinshead, W. H. (1962). "Textbook of Anatomy," pp. 182–185. Harper, New York.

Hollmann, K. H. (1968). A morphometric study of sub-cellular organization in mouse mammary cancers and normal lactating tissue. *Z. Zellforsch. Mikrosk. Anat.* **87**, 266–277.

Hughes, P., Gillespie, A., and Dewhurst, C. J. (1972). Amenorrhea and galactorrhea. *Obstet. Gynecol.* **40**, 147–151.

Hume, M., Sevitt, S., and Thomas, D. P. (1970). "Venous Thrombosis and Pulmonary Embolism," pp. 54–84. Harvard Univ. Press, Cambridge Massachusetts.

Hunter, W. M., and Greenwood, F. C. (1964). A radio-immunoelectrophoretic assay for human growth hormone. *Biochem. J.* **91**, 43–56.

Hwang, P., Guyda, H., and Friesen, H. (1971a). A radioimmunoassay for human prolactin. *Proc. Nat. Acad. Sci. U.S.* **68**, 1902–1906.

Hwang, P., Guyda, H., and Friesen, H. (1971b). Human prolactin (HPr): purification and clinical studies. *Clin. Res.* **19**, 772.

Hwang, P., Guyda, H., and Friesen, H. (1972). Purification of human prolactin. *J. Biol. Chem.* **247**, 1955–1958.

Jacobs, L. S., Mariz, I. K., and Daughaday, W. H. (1972). A mixed heterologous radioimmunoassay for human prolactin. *J. Clin. Endocrinol. Metab.* **34**, 484–490.

Jaffe, R. B. (1971). Comment in Editor's note on, Injected progestogen and lactation. *Obstet. Gynecol. Surv.* **26**, 654.

Jaffe, R. B., Lee, P. A., and Midgley, A. R., Jr. (1969). Serum gonadotropins before, at the inception of, and following human pregnancy. *J. Clin. Endocrinol. Metab.* **29**, 1281–1283.

Johke, T. (1969). Prolactin release in response to milking stimulus in the cow and goat estimated by radioimmunoassay. *Endocrinol. Jap.* **16**, 179–185.

Josimovich, J. B., Kosor, B., and Mintz, D. H. (1969). Roles of placental lactogen in foetal-maternal relations. *Foetal Autonomy, Ciba Found. Symp., 1969* pp. 117–131.

Jubelin, J., and Boyer, J. (1972). The lipolytic activity of human milk. *Eur. J. Clin. Invest.* **2**, 417–421.

Kaern, T. (1967). Effect of an oral contraceptive immediately post partum on initiation of lactation. *Brit. Med. J.* **3**, 644–645.

Kaiser, R. (1969). Schwangerschaft und Laktationsperiode als Phasen der Proliferationsruhe an den weiblichen Fortpflanzungsorganen. *Geburtsh. Frauenheilk.* **29**, 420–430.

Kamal, I., Hefnawi, F., Ghoneim, M., Talaat, M., Younis, N., Tagui, A., and Abdalla, M. (1969a). Clinical, biochemical, and experimental studies on lactation. I. Lactation pattern in Egyptian women. *Amer. J. Obstet. Gynecol.* **105**, 314–323.

Kamal, I., Hefnawi, F., Ghoneim, M., Talaat, M., Younis, N., Tagui, A., and Abdalla, M. (1969b). Clinical, biochemical, and experimental studies on lactation. II. Clinical effects of gestagens on lactation. *Amer. J. Obstet. Gynecol.* **105**, 324–334.

Kamberi, I. A., Mical, R. S., and Porter, J. C. (1971). Effects of melatonin and serotonin on the release of FSH and prolactin. *Endocrinology* **88**, 1288–1293.

Kantor, H. I., Leib, L., and Kamholz, J. H. (1963). A new formulation with dual purpose: To inhibit lactation or to permit nursing. *Amer. J. Obstet. Gynecol.* **85**, 865–869.

Kaplan, S. L., and Grumbach, M. M. (1965). Serum chorionic "growth hormone-prolactin" and serum pituitary growth hormone in mother and fetus at term. *J. Clin. Endocrinol. Metab.* **25**, 1370–1374.

Kaplan, S. L., Grumbach, M. M., Friesen, H. G., and Costom, B. H. (1972). Thyrotropin-releasing factor (TRF) effect on secretion of human pituitary prolactin and thyrotropin in children and in idiopathic hypopituitary dwarfism: Further evidence for hypophysiotropic hormone deficiencies. *J. Clin. Endocrinol. Metab.* **35**, 825–830.

Karim, M., Ammar, R., El Mahgoub, S., El Ganzoury, B., Fikri, F., and Abdou, I. (1971). Injected progestogen and lactation. *Brit. Med. J.* **1**, 200–203.

Kates, R. B., Goss, D. A., and Townes, P. J. (1969). Immediate postpartum use of sequential oral contraceptive therapy. *S. Med. J.* **62**, 694–696.

Keller, P. J. (1968). Excretion of follicle-stimulating and luteinizing hormone during lactation. *Acta Endocrinol. (Copenhagen)* **57**, 529–535.

Kessler, E. (1968). Stillen und Brustkrebs. *Deut. Med. Wochenschr.* **93**, 639–641.

Keydar, J., Gilead, Z., Karby, S., and Harel, E. (1973). Production of virus by embryonic cultures co-cultivated with breast tumour cells or infected with milk from breast cancer patients. *Nature, New Biol. (London),* **106,** 49–52.

Khojandi, M., and Tyson, J. E. (1973). Non-puerperal galactorrhea: Clinical and experimental. *Clin. Res.* **21**, 495.

Kinch, R. A. H., Plunkett, E. R., and Devlin, M. C. (1969). Postpartum amenorrhea-galactorrhea of hypothyroidism. *Amer. J. Obstet. Gynecol.* **105**, 766–772.

Kleeman, C. R., and Vorherr, H. (1969). Water metabolism and the neurohypophysial hormones. *In* "Duncan's Diseases of Metabolism, Endocrinology and Nutrition" (P. Bondy, ed.), pp. 1103–1149. Saunders, Philadelphia, Pennsylvania.

Kleinberg, D. L., and Frantz, A. G. (1971). Human prolactin: Measurement in plasma by in vitro bioassay. *J. Clin. Invest.* **50**, 1557–1568.

Knowles, J. A. (1965). Excretion of drugs in milk—a review. *J. Pediat.* **66,** 1068–1082.

Koetsawang, S., Bhiraleus, P., and Chiemprajert, T. (1972). Effects of oral contraceptives on lactation. *Fert. Steril.* **23**, 24–28.

Kolodny, R. C., Jacobs, L. S., and Daughaday, W. H. (1972). Mammary stimulation causes prolactin secretion in non-lactating women. *Nature (London)* **238,** 284–286.

Kon, S. K., and Cowie, A. T. (1961). "Milk: The Mammary Gland and its Secretion," Vol. 1. Academic Press, New York.

Kopsch, Fr. (1955). "Rauber-Kopsch Lehrbuch und Atlas der Anatomie des Menschen," Vol. II, pp. 710–715. Thieme, Stuttgart.

Kora, S. J. (1969). Effect of oral contraceptives on lactation. *Fert. Steril.* **20,** 419–423.

Krieger, D. T. (1971). The hypothalamus and neuroendocrine pathology. *Hosp. Pract.* **6,** (November), 127–138.

Kurcz, M., Nagy, I., Kiss, Cs., and Halmy, L. (1969). Ergebnisse der Prolactinbestimmung im Blut. *Arch. Gynaekol.* **208**, 19–32.

Kurosumi, K., Kobayashi, Y., and Baba, N. (1968). The fine structure of mammary glands of lactating rats, with special reference to the apocrine secretion. *Exp. Cell Res.* **50**, 177–192.

Langer, E., and Huhn, S. (1958). Der submikroskopische Bau der Myoepithelzelle. *Z. Zellforsch. Mikrosk. Anat.* **47**, 507–516.

Langman, J. (1969). Integumentary system. *In* "Medical Embryology," pp. 367–368. Williams & Wilkins, Baltimore, Maryland.

Lassmann, G. (1964). Beitrag zur Innervation der menschlichen Brustdrüse. *Acta Anat.* **58**, 131–159.

Laumas, K. R., Malkani, P. K., Bhatnagar, S., and Laumas, V. (1967). Radioactivity in the breast milk of lactating women after oral administration of ^3H-norethynodrel. *Amer. J. Obstet. Gynecol.* **98**, 411–413.

Leis, H. P. (1970). Anatomy. *In* "Diagnosis and treatment of breast lesions," pp. 15–26. Medical Examination Publ., Flushing, New York.

LeMaire, W. J., Conly, P. W., Moffett, A., Spellacy, W. N., Cleveland, W. W., and Savard, K. (1971). Function of the human corpus luteum during the puerperium: Its maintenance by exogenous human chorionic gonadotropin. *Amer. J. Obstet. Gynecol.* **110**, 612–618.

Levine, H. J., Bergenstal, D. M., and Thomas, L. B. (1962). Persistent lactation: Endocrine and histologic studies in 5 cases. *Amer. J. Med. Sci.* **243**, 64–74.

L'Hermite, M., Copinschi, G., Golstein, J., Vanhaelst, L., Leclercq, R., and Bruno, O. D. (1972). Prolactin release after injection of thyrotrophin-releasing hormone in man. *Lancet* **1**, 763–765.

Li, C. H., Dixon, J. S., Schmidt, K. D., Pankov, Y. A., and Lo, T. B. (1969). Amino and carboxy-terminal sequences of ovine lactogenic hormone. *Nature (London)* **222**, 1268–1269.

Lieser, H., and Bässler, R. (1969). Morphologie der Mamma bei hormonaler Lactationshemmung. *Arch. Gynaekol.* **207**, 423–437.

Ling, E. R., Kon, S. K., and Porter, J. W. G. (1961). III. Milk proteins and other nitrogenous constituents. *In* "Milk: The Mammary Gland and Its Secretion" (S. K. Kon, and A. T. Cowie, eds.), Vol. 2, pp. 204–215. Academic Press, New York.

Linzell, J. L. (1961). Recent advances in the physiology of the udder. *In* "Veterinary Annual" (W. A. Pool, ed.), pp. 44–53. Wright, Bristol.

Linzell, J. L. (1971). The role of the mammary glands in reproduction. *Res. in Reprod.* **3**, No. 6 (November), 2–3.

Linzell, J. L., and Peaker, M. (1971). Mechanism of milk secretion. *Physiol. Rev.* **51**, 564–597.

Lo Presto, B., and Caypinar, E. Y. (1959). Prevention of postpartum lactation by administration of Deladumone during labor. *J. Amer. Med. Ass.* **169**, 250–252.

Lovell, R., and Rees, T. A. (1961). Immunological aspects of colostrum. *In* "Milk: The Mammary Gland and Its Secretion" (S. K. Kon and A. T. Cowie, eds.), Vol. 2, pp. 363–381. Academic Press, New York.

Lu, K. H., Koch, Y., and Meites, J. (1971). Direct inhibition by ergocornine of pituitary prolactin release. *Endocrinology* **89**, 229–233.

Lutterbeck, P. M., Pryor, J. S., Varga, L., and Wenner, R. (1971). Treatment of non-puerperal galactorrhoea with an ergot alkaloid. *Brit. Med. J.* **3**, 228–229.

Lyons, W. R. (1958). Hormonal synergism in mammary growth. *Proc. Roy. Soc., Ser. B.* **149**, 303–325.

McDevitt, E., and Smith, B. (1969). Thrombophlebitis during pregnancy and the puerperium. *In* "Thrombosis" (S. Sherry *et al.*, eds.), pp. 55–66. Nat. Acad. Sci., Washington, D.C.

MacDonald, D., and O'Driscoll, K. (1965). Suppression of lactation. A double-blind trial. *Lancet* **2**, 623.

McGlone, J. (1969). An assessment of quinestrol in the inhibition of lactation. *Practitioner* **203**, 187–188.

MacLeod, R. M. (1969). Influence of norepinephrine and catecholamine-depleting agents on the synthesis and release of prolactin and growth hormone. *Endocrinology* **85**, 916–923.

MacLeod, R. M., Abad, A., and Eidson, L. L. (1969). *In vivo* effect of sex hormones on the *in vitro* synthesis of prolactin and growth hormone in normal and pituitary tumor-bearing rats. *Endocrinology* **84**, 1475–1483.

Macy, I. G., and Kelly, H. J. (1961). Human milk and cow's milk in infant nutrition. *In* "Milk: The Mammary Gland and Its Secretion" (S. K. Kon and A. T. Cowie, eds.), Vol. 2, pp. 265–304. Academic Press, New York.

Malarkey, W. B., Jacobs, L. S., and Daughaday, W. H. (1971). Levodopa suppression of prolactin in nonpuerperal galactorrhea. *N. Engl. J. Med.* **285**, 1160–1163.

Manaro, J. M., De Spremolla, I. M., Michelotti, M., and Maggiolo, J. (1971). Determination of prolactin in urine in normal subjects (of both sexes in different stages of life) using a specific cytologic test in castrated pigeons. *Acta Endocrinol. Panam.* **2**, 117–125.

Mann, C. W. (1971). Lactation inhibition in the outer hebrides. *Practitioner* **206,** 246–247.

Marcovitz, S., and Friesen, H. (1971). Regulation of prolactin secretion in man. *Clin. Res.* **19**, 773.

Martin, J. E., MacDonald, P. C., and Kaplan, N. M. (1970). Successful pregnancy in a patient with Sheehan's syndrome. *N. Engl. J. Med.* **282**, 425–427.

Mata, L. J., and Wyatt, R. G. (1971). Host resistance to infection. *Amer. J. Clin. Nutr.* **24**, 976–986.

Meites, J. (1973). Breast Ca may be curbed by reducing prolactin secretion. *Ob. Gyn. News* **8,** No. 9, 4.

Meldolesi, J., Marini, D., and Demonte Marini, M. L. (1972). Studies on *in vitro* synthesis and secretion of growth hormone and prolactin. I. Hormone pulse labeling with radioactive leucine. *Endocrinology* **91**, 802–808.

Meyer, H. F. (1968). Breast feeding in the United States. Report of a 1966 national survey with comparable 1946 and 1956 data. *Clin. Pediat.* **7**, 708–715.

Michael, J. G., Ringenback, R., and Hottenstein, S. (1971). The antimicrobial activity of human colostral antibody in the newborn. *J. Infec. Dis.* **124**, 445–448.

Millar, D. G. (1969). The lactating breast. *Practitioner* **203**, 158–165.

Miller, G. H., and Hughes, L. R. (1970). Lactation and genital involution effects of a new low-dose oral contraceptive on breast-feeding mothers and their infants. *Obstet. Gynecol.* **35**, 44–50.

Miller, M. R., and Kasahara, M. (1959). The cutaneous innervation of the human female breast. *Anat. Rec.* **135**, 153–157.

Miller, R. W., and Fraumeni, J. F. (1972). Does breast-feeding increase the child's risk of breast cancer? *Pediatrics* **49**, 645–646.

Miyamoto, J., Gomez, L., and Gold, J. J. (1963). Effect of MRL-41 on postpartum breast manifestations. *Amer. J. Obstet. Gynecol.* **85**, 870–872.

Monroe, B. G., and Scott, D. E. (1966). Ultrastructural changes in the neural lobe of the hypophysis of the rat during lactation and suckling. *J. Ultrastruct. Res.* **14**, 497–517.

Moore, D. H., Charney, J., Kramarsky, B., Lasfargues, E. Y., Sarkar, N. H., Brennan, M. J., Burrows, J. H., Sirsat, S. M., Paymaster, J. C., and Vaidya, A. B. (1971). Search for a human breast cancer virus. *Nature (London)* **229,** 611–615.

Morris, J. A., Creasy, R. K., and Hohe, P. T. (1970). Inhibition of puerperal lactation. *Obstet. Gynecol.* **36**, 107–114.

Nabarro, J. D. N. (1972). Pituitary tumours and hypopituitarism. *Brit. Med. J.* **1,** 492–495.

Nagai, S., and Yamada, C. (1971). Sucking ability of newborns. *Tohoku J. Exp. Med.* **103**, 231–245.

Nasr, H., Mozaffarian, G., Pensky, J., and Pearson, O. H. (1972). Prolactin-secreting pituitary tumors in women. *J. Clin. Endocrinol. Metab.* **35**, 505–512.

Netter, F. H. (1965). Section XIII. Anatomy and pathology of the mammary gland. *In* "The Ciba Collection of Medical Illustrations, Reproductive System" (E. Oppenheimer, ed.), Vol. 2, pp. 245–249. Ciba Pharm. Inc., Summit, New Jersey.

Newton, M. (1961). Human lactation. *In* "Milk: The Mammary Gland and Its Secretion" (S. K. Kon and A. T. Cowie, eds.,) Vol. 1, pp. 281–320. Academic Press, New York.

Noel, G. L., Suh, H. K., and Frantz, A. G. (1971). Stimulation of prolactin release by stress in humans. *Clin. Res.* **19**, 718.

Noel, G. L., Suh, H. K., Stone, G., and Frantz, A. G. (1972). Human prolactin and growth hormone release during surgery and other conditions of stress. *J. Clin. Endocrinol. Metab.* **35**, 840–851.

Nokin, J., Vekemans, M., L'Hermite, M., and Robyn, C. (1972). Circadian periodicity of serum prolactin concentration in man. *Brit. Med. J.* **3**, 561–562.

Ordway, N. K. (1970). Formula feeding of infants in the first months. *Postgrad. Med.* **48,** (August) 167–172.

Oser, B. L. (1965). Milk. *In* "Hawk's Physiological Chemistry," 14th ed., pp. 368–384. McGraw-Hill (Blakiston), New York.

Paige, D. M., Bayless, T. M., Ferry, G. D., and Graham, G. G. (1971). Lactose malabsorption and milk rejection in negro children. *Johns Hopkins Med. J.* **129,** 163–169.

Parsons, J. A., and Nicoll, C. S. (1971). Mechanism of action of prolactin-inhibiting factor. *Neuroendocrinology* **8**, 213–227.

Pasteels, J. L. (1970). General discussion on prolactin. *In* "Ovo-Implantation Human Gonadotropins and Prolactin," pp. 306–307. Karger, Basel.

Pasteels, J. L., Gausset, P., Danguy, A., Ectors, F., Nicoll, C. S., and Varavudhi, P. (1972). Morphology of the lactotropes and somatotropes in man and rhesus monkeys. *J. Clin. Endocrinol. Metab.* **34**, 959–967.

Peake, G. T., McKeel, D. W., Jarett, L., and Daughaday, W. H. (1969). Ultrastructural, histologic and hormonal characterization of a prolactin-rich human pituitary tumor. *J. Clin. Endocrinol. Metab.* **29**, 1383–1393.

Pearson, O. H. (1969). Prolactin may be key to breast cancer control. *Ob. Gyn. News.* **4**, 32.

Peckham, C. H. (1934). An investigation of some effects of pregnancy noted six weeks and one year after delivery. *Bull. Johns Hopkins Hosp.* **54**, 186–207.

Pedersen, V. B., Keil-Dlouhá, V., and Keil, B. (1971). On the properties of trypsin inhibitors from human and bovine colostrum. *FEBS Lett.* **17**, 23–26.

Permutt, M. A., Parker, C. W., and Utiger, R. D. (1966). Immunochemical studies with lysine vasopressin. *Endocrinology* **78**, 809–814.

Pernkopf, E. (1964). The thorax. *In* "Atlas of Topographical and Applied Human Anatomy" (H. Ferner, ed.), Vol. II, pp. 6–7. Saunders, Philadelphia, Pennsylvania.

Pernoll, M. L. (1971). Diagnosis and treatment of galactorrhea. *Postgrad. Med.* **49,** 76–82.

Pincus, G. (1965). Mammary glands and lactation. "The Control of Fertility," pp. 260–263. Academic Press, New York.

Ramadan, M. A., Salah, M. M., Eid, S. Z., and Sammour, M. B. (1972). The effect of the oral contraceptive Ovosiston on the composition of human milk. *J. Reprod. Med.* **9**, 81–83.

Ramberg, C. F., Jr., Mayer, G. P., Kronfeld, D. S., Phang. J. M., and Berman, M. (1970). Calcium kinetics in cows during late pregnancy, parturition, and early lactation. *Amer. J. Physiol.* **219**, 1166–1177.

Rankin, J. S., Goldfarb, A. F., and Rakoff, A. E. (1969). Galactorrhea-amenorrhea syndromes: Postpartum galactorrhea-amenorrhea in the absence of intracranial neoplasm. *Obstet. Gynecol.* **33**, 1–10.

Rapoport, B., Refetoff, S., Fang, V. S., and Friesen, H. G. (1973). Suppression of serum thyrotropin (TSH) by L-dopa in chronic hypothyroidism: Interrelationships in the regulation of TSH and prolactin secretion. *J. Clin. Endocrinol. Metab.* **36**, 256–262.

Refetoff, S., Block, M. B., Ehrlich, E. N., and Friesen, H. G. (1972). Chiari-Frommel syndrome in a patient with primary adrenocortical insufficiency. *N. Engl. J. Med.* **287**, 1326–1328.

Relkin, R. (1965). Galactorrhea: A review. *N.Y. State J. Med.* **65**, 2800–2807.

Rhodes, P. (1971). Antenatal and postnatal physiotherapy. *Practitioner* **206,** 758–764.

Rice, B. F., Schneider, G., and Weed, J. (1969). Serum calcium and magnesium concentration during early labor and the postpartum period. *Amer. J. Obstet. Gynecol.* **104**, 1159–1162.

Rice-Wray, E. (1963). Oral contraception in Latin America. *Proc. Conf. Int. Planned Parenthood Fed., 7th, 1963* Excerpta Med. Int. Cong. Ser. No. 72, pp. 358–368.

Rice-Wray, E., Goldzieher, J. W., and Aranda-Rosell, A. (1963). Oral progestins in fertility control: A comparative study. *Fert. Steril.* **14**, 402–409.

Richard, Ph. (1970). An electrophysiological study in the ewe of the tracts which transmit impulses from the mammary glands to the pituitary stalk. *J. Endocrinol.* **47**, 37–44.

Richardson, G. S. (1970). Reflex lactation (thoracotomy) and reflex ovulation (intercostal block): Case report, review of the literature, and discussion of mechanisms. *Obstet. Gynecol. Surv.* **25**, 1021–1036.

Rifkind, A. B., Kulin, H. E., and Ross, G. T. (1967). Follicle-stimulating hormone (FSH) and luteinizing hormone (LH) in the urine of prepubertal children. *J. Clin. Invest.* **46,** 1925–1931.

Rolland, R., and Schellekens, L. (1973). A new approach to the inhibition of puerperal lactation. *J. Obstet. Gynaecol. Brit. Commonw.* **80,** 945–951.

Rorie, D. K., and Newton, M. (1964). Oxytocic factors in the plasma of the human male. *Fert. Steril.* **15**, 135–142.

Ross, F., and Nusynowitz, M. L. (1968). A syndrome of primary hypothyroidism, amenorrhea and galactorrhea. *J. Clin. Endocrinol. Metab.* **28**, 591–595.

Roth, J., Glick, S. M., Klein, L. A., and Petersen, M. J. (1966). Specific antibody to vasopressin in man. *J. Clin. Endocrinol. Metab.* **26**, 671–675.

Rushworth, R. G. (1971). Pituitary apoplexy. *Med. J. Aust.* **1**, 251–254.

Sachs, H. (1967). Biosynthesis and release of vasopressin. *Amer. J. Med.* **42**, 687–700.

Sachson, R., Rosen, S. W., Cuatrecasas, P., Roth, J., and Frantz, A. G. (1972). Prolactin stimulation by thyrotropin-releasing hormone in a patient with isolated thyrotropin deficiency. *N. Engl. J. Med.* **287**, 972–973.

Saito, S., Abe, K., Nagata, N., Nakamura, E., and Tanaka, K. (1972). Effect of L-dopa on anterior pituitary hormone release in man. *Endocrinol. Jap.* **19**, 435–442.

Sar, M., and Meites, J. (1969). Effects of suckling on pituitary release of prolactin, GH, and TSH in postpartum lactating rats. *Neuroendocrinology* **4**, 25–31.

Sassin, J. F., Frantz, A. G., Weitzman, E. D., and Kapen, S. (1972). Human prolactin: 24-hour pattern with increased release during sleep. *Science* **177**, 1205–1207.

Schäfer, A., and Bässler, R. (1969). Vergleichende elektronenmikroskopische Untersuchungen am Drüsenepithel und am sog. lobulären Carcinom der Mamma. *Virchows Arch., A* **346**, 269–286.

Schally, A. V., Redding, T. W., Bowers, C. Y., and Barrett, J. F. (1969). Isolation and properties of porcine thyrotropin-releasing hormone. *J. Biol. Chem.* **244**, 4077–4088.

Schally, A. V., Arimura, A., Kastin, A. J., Matsuo, H., Baba, Y., Redding, T. W., Nair, R. M. G., Debeljuk, L., and White, W. F. (1971). Gonadotropin-releasing hormone: One polypeptide regulates secretion of luteinizing and follicle-stimulating hormones. *Science* **173**, 1036–1038.

Schams, D. (1972). Prolactin levels in bovine blood, influenced by milking manipulation, genital stimulation and oxytocin administration with specific consideration of the seasonal variations. *Acta Endocrinol. (Copenhagen)* **71**, 684–696.

Scharrer, B. (1959). The role of neurosecretion in neuroendocrine integration. *In* "Comparative Endocrinology" (A. Gorbman, ed.), pp. 134–148. Wiley, New York.

Schelin, U., and Lundin, P. M. (1971). An electron microscopic study of normal and neoplastic acidophil cells of the rat pituitary. *Acta Endocrinol.* (*Copenhagen*) **67**, 29–39.

Schwenk, A. (1969). Die Vorbereitung der Fortpflanzungsfunktionen von der Kindheit bis zur Pubertät und ihre Störungen. *In* "Gynäkologie und Geburtshilfe" (O. Käser *et al.*, eds.), Vol. I, pp. 209–249. Thieme, Stuttgart.

Semm, K. (1966). Contraception and lactation. *In* "Social and Medical Aspects of Oral Contraception" (M. N. G. Dukes, ed.), Int. Congr. Ser. No. 130, pp. 98–101. Excerpta Med. Found., Amsterdam.

Semm, K., and Waidl, E. (1962). Histochemische Untersuchungen ueber die Serum-Oxytocinase-Bildung in menschlichen Trophoblasten. *Z. Geburtsh. Gynaekol.* **158**, 165–171.

Sharman, A. (1951). Ovulation after pregnancy. *Fert. Steril.* **2**, 371–393.

Sharman, A. (1966). "Reproductive Physiology of the Post-Partum Period." Livingstone, Edinburgh.

Shearman, R. P., and Turtle, J. R. (1970). Secondary amenorrhea with inappropriate lactation. *Amer. J. Obstet. Gynecol.* **106**, 818–827.

Sheehan, H. L., and Davis, J. C. (1968). Pituitary necrosis. *Brit. Med. Bull.* **24,** 59–70.

Sheld, H. H. (1968). Non-puerperal galactorrhea following hysterectomy. *Rocky Mt. Med. J.* **65,** (March), 57–59.

Sheld, H. H., and Charme, L. S. (1969). Nonpuerperal galactorrhea following hysterectomy. *N.Y. State J. Med.* **69**, 590–593.

Sherman, L., and Kolodny, H. D. (1971). The hypothalamus, brain-catecholamines, and drug therapy for gigantism and acromegaly. *Lancet* **1**, 682–685.

Sherwood, L. M. (1971). Current concepts. Human prolactin. *N. Engl. J. Med.* **284,** 774–777.

Shevach, A. B., and Spellacy, W. N. (1971). Galactorrhea and contraceptive practices. *Obstet. Gynecol.* **38**, 286–289.

Shiino, M., Williams, G., and Rennels, E. G. (1972). Ultrastructural observation of pituitary release of prolactin in the rat by suckling stimulus. *Endocrinology* **90,** 176–187.

Silverman, W. A. (1961). The feeding and nutrition of premature infants. *In* "Dunham's Premature Infants," pp. 151–179. Harper (Hoeber), New York.

Simkin, B., and Arce, R. (1963). Prolactin activity in blood during the normal human menstrual cycle. *Proc. Soc. Exp. Biol. Med.* **113**, 485–488.

Simkin, B., and Goodart, D. (1960). Preliminary observations on prolactin activity in human blood. *J. Clin. Endocrinol. Metab.* **20**, 1095–1106.

Sinha, Y. N., Lewis, U. J., and VanderLaan, W. P. (1973a). A homologous radioimmunoassay for human prolactin. *Clin. Res.* **21**, 189.

Sinha, Y. N., Selby, F. W., and VanderLaan, W. P. (1973b). Radioimmunoassay of prolactin in the urine of mouse and man. *J. Clin. Endocrinol. Metab.* **36,** 1039–1042.

Sinha, Y. N., Selby, F. W., Lewis, U. J., and VanderLaan, W. P. (1973c). A homologous radioimmunoassay for human prolactin. *J. Clin. Endocrinol. Metab.* **36,** 509–516.

Smith, R. E., and Farquhar, M. G. (1966). Lysosome function in the regulation of the secretory process in cells of the anterior pituitary gland. *J. Cell Biol.* **31**, 319–347.

Speert, H. (1958). Johann Chiari, Richard Frommel, and the Chiari-Frommel syndrome. *In* "Obstetric and Gynecologic Milestones," pp. 385–391. Macmillan, New York.

Spellacy, W. N., and Buhi, W. C. (1969). Pituitary growth hormone and placental lactogen levels measured in normal term pregnancy and at the early and late postpartum periods. *Amer. J. Obstet. Gynecol.* **105**, 888–896.

Spellacy, W. N., Carlson, K. L., and Schade, S. L. (1968). Human growth hormone studies in patients with galactorrhea (Ahumada-Del Castillo syndrome). *Amer. J. Obstet. Gynecol.* **100**, 84–89.

Spellacy, W. N., Buhi, W. C., and Birk, S. A. (1970). Normal lactation and blood growth hormone studies. *Amer. J. Obstet. Gynecol.* **107**, 244–249.

Spies, H. G., and Clegg, M. T. (1971). Pituitary as a possible site of prolactin feedback in autoregulation. *Neuroendocrinology* **8**, 205–212.

Spona, J., and Janisch, H. (1971). Serum placental lactogen (HPL) as index of placental function. *Acta Endocrinol.* (*Copenhagen*) **68**, 401–412.

Stearns, E., Winter, J. S. D., and Faiman. C. (1972). The effect of coitus on gonadotropin, prolactin and sex steroid levels in man. *Clin. Res.* **20**, 923.

Stewart, P. S., Puppione, D. L., and Patton, S. (1972). The presence of microvilli and other membrane fragments in the non-fat phase of bovine milk. *Z. Zellforsch. Mikrosk. Anat.* **123**, 161–167.

Stockley, I. H. (1970). Mechanisms of drug interaction in the pharmacokinetic phase of drug activity. *Amer. J. Hosp. Pharm.* **27**, 977–985.

Sud, S. C. (1971). Suppression of prolactin release from adenohypophysis by prolactin: An *in vitro* study. *Indian J. Exp. Biol.* **9**, 260–261.

Sulman, F. G. (1970). Chapter 3: Prolactin. *In* "Hypothalamic Control of Lactation" (F. Gross *et al.*, eds.), Vol. 3, pp. 29–37. Springer-Verlag, Berlin and New York.

Tanner, J. M. (1962). "Wachstum und Reifung des Menschen," pp. 45–46. Thieme, Stuttgart.

Taylor, E. S. (1966). Lactation. *In* "Beck's Obstetrical Practice," pp. 216–222. Williams & Wilkins, Baltimore, Maryland.

Thompson, J. P., and Kempers, R. D. (1965). Amenorrhea and galactorrhea. *Amer. J. Obstet. Gynecol.* **93**, 65–71.

Tindal, J. S. (1972). Reflex pathways controlling lactation. *Proc. Roy. Soc. Med.* **65**, 1085–1086.

Tindal, J. S., and Yokoyama, A. (1962). Assay of oxytocin by the milk-ejection response in the anesthetized lactating guinea pig. *Endocrinology* **71**, 196–202.

Tindal, J. S., Knaggs, G. S., and Turvey, A. (1969). The afferent path of the milk-ejection reflex in the brain of the rabbit. *J. Endocrinol.* **43**, 663–671.

Tindall, V. R. (1968). Factors influencing puerperal thrombo-embolism. *J. Obstet. Gynaecol. Brit. Commonw.* **75**, 1324–1327.

Toker, C. (1967). Observations on the ultrastructure of a mammary ductule. *J. Ultrastruct. Res.* **21**, 9–25.

Tokuhata, G. K. (1969). Morbidity and mortality among offspring of breast cancer mothers. *Amer. J. Epidemiol.* **89**, 139–153.

Traeger, A. (1970). Die Ausscheidung von Arzneimitteln mit der Muttermilch. *Z. Aerztl. Fortbild.* **64**, 724–728.

Tucker, H. A., Convey, E. M., and Koprowski, J. A. (1973). Milking-induced release of endogenous prolactin in cows infused with exogenous prolactin (36961). *Proc. Soc. Exp. Biol. Med.* **142**, 72–75.

Tuppy, H. (1960). Enzymic inactivation and degradation of oxytocin and vasopressin. *In* "Polypeptides which Affect Smooth Muscles and Blood Vessels" (M. Schachter, ed.), pp. 49–58. Pergamon, Oxford.

Turkington, R. W. (1968a). Hormone-induced synthesis of DNA by mammary gland *in vitro. Endocrinology* **82**, 540–546.

Turkington, R. W. (1968b). Induction of milk protein synthesis by placental lactogen and prolactin *in vitro. Endocrinology* **82**, 575–583.

Turkington, R. W. (1972a). Secretion of prolactin by patients with pituitary and hypothalamic tumors. *J. Clin. Endocrinol. Metab.* **34**, 159–164.

Turkington R. W. (1972b). Phenothiazine stimulation test for prolactin reserve: The syndrome of isolated prolactin deficiency. *J. Clin. Endocrinol. Metab.* **34**, 247–249.

Turkington, R. W. (1972c). Inhibition of prolactin secretion and successful therapy of the Forbes-Albright syndrome with L-dopa. *J. Clin. Endocrinol. Metab.* **34,** 306–311.

Turkington, R. W. (1972d). Human prolactin. An ancient molecule provides new insights for clinical medicine. *Amer. J. Med.* **53**, 389–394.

Turkington, R. W. (1972e). Prolactin secretion in patients treated with various drugs. *Arch. Intern. Med.* **130**, 349–354.

Turkington, R. W. (1972f). Multiple hormonal interactions. The mammary gland. *In* "Biochemical Actions of Hormones" (G. Litwack, ed.), Vol. 2, pp. 55–80. Academic Press, New York.

Turnbull, A. C. (1968). Puerperal thrombo-embolism and suppression of lactation. *J. Obstet. Gynaecol. Brit. Commonw.* **75**, 1321–1323.

Turnbull, A. C., Daniel, D. G., and McGarry, J. M. (1971). Antenatal and postnatal thrombo-embolism. *Practitioner* **206**, 727–735.

Turner, C. W. (1952). "The Mammary Gland," p. 130. Lucas Brothers, Columbia, Missouri.

Tyson, J. E., and Blizzard, R. M. (1972). Prolactin and thyrotropin secretion in human pregnancy. *Clin. Res.* **20**, 443.

Tyson, J. E., Hwang, P., Guyda, H., and Friesen, H. G. (1972a). Studies of prolactin secretion in human pregnancy. *Amer. J. Obstet. Gynecol.* **113**, 14–20.

Tyson, J. E., Friesen, H., Guyda, H., and Hwang, P. (1972b). Secretion of human prolactin in pregnancy and the puerperium. *19th Ann. Meet., Soc. Clin. Invest. 1972. Proc. Soc. Gynecol. Invest.* Abstract No. 73.

Tyson, J. E., Huth, J., Smith, B., and Thomas, P. (1973). Prolactin induced alterations in human milk. *Clin. Res.* **21**, 641.

Urban, I., Moss, R. L., and Cross, B. A. (1971). Problems in electrical stimulation of afferent pathways for oxytocin release. *J. Endocrinol.* **51**, 347–358.

van der Molen, H. J., Hart, P. G., and Wijmenga, H. G. (1969). Studies with 4-^{14}C-lynestrenol in normal and lactating women. *Acta Endocrinol. (Copenhagen)* **61**, 255–274.

Van Wyk, J. J., and Grumbach, M. M. (1960). Syndrome of precocious menstruation and galactorrhea in juvenile hypothyroidism: An example of hormonal overlap in pituitary feedback. *J. Pediat.* **57**, 416–435.

Varga, L., Lutterbeck, P. M., Pryor, J. S., Wenner, R., and Erb, H. (1972a). Suppression of puerperal lactation with an ergot alkaloid: A double-blind study. *Brit. Med. J.* **2**, 743–744.

Varga, L., Lutterbeck, P. M., Pryor, J. S., Wenner, R., and Erb, H. (1972b). Unterdrückung der puerperalen Laktation mit einem Mutterkornalkaloid. *Schweiz. Med. Wochenschr.* **102**, 1284–1285.

Volpé, R., Killinger, D., Bird, C., Clark, A. F., and Friesen, H. (1972). Idiopathic galactorrhea and mild hypogonadism in a young adult male. *J. Clin. Endocrinol. Metab.* **35**, 684–692.

von Berswordt-Wallrabe, I., Herlyn, U., Flaskamp, D., and Hellige, G. (1971). Die Biologische Bestimmung von Lactotropem Hormon (LTH) und ihre Anwendbarkeit beim Menschen. *Arch. Gynaekol.* **209,** 380–395.

Vorherr, H. (1968). The pregnant uterus: Process of labor, puerperium, and lactation. *In* "Biology of Gestation" (N. S. Assali, ed.), Vol. 1, pp. 426–448. Academic Press, New York.

Vorherr, H. (1971). Catecholamine antagonism to oxytocin-induced milk-ejection. *Acta Endocrinol. (Copenhagen)* **67**, Suppl. 154, 5–38.

Vorherr, H. (1972a). Disorders of uterine functions during pregnancy, labor, and puerperium. *In* "Pathophysiology of Gestation" (N. S. Assali, ed.), Vol. 1, pp. 145–268. Academic Press, New York.

Vorherr, H. (1972b). Puerperium: Maternal involutional changes and lactation. *In* "Davis' Gynecology and Obstetrics" (J. J. Rovinsky, ed.), Vol. 1, Chapter 20, pp. 1–46. Harper, New York.

Vorherr, H. (1972c). ADH and oxytocin in blood and urine of gravidas and parturients. *Proc. Soc. Gynecol. Invest.* Abstr. Pap., p. 30.

Vorherr, H. (1972d). To breast-feed or not to breast-feed? *Postgrad. Med.* **51,** 127–134.

Vorherr, H. (1972e). Suppression of postpartum lactation. *Postgrad. Med.* **52,** 145–152.

Vorherr, H. (1973). Contraception after abortion and post partum. *Amer. J. Obstet. Gynecol.* **117**, 1002–1025.

Vorherr, H., and Friedberg, V. (1964). Über die Wirkung von Adiuretin auf die Diurese in der Schwangerschaft. *Klin. Wochenschr.* **42**, 201–204.

Vorherr, H., and Munsick, R. A. (1970). Identification of neurohypophysial hormones with their antisera. *J. Clin. Invest.* **49**, 828–836.

Vorherr, H., Bradbury, M. W. B., Hoghoughi, M., and Kleeman, C. R. (1968). Antidiuretic hormone in cerebrospinal fluid during endogenous and exogenous changes in its blood level. *Endocrinology* **83**, 246–250.

Wang, R. I. H., and Ober, K. F. (1971). A survey of drug interactions for the practicing physician. *Drug Ther.* **1,** No. 10, 48–55.

Watrous, J. B., Ahearn, R. E., and Carvalho, M. A. (1959). Lactation inhibition by Deladumone injected during labor or just after delivery. *J. Amer. Med. Ass.* **169**, 246–249.

Watson, P. S. (1969). 'Instant' inhibition of lactation. *Practitioner* **203,** 184–186.

Weitzel, D., and Bässler, R. (1971). Beiträge zur Angioarchitektur der weiblichen Brustdrüse. Dargestellt an Injektionspräparaten und Angiographien. *Z. Anat. Entwicklungsgesch.* **133**, 73–88.

Wenner, R. (1967). Physiologische und pathologische Lactation. *Arch. Gynaekol.* **204**,171–206.

White, A., Handler, P., and Smith, E. L. (1964). Specialized extracellular fluids. *In* "Principles of Biochemistry," 3rd ed., pp. 717–722. McGraw-Hill (Blakiston), New York.

White, W. F. (1970). On the identity of the LH- and FSH-releasing hormones. *Colloq. Ges. Biol. Chem.* **21**, 84–87.

Winberg, J., and Wessner, G. (1971). Does breast milk protect against septicaemia in the newborn? *Lancet* **1**, 1091–1094.

Winkelmann, W. (1971). Hypophyseotrope Hypothalamushormone. *Internist* **12,** 193–198.

Wong, Y. K., and Wood, B. S. B. (1971). Breast-milk jaundice and oral contraceptives. *Brit. Med. J.* **4**, 403–404.

Wuttke, W., Cassell, E., and Meites, J. (1971). Effects of ergocornine on serum prolactin and LH, and on hypothalamic content of PIF and LRF. *Endocrinology* **88**, 737–741.

Yen, S. S. C., Vicic, W. J., and Kearchner, D. V. (1969). Gonadotropin levels in puberty. I. Serum luteinizing hormone. *J. Clin. Endocrinol. Metab.* **29**, 382–385.

Young, R. L., Bradley, E. M., Goldzieher, J. W., Myers, P. W., and Lecocq, F. R. (1967). Spectrum of nonpuerperal galactorrhea: Report of two cases evolving through the various syndromes. *J. Clin. Endocrinol. Metab.* **27**, 461–466.

Zárate, A., Villalobos, M., and Valenzuela, S. (1968). A case of Chiari-Frommel syndrome associated with polycystic ovaries. *Amer. J. Obstet. Gynecol.* **101,** 1131–1132.

Zárate, A., Canales, E. S., Soria, J., Ruiz, F., and MacGregor, C. (1972). Ovarian refractoriness during lactation in women: Effect of gonadotropin stimulation. *Amer. J. Obstet. Gynecol.* **112**, 1130–1132.

Zeppa, R. (1969). Vascular response of the breast to estrogen. *J. Clin. Endocrinol. Metab.* **29**, 695–700.

Zilliacus, H. (1967). Physiologie und Pathologie des Wochenbettes. *In* "Gynäkologie und Geburtshilfe" (O. Käser, *et al.,* eds.), Vol. II, pp. 966–997. Thieme, Stuttgart.

Zondek, B., Bromberg, Y. M., and Rozin, S. (1951). An anterior pituitary hyperhormonotrophic syndrome (excessive uterine bleeding, galactorrhoea, hyperthyroidism). *J. Obstet. Gynaecol. Brit. Emp.* **58**, 525–537.

Subject Index

B

C

D

G

H

I

L

M

N

A 4
B 5
C 6
D 7
E 8
F 9
G 0
H 1
I 2
J 3

ST. LOUIS COLLEGE OF PHARMACY
3 2201 90026 6442

WITHDRAWN